Serge Bouc

Biset Functors for Finite Groups

Springer

Serge Bouc
Université de Picardie
Fac. Mathématiques (LAMFA-CNRS)
33 rue St. Leu
80039 Amiens
France
serge.bouc@u-picardie.fr
http://www.lamfa.u-picardie.fr/bouc/

ISBN: 978-3-642-11296-6 e-ISBN: 978-3-642-11297-3
DOI: 10.1007/978-3-642-11297-3
Springer Heidelberg Dordrecht London New York

Lecture Notes in Mathematics ISSN print edition: 0075-8434
ISSN electronic edition: 1617-9692

Library of Congress Control Number: 2010920971

Mathematics Subject Classification (2000): 20J15, 19A22, 20C15, 20G05

Cover design: SPi Publisher Services

Printed on acid-free paper

springer.com

Merci à Cathy Ricordeau-Hamer,
pour son amitié et son soutien indéfectibles.

Contents

Chapter 1
Examples

This chapter illustrates by examples some situations where the use of biset functors is natural. It is meant to be an informal introduction, most of the technical details being omitted, and postponed to subsequent chapters.

1.1. Representation Groups

1.1.1. Let $\mathbb{F}$ be a field. If G is a finite group, denote by $R_{\mathbb{F}}(G)$ the representation group of G over $\mathbb{F}$, i.e. the Grothendieck group of the category of finitely generated $\mathbb{F}G$-modules. It can be computed as the quotient of the free abelian group on the set of isomorphism classes of finitely generated $\mathbb{F}G$-modules, by the subgroup generated by all the elements of the form $[U]-[V]-[W]$ where U, V and W are finitely generated $\mathbb{F}G$-modules appearing in a short exact sequence of $\mathbb{F}G$-modules $0 \to V \to U \to W \to 0$, and $[U]$ denotes the isomorphism class of U.

1.1.2. Operations. There are various natural operations connecting the groups $R_{\mathbb{F}}(G)$ and $R_{\mathbb{F}}(H)$ for finite groups G and H:

- If H is a subgroup of G, then restriction of modules from $\mathbb{F}G$ to $\mathbb{F}H$ induces a *restriction map* $\mathrm{Res}_H^G : R_{\mathbb{F}}(G) \to R_{\mathbb{F}}(H)$.
- In the same situation, induction of modules from $\mathbb{F}H$ to $\mathbb{F}G$ yields an *induction map* $\mathrm{Ind}_H^G : R_{\mathbb{F}}(H) \to R_{\mathbb{F}}(G)$.
- If $\varphi : G \to H$ is a *group isomorphism*, there is an obvious associated linear map $\mathrm{Iso}(\varphi) : R_{\mathbb{F}}(G) \to R_{\mathbb{F}}(H)$.
- If N is a normal subgroup of G, and $H = G/N$, then inflation of modules from $\mathbb{F}H$ to $\mathbb{F}G$ yields an *inflation map* $\mathrm{Inf}_{G/N}^G : R_{\mathbb{F}}(G/N) \to R_{\mathbb{F}}(G)$.
- Another operation can be defined in the same situation, with the additional hypothesis that the characteristic of $\mathbb{F}$ is coprime to the order of N: starting with an $\mathbb{F}G$-module V, one can consider the module V_N of coinvariants of N on V, i.e. the largest quotient vector space of V on which N acts trivially. Then V_N is a $\mathbb{F}H$-module, and the construction $V \mapsto V_N$ is an exact functor from the category of $\mathbb{F}G$-modules to the category of $\mathbb{F}H$-modules, because of the hypothesis on

S. Bouc, *Biset Functors for Finite Groups*, Lecture Notes in Mathematics 1990,
DOI 10.1007/978-3-642-11297-3_1, © Springer-Verlag Berlin Heidelberg 2010

the characteristic of $\mathbb{F}$: indeed $V_N \cong \mathbb{F} \otimes_{\mathbb{F}N} V$, and the $\mathbb{F}N$-module $\mathbb{F}$ is projective, hence flat. So the correspondence $V \mapsto V_N$ induces a *deflation map* $\mathrm{Def}^G_{G/N} : R_{\mathbb{F}}(G) \to R_{\mathbb{F}}(G/N)$.

1.1.3. Relations. All these operations are subject to various compatibility conditions (in the following list, all the conditions involving a deflation map hold whenever they are defined, i.e. when the orders of the corresponding normal subgroups are not divisible by the characteristic of $\mathbb{F}$):

1. *Transitivity conditions:*

   a. If K and H are subgroups of G with $K \leq H \leq G$, then

   $$\mathrm{Res}^H_K \circ \mathrm{Res}^G_H = \mathrm{Res}^G_K\ , \qquad \mathrm{Ind}^G_H \circ \mathrm{Ind}^H_K = \mathrm{Ind}^G_K\ .$$

   b. If $\varphi : G \to H$ and $\psi : H \to K$ are group isomorphisms, then

   $$\mathrm{Iso}(\psi) \circ \mathrm{Iso}(\varphi) = \mathrm{Iso}(\psi\varphi)\ .$$

   c. If N and M are normal subgroups of G with $N \leq M$, then

   $$\mathrm{Inf}^G_{G/N} \circ \mathrm{Inf}^{G/N}_{G/M} = \mathrm{Inf}^G_{G/M}\ , \qquad \mathrm{Def}^{G/N}_{G/M} \circ \mathrm{Def}^G_{G/N} = \mathrm{Def}^G_{G/M}\ .$$

2. *Commutation conditions:*

   a. If $\varphi : G \to H$ is a group isomorphism, and K is a subgroup of G, then

   $$\begin{aligned}\mathrm{Iso}(\varphi') \circ \mathrm{Res}^G_K &= \mathrm{Res}^H_{\varphi(K)} \circ \mathrm{Iso}(\varphi)\\ \mathrm{Iso}(\varphi) \circ \mathrm{Ind}^G_K &= \mathrm{Ind}^H_{\varphi(K)} \circ \mathrm{Iso}(\varphi')\ ,\end{aligned}$$

   where $\varphi' : K \to \varphi(K)$ is the restriction of φ.

   b. If $\varphi : G \to H$ is a group isomorphism, and N is a normal subgroup of G, then

   $$\begin{aligned}\mathrm{Iso}(\varphi") \circ \mathrm{Def}^G_{G/N} &= \mathrm{Def}^H_{H/\varphi(N)} \circ \mathrm{Iso}(\varphi)\\ \mathrm{Iso}(\varphi) \circ \mathrm{Inf}^G_{G/N} &= \mathrm{Inf}^H_{H/\varphi(N)} \circ \mathrm{Iso}(\varphi")\ ,\end{aligned}$$

   where $\varphi" : G/N \to H/\varphi(N)$ is the group isomorphism induced by φ.

   c. *(Mackey formula)* If H and K are subgroups of G, then

   $$\mathrm{Res}^G_H \circ \mathrm{Ind}^G_K = \sum_{x \in [H\backslash G/K]} \mathrm{Ind}^H_{H \cap {}^xK} \circ \mathrm{Iso}(\gamma_x) \circ \mathrm{Res}^K_{H^x \cap K}\ ,$$

   where $[H\backslash G/K]$ is a set of representatives of (H,K)-double cosets in G, and $\gamma_x : H^x \cap K \to H \cap {}^xK$ is the group isomorphism induced by conjugation by x.

d. If N and M are normal subgroups of G, then

$$\mathrm{Def}_{G/N}^{G} \circ \mathrm{Inf}_{G/M}^{G} = \mathrm{Inf}_{G/NM}^{G/N} \circ \mathrm{Def}_{G/NM}^{G/M} .$$

e. If H is a subgroup of G, and if N is a normal subgroup of G, then

$$\mathrm{Def}_{G/N}^{G} \circ \mathrm{Ind}_{H}^{G} = \mathrm{Ind}_{HN/N}^{G/N} \circ \mathrm{Iso}(\varphi) \circ \mathrm{Def}_{H/H\cap N}^{H} ,$$
$$\mathrm{Res}_{H}^{G} \circ \mathrm{Inf}_{G/N}^{G} = \mathrm{Inf}_{H/H\cap N}^{H} \circ \mathrm{Iso}(\varphi^{-1}) \circ \mathrm{Res}_{HN/N}^{G/N} ,$$

where $\varphi : H/H \cap N \to HN/N$ is the canonical group isomorphism.

f. If H is a subgroup of G, if N is a normal subgroup of G, and if $N \leq H$, then

$$\mathrm{Res}_{H/N}^{G/N} \circ \mathrm{Def}_{G/N}^{G} = \mathrm{Def}_{H/N}^{H} \circ \mathrm{Res}_{H}^{G}$$
$$\mathrm{Ind}_{H}^{G} \circ \mathrm{Inf}_{H/N}^{H} = \mathrm{Inf}_{G/N}^{G} \circ \mathrm{Ind}_{H/N}^{G/N} .$$

3. *Triviality conditions:* If G is a group, then

$$\mathrm{Res}_{G}^{G} = \mathrm{Id} , \quad \mathrm{Ind}_{G}^{G} = \mathrm{Id} , \quad \mathrm{Def}_{G/\mathbf{1}}^{G} = \mathrm{Id} , \quad \mathrm{Inf}_{G/\mathbf{1}}^{G} = \mathrm{Id} ,$$

$$\mathrm{Iso}(\varphi) = \mathrm{Id} , \quad \text{if } \varphi \text{ is an inner automorphism} .$$

1.1.4. Simplifications. So at this point, there is a rather complicate formalism involving the natural operations introduced in 1.1.2 and relations between them. The first observation that allows for a simplification, is that for each of the operations of 1.1.2, the map $R_{\mathbb{F}}(G) \to R_{\mathbb{F}}(H)$ is induced by a functor sending an $\mathbb{F}G$-module M to the $\mathbb{F}H$-module $L \otimes_{\mathbb{F}G} M$, where L is some finite dimensional $(\mathbb{F}H, \mathbb{F}G)$-bimodule:

- When H is a subgroup of G, and M is an $\mathbb{F}G$-module, then $\mathrm{Res}_{H}^{G} M \cong \mathbb{F}G \otimes_{\mathbb{F}G} M$, so $L = \mathbb{F}G$ in this case, for the $(\mathbb{F}H, \mathbb{F}G)$-bimodule structure given by left multiplication by elements of $\mathbb{F}H$ and right multiplication by elements of $\mathbb{F}G$.
- In the same situation, if N is an $\mathbb{F}H$-module, then $\mathrm{Ind}_{H}^{G} N \cong \mathbb{F}G \otimes_{\mathbb{F}H} N$, so $L = \mathbb{F}G$ again, but with its $(\mathbb{F}G, \mathbb{F}H)$-bimodule structure given by left multiplication by $\mathbb{F}G$ and right multiplication by $\mathbb{F}H$.
- If $\varphi : G \to H$ is a group isomorphism, and M is an $\mathbb{F}G$-module, then the image of M by $\mathrm{Iso}(\varphi)$ is the $\mathbb{F}H$-module $\mathrm{Iso}(\varphi)(M) \cong \mathbb{F}H \otimes_{\mathbb{F}G} M$, so $L = \mathbb{F}H$ in this case, for the $(\mathbb{F}H, \mathbb{F}G)$-bimodule structure given by left multiplication by $\mathbb{F}H$, and by first taking images by φ of elements of $\mathbb{F}G$, and then multiplying on the right.
- If N is a normal subgroup of G, and $H = G/N$, then the inflated module from $\mathbb{F}H$ to $\mathbb{F}G$ of the $\mathbb{F}H$-module V is isomorphic to $\mathbb{F}H \otimes_{\mathbb{F}H} V$, so

$L = \mathbb{F}H$ in this case, with the $(\mathbb{F}G, \mathbb{F}H)$-bimodule structure given by multiplication on the right by $\mathbb{F}H$, and projection from $\mathbb{F}G$ onto $\mathbb{F}H$, followed by left multiplication.

- In the same situation, the module of coinvariants by N on the $\mathbb{F}G$-module M is isomorphic to $\mathbb{F}H \otimes_{\mathbb{F}G} M$, so $L = \mathbb{F}H$ in this case, with the $(\mathbb{F}H, \mathbb{F}G)$-bimodule structure obtained by reversing the actions in the previous case.

The second observation is that in each case, the $(\mathbb{F}H, \mathbb{F}G)$-bimodule L is actually a permutation bimodule: there exist an $\mathbb{F}$-basis U of L which is globally invariant under the action of both H and G, i.e. such that $hUg = U$, for any $h \in H$ and $g \in G$. In particular, the set U is endowed with a left H-action, and a right G-action, which commute, i.e. such that $(hu)g = h(ug)$ for any $h \in H$, $u \in U$ and $g \in G$. Such a set is called *an (H, G)-biset.*

Conversely, if U is a finite (H, G)-biset, then the finite dimensional $\mathbb{F}$-vector space $\mathbb{F}U$ with basis U inherits a natural structure of $(\mathbb{F}H, \mathbb{F}G)$-bimodule. If the functor $M \mapsto L \otimes_{\mathbb{F}G} M$ from $\mathbb{F}G$-modules to $\mathbb{F}H$-modules is exact, it induces a group homomorphism $R_{\mathbb{F}}(G) \to R_{\mathbb{F}}(H)$, that will be denoted by $R_{\mathbb{F}}(U)$. The exactness condition means that the module $\mathbb{F}U$ is flat as a right $\mathbb{F}G$-module. Equivalently, since it is finitely generated, it is a projective $\mathbb{F}G$-module. Using Higman's criterion (see [45] III 14.4 Lemme 20), it is easy to see that this is equivalent to say that for each $u \in U$, the order of the stabilizer of u in G is not divisible by the characteristic l of $\mathbb{F}$. In this case, the biset U will be called *right l-free*. Note that if $l = 0$, this conditions is always fulfilled.

1.1.5. Formalism. Now the situation is the following: to each finite group G is associated an abelian group $R_{\mathbb{F}}(G)$. If G and H are finite groups, then to any a finite right l-free (H, G)-biset U corresponds a group homomorphism $R_{\mathbb{F}}(U) : R_{\mathbb{F}}(G) \to R_{\mathbb{F}}(H)$, with the following properties:

1. Let G and H be finite groups, and let U_1 and U_2 be finite right l-free (H, G)-bisets. If U_1 and U_2 are isomorphic as bisets, i.e. if there exists a bijection $f : U_1 \to U_2$ such that $f(hug) = hf(u)g$ for any $h \in H$, $u \in U_1$ and $g \in G$, then $R_{\mathbb{F}}(U_1) = R_{\mathbb{F}}(U_2)$. This is because the $(\mathbb{F}H, \mathbb{F}G)$-bimodules $\mathbb{F}U_1$ and $\mathbb{F}U_2$ are isomorphic in this case. This first property can be summarized as

   (B1) $$U_1 \cong U_2 \Rightarrow R_{\mathbb{F}}(U_1) = R_{\mathbb{F}}(U_2) \ .$$

2. If G and H are finite groups, and if U and U' are finite right l-free (H, G)-bisets, then $R_{\mathbb{F}}(U \sqcup U') = R_{\mathbb{F}}(U) + R_{\mathbb{F}}(U')$, where $U \sqcup U'$ is the disjoint union of U and U', endowed with the obvious (H, G)-biset structure. Indeed, the $(\mathbb{F}H, \mathbb{F}G)$-bimodules $\mathbb{F}(U \sqcup U')$ and $\mathbb{F}U \oplus \mathbb{F}U'$ are isomorphic. This property can be recorded as

   (B2) $$R_{\mathbb{F}}(U \sqcup U') = R_{\mathbb{F}}(U) + R_{\mathbb{F}}(U') \ .$$

3. Let G, H, and K be finite groups, let U be a finite right l-free (H,G)-biset, and let V be a finite right l-free (K,H)-biset. Then consider the set
$$V \times_H U = (V \times U)/H \ ,$$
where the right action of H on $V \times U$ is given by $(v,u)h = (vh, h^{-1}u)$, for $v \in V$, $u \in U$, and $h \in H$. The set $V \times_H U$ is a (K,G)-biset for the action induced by $k(v,u)g = (kv, ug)$, for $k \in K$, $v \in V$, $u \in U$, and $g \in G$. Moreover, it is right l-free if V and U are: the pair (v,u) of $V \times_H U$ is invariant by $g \in G$ if and only if there exists $h \in H$ such that $vh = v$ and $hu = ug$. Since V is l-free, there is an integer m, not divisible by l, such that $h^m = 1$. But then $h^m u = ug^m = u$, and since U is l-free, there is an integer m', not divisible by l, such that $(g^m)^{m'} = g^{mm'} = 1$. In this situation, one checks easily that

   (B3) $$R_{\mathbb{F}}(V) \circ R_{\mathbb{F}}(U) = R_{\mathbb{F}}(V \times_H U) \ .$$

   This is because the $(\mathbb{F}K, \mathbb{F}G)$-bimodules $\mathbb{F}V \otimes_{\mathbb{F}H} \mathbb{F}U$ and $\mathbb{F}(V \times_H U)$ are isomorphic.
4. Finally, if G is a finite group, and if Id_G is the set G, viewed as a (G,G)-biset for left and right multiplication, then Id_G is left and right free, hence right l-free, and $R_{\mathbb{F}}(\mathrm{Id}_G)$ is the identity map: this is because the functor $\mathbb{F}G \otimes_{\mathbb{F}G} -$ is isomorphic to the identity functor on the category of $\mathbb{F}G$-modules. Thus:

   (B4) $$R_{\mathbb{F}}(\mathrm{Id}_G) = \mathrm{Id}_{R_{\mathbb{F}}(G)} \ .$$

This formalism of maps associated to bisets yields a nice way to encode all the relations listed in Sect. 1.1.3: more precisely, the triviality conditions follow from Properties (B4) and (B1). For transitivity and commutation conditions, the left hand side can always be expressed as $R_{\mathbb{F}}(V) \circ R_{\mathbb{F}}(U)$, where V is a (K,H)-biset and U is an (H,G)-biset, for suitable finite groups K, H, and G. By property (B3), this is equal to $R_{\mathbb{F}}(V \times_H U)$.

Now the right hand side of the transitivity conditions is of the form $R_{\mathbb{F}}(W)$, where W is some (K,G)-biset, and one checks easily in this case that the (K,G)-bisets $V \times_H U$ and W are isomorphic. So the transitivity conditions follow from property (B1).

Similarly, the right hand side of the commutation conditions can always be written as a composition of two or three maps of the form $R_{\mathbb{F}}(T_i)$, for suitable bisets T_i, or a sum of such compositions in the case of the Mackey formula. In any case, using properties (B2) and (B3), this right hand side can always we written as $R_{\mathbb{F}}(W)$, for a suitable (K,G)-biset W, and the corresponding relation follows from a biset isomorphism $V \times_H U \cong W$, using property (B1). For example, the Mackey formula

$$\mathrm{Res}_H^G \circ \mathrm{Ind}_K^G = \sum_{x \in [H\backslash G/K]} \mathrm{Ind}_{H\cap {}^xK}^H \circ \mathrm{Iso}(\gamma_x) \circ \mathrm{Res}_{H^x \cap K}^K ,$$

when H and K are subgroups of the group G, can be seen as a translation of the isomorphism of (H, K)-bisets

$$G \times_G G \cong \bigsqcup_{x \in [H\backslash G/K]} H \times_{H\cap {}^xK} \Gamma_x \times_{H^x\cap K} K ,$$

where Γ_x is the $(H \cap {}^xK, H^x \cap K)$-biset associated to the group isomorphism $H^x \cap K \to H \cap {}^xK$ given by conjugation by x on the left. This biset isomorphism is nothing but the decomposition of the (H, K)-biset $G \times_G G \cong G$ as a disjoint union

$$G = \bigsqcup_{x \in [H\backslash G/K]} HxK$$

of its (H, K) double cosets, keeping track of the (H, K)-biset structure of each orbit.

1.2. Other Examples

1.2.1. Groups of Projective Modules. Let $\mathbb{F}$ be a field. If G is a finite group, denote by $P_{\mathbb{F}}(G)$ *the group of finitely generated projective* $\mathbb{F}G$-*modules*. Recall that $P_{\mathbb{F}}(G)$ is the quotient of the free abelian group on the set of isomorphism classes of finitely generated projective $\mathbb{F}G$-modules, by the subgroup generated by the elements of the form $[P \oplus Q] - [P] - [Q]$, where P and Q are two such modules, and $[P]$ denotes the isomorphism class of P.

If H is another finite group, and U is a finite (H, G)-biset, a natural question is to ask if the functor $M \mapsto \mathbb{F}U \otimes_{\mathbb{F}G} M$ maps a projective $\mathbb{F}G$-module M to a projective $\mathbb{F}H$-module. In this case in particular, it maps the module $\mathbb{F}G$ to a projective $\mathbb{F}H$-module, hence $\mathbb{F}U$ is a projective $\mathbb{F}H$-module. Conversely, if $\mathbb{F}U$ is a projective $\mathbb{F}H$-module, and if M is a projective $\mathbb{F}G$-module, then $\mathbb{F}U \otimes_{\mathbb{F}G} M$ is a projective $\mathbb{F}H$-module: indeed M is a direct summand of some free module $(\mathbb{F}G)^{(I)}$, where I is some set. Thus $\mathbb{F}U \otimes_{\mathbb{F}G} M$ is a direct summand of $\mathbb{F}U \otimes_{\mathbb{F}G} (\mathbb{F}G)^{(I)} \cong (\mathbb{F}U)^{(I)}$, which is a projective $\mathbb{F}H$-module.

Using Higman's criterion, it is easy to see that $\mathbb{F}U$ is a projective $\mathbb{F}H$-module if and only if the biset U is *left l-free*, i.e. if for any $u \in U$, the order of the stabilizer of u in H is not divisible by l. So if U is a finite left l-free (H, G)-biset, the functor $P \mapsto \mathbb{F}U \otimes_{\mathbb{F}G} P$ maps a finitely projective $\mathbb{F}G$-module to a finitely generated projective $\mathbb{F}H$-module. Since this functor preserves direct sums, it induces a map $P_{\mathbb{F}}(U) : P_{\mathbb{F}}(G) \to P_{\mathbb{F}}(H)$.

Now the situation is similar to the case of representation groups, except that the condition "right l-free" is replaced by "left l-free": to each finite group G is associated an abelian group $P_{\mathbb{F}}(G)$, and to any finite *left l-free* (H,G)-biset U is associated a group homomorphism $P_{\mathbb{F}}(U)$ from $P_{\mathbb{F}}(G)$ to $P_{\mathbb{F}}(H)$. These operations are easily seen to have properties (B1)–(B4).

Considering the formalization procedure of Sect. 1.1 backwards, it means that for the correspondence $G \mapsto P_{\mathbb{F}}(G)$, there are natural operations of restriction, induction, deflation, and transport by group isomorphism, and also an inflation operation $\mathrm{Inf}_{G/N}^{G} : P_{\mathbb{F}}(G/N) \to P_{\mathbb{F}}(G)$, which is only defined when N is a normal subgroup of G whose order is not divisible by l. All these operations satisfy the relations of Sect. 1.1.3, whenever they are defined.

1.2.2. Burnside Groups. If G is a finite group, denote by $B(G)$ the Burnside group of G, i.e. the Grothendieck group of the category of finite G-sets. Recall that $B(G)$ is the quotient of the free abelian group on the set of isomorphism classes of finite (left) G-sets, by the subgroup generated by the elements of the form $[X \sqcup Y]-[X]-[Y]$, where X and Y are finite G-sets, and $X \sqcup Y$ is their disjoint union, and $[X]$ denotes the isomorphism class of X (see Sect. 2.4 for details).

If G and H are finite groups, and if U is a finite (H,G)-biset, then the correspondence $X \mapsto U \times_G X$ from G-sets to H-sets induces a map $B(U) : B(G) \to B(H)$. One can check easily that these maps have the properties (B1)–(B4), with $R_{\mathbb{F}}$ replaced by B.

1.2.3. Remark : Let $\mathbb{F}$ be a field, of characteristic l. If X is a finite G-set, the $\mathbb{F}$-vector space $\mathbb{F}X$ with basis X has a natural $\mathbb{F}G$-module structure, induced by the action of G on X. The construction $X \mapsto \mathbb{F}X$ maps disjoint unions of G-sets to direct sums of $\mathbb{F}G$-modules, so it induces a map

$$\chi_{\mathbb{F},G} : B(G) \to R_{\mathbb{F}}(G) \; ,$$

called the *linearization morphism* at the group G. These maps are compatible with the maps $B(U)$ and $R_{\mathbb{F}}(U)$ corresponding to bisets, in the following sense: if G and H are finite groups, and if U is a finite right l-free (H,G)-biset, then the diagram

$$\begin{array}{ccc} B(G) & \xrightarrow{\chi_{\mathbb{F},G}} & R_{\mathbb{F}}(G) \\ {\scriptstyle B(U)}\downarrow & & \downarrow{\scriptstyle R_{\mathbb{F}}(U)} \\ B(H) & \xrightarrow[\chi_{\mathbb{F},H}]{} & R_{\mathbb{F}}(H) \end{array}$$

is commutative.

1.2.4. Cohomology and Inflation Functors. Let G be a finite group, and R be a commutative ring with identity. When k is a non negative integer, *the k-th cohomology group* $H^k(G,R)$ of G with values in R is defined as the extension group $\mathrm{Ext}_{RG}^k(R,R)$. In other words

$$H^k(G,R) = \mathrm{Hom}_{D(RG)}(R, R[k]) \; ,$$

where $D(RG)$ is the derived category of RG-modules.

Now suppose that H is another finite group, and that U is a finite *right free* (H,G)-biset. This hypothesis implies that RU is free as a right RG-module. It follows that the total tensor product $RU \otimes^{\mathbb{L}}_{RG} -$ coincides with the ordinary tensor product $RU \otimes_{RG} -$. This yields a map

$$\Theta_U : \mathrm{Hom}_{D(RG)}(R, R[k]) \to \mathrm{Hom}_{D(RH)}(RU \otimes_{RG} R, RU \otimes_{RG} R[k]) \; .$$

Now $RU \otimes_{RG} R \cong R(U/G)$, and there are two maps of RH-modules

$$\varepsilon_U : R \to R(U/G) \qquad\qquad \eta_U : R(U/G) \to R$$

defined by $\varepsilon_U(1) = \sum_{uG \in U/G} uG$, and $\eta_U(uG) = 1$ for any $uG \in U/G$. By composition, this gives a map

$$H^k(U) : H^k(G,R) = \mathrm{Hom}_{D(RG)}(R, R[k]) \to \mathrm{Hom}_{D(RH)}(R, R[k]) = H^k(H,R)$$

defined by $H^k(U)(\varphi) = \eta_U[k] \circ (RU \otimes_{RG} \varphi) \circ \varepsilon_U$.

If K is another finite group, and V is a finite right free (K,H)-biset, one can check that $H^k(V) \circ H^k(U) = H^k(V \times_H U)$: this follows from the fact that since U and V are right free, the set $(V \times_H U)/G$ is in one to one correspondence with $(V/H) \times (U/G)$.

It follows easily that the maps $H^k(U)$ between cohomology groups, for finite right free bisets, have the properties (B1)–(B4), with $R_{\mathbb{F}}$ replaced by H^k.

In this case, the formalism of maps associated to bisets encodes the usual operations of restriction, transfer, transport by isomorphism, and inflation on group cohomology. The condition imposed on bisets to be right free expresses the fact that there is no natural deflation map for group cohomology, that would be compatible with the other operations in the sense of relations of Sect. 1.1.3. Group cohomology is an example of *inflation functor*. These functors have been considered by P. Symonds [47], and also by P. Webb [54], who gave their name. More recently, E. Yaraneri [59] studied the composition factors of the inflation functor $R_{\mathbb{F}}$.

1.2.5. Global Mackey Functors. It may happen that for some construction similar to the previous examples, the only operations that are naturally defined are those of restriction to a subgroup, induction from a subgroup, and transport by group isomorphism. Functors of this type are called *global Mackey functors*, as opposed to the Mackey functors for a fixed finite group G (see [51]). These global Mackey functors have also been considered by P. Webb [54]. They can be included in the general formalism of biset functors, by restricting bisets to be *left and right free*.

1.3. Biset Functors

The above examples lead to the following informal definition: let $\mathcal{D}$ be a class of finite groups, and for each G and H in $\mathcal{D}$, let $\beta(H,G)$ be a class of finite (H,G)-bisets. *A biset functor* F (for $\mathcal{D}$ and β) with values in the category R-Mod of R-modules, where R is a commutative ring with identity, consists of the following data:

1. For each $G \in \mathcal{D}$, an R-module $F(G)$.
2. For each G and H in $\mathcal{D}$, and for each finite (H,G)-biset U in $\beta(H,G)$, a map of R-modules $F(U) : F(G) \to F(H)$.

These data are subject to the following conditions:

(B1) If G and H are in $\mathcal{D}$, then $\beta(H,G)$ is closed by isomorphism of (H,G)-bisets, and

$$\forall U_1, U_2 \in \beta(H,G),\ \ U_1 \cong U_2 \Rightarrow F(U_1) = F(U_2)\ .$$

(B2) If G and H are in $\mathcal{D}$, then $\beta(H,G)$ is closed by disjoint union of (H,G)-bisets, and

$$\forall U, U' \in \beta(H,G),\ \ F(U \sqcup U') = F(U) + F(U')\ .$$

(B3) If G, H, and K are in $\mathcal{D}$, then $\beta(K,H) \times_H \beta(H,G) \subseteq \beta(K,G)$, and

$$\forall V \in \beta(K,H),\ \forall U \in \beta(H,G),\ \ F(V) \circ F(U) = F(V \times_H U)\ .$$

(B4) If $G \in \mathcal{D}$, then the (G,G)-biset Id_G is in $\beta(G,G)$, and

$$F(\mathrm{Id}_G) = \mathrm{Id}_{F(G)}\ .$$

If F and F' are such biset functors, then *a morphism of biset functors* $f : F \to F'$ is a collection of maps $f_G : F(G) \to F'(G)$, for $G \in \mathcal{D}$, such that all the diagrams

$$\begin{array}{ccc} F(G) & \xrightarrow{f_G} & F'(G) \\ {\scriptstyle F(U)}\downarrow & & \downarrow{\scriptstyle F'(U)} \\ F(H) & \xrightarrow[f_H]{} & F'(H) \end{array}$$

are commutative, where G and H are in $\mathcal{D}$, and $U \in \beta(H,G)$.

Morphisms of biset functors can be composed in the obvious way, so biset functors (for $\mathcal{D}$ and β) with values in R-Mod form a category.

Equivalently, biset functors (for $\mathcal{D}$ and β) can be seen as additive functors from some additive subcategory of the *biset category* (see Definition 3.1.1), depending on $\mathcal{D}$ and β, to the category of R-modules.

1.4. Historical Notes

The notion of biset functor and related categories has been considered by various authors, under different names: Burnside functors, global Mackey functors, globally defined Mackey functors, functors with a Mackey structure, inflation functors. I wish to thank P. Webb for the major part of the following list of references [56]:

- 1981: Haynes Miller wrote to Frank Adams, describing *the Burnside category*, which is the (non full) subcategory of the biset category in which morphisms are provided by right-free bisets.
- 1985: The Burnside category appears in print: J.F. Adams, J.H. Gunawardena, H. Miller [1].
- 1987: T. tom Dieck [53, page 278] uses the term *global Mackey functor* for functors on this category.
- 1990: I. Hambleton, L. Taylor, B. Williams [37] consider very similar categories and functors (the main difference being that morphisms in their category RG-Morita are *permutation bimodules* instead of bisets).
- 1991: P. Symonds [47] also considers Mackey functors with inflation, that he calls *functors with Mackey structure.*
- 1993: P. Webb [54] considers *global Mackey functors* and *inflation functors.*
- 1996: I consider *foncteurs d'ensembles munis d'une double action* (in french [6]), i.e. functors on specific subcategories of the (yet unnamed) biset category.
- 2000: In Sect. 6 of [8], I use the name *biset-functor* for the Burnside functor.
- 2000: P. Webb [55] uses the name *globally defined Mackey functors* for biset functors.
- 2000: J. Thévenaz and I use *a functorial approach* to study the Dade group of a finite p-group.
- 2005: In Sect. 7 of [14], I define *rational biset functors*, now called rational p-biset functors.
- 2007: E. Yalçın and I [22] use the name of *biset category.*

1.5. About This Book

This book is organized as follows: Part I exposes a few generalities on bisets and biset functors, in a rather general framework. Some details on simple biset functors can be found in Chap. 4. Part II focuses on biset functors defined on *replete subcategories* of the biset category (see Definition 4.1.7), i.e. functors with the above five type of operations, but possibly defined over some particular class of finite groups, closed under taking subquotients (see Definition 4.1.7). The special case of p-biset functors is handled in Part III, and some important applications are detailed in Chaps. 11 and 12.

It follows from this rough summary that many interesting aspects and applications of biset functors, such as connections to homotopy theory, or highest weight categories, are not handled here: the main reason for these omissions is my lack of sufficient knowledge in those fields to treat them properly. I still hope this book will be useful to anyone interested in biset functors and their applications.

1.5.1. Notation and Conventions.

- The trivial group is denoted by $\mathbf{1}$.
- If G is a group, then $H \leq G$ means that H is a subgroup of G. If H and K are subgroups of G, then $H =_G K$ means that H and K are conjugate in G, and $H \leq_G K$ means that H is conjugate in G to some subgroup of K.
- If G is a group, if $g \in G$, and $H \leq G$, then $H^g = g^{-1}Hg$ and ${}^gH = gHg^{-1}$. The normalizer of H in G is denoted by $N_G(H)$, the centralizer of G in H by $C_G(H)$, and the center of G by $Z(G)$.
- If G is a group, then s_G is the set of subgroups of G, and $[s_G]$ is a set of representatives of conjugacy classes of G on s_G.
- The cardinality of a set X is denoted by $|X|$.
- If R is a ring, then R-Mod is the category of left R-modules. If B is a $\mathbb{Z}$-module, then RB denotes the R-module $R \otimes_{\mathbb{Z}} B$.
- If $\mathcal{C}$ is a category, then $\mathrm{Ob}(\mathcal{C})$ is its class of objects, and $x \in \mathrm{Ob}(\mathcal{C})$ is often abbreviated to $x \in \mathcal{C}$, to denote that x is an object of $\mathcal{C}$.
- If $\mathcal{C}$ is a category, then $\mathcal{D} \subseteq \mathcal{C}$ means that $\mathcal{D}$ is a subcategory of $\mathcal{C}$. Subcategories need not be full, but are assumed non empty.
- If $\mathcal{C}$ is a category, then $\mathcal{C}^{op}$ denotes the opposite category of $\mathcal{C}$, i.e. the category with the same objects as $\mathcal{C}$, and morphisms reversed.

Acknowledgements: I would like to thank Laurence Barker, Mélanie Baumann, Robert Boltje, Olcay Coşkun, Zongzhu Lin, Nadia Mazza, Fumihito Oda, Kári Ragnarsson, Cathy Ricordeau-Hamer, Radu Stancu, Peter Symonds, Yugen Takegahara, Jacques Thévenaz, Peter Webb, Ergün Yalçın, Tomoyuki Yoshida, Alexander Zimmermann, for stimulating conversations and valuable comments. I also thank the MSRI for its hospitality while the last chapter of this book was written, in spring 2008.

Part I
General Properties

Chapter 2
G-Sets and (H, G)-Bisets

This chapter is devoted to recalling the basic definitions and properties of G-sets and (H, G)-bisets, and the associated Burnside groups.

2.1. Left G-Sets and Right G-Sets

2.1.1. Definition : *If G is a group, denote by G^{op}* the opposite group*: the underlying set of G^{op} is the set G, and the group law in G^{op} is defined by*

$$\forall g, h \in G, \quad gh \text{ (in } G^{op}) = hg \text{ (in } G) \ .$$

2.1.2. Definition : *Let G be a group. A* left G-set *(resp.* a right G-set*) is a pair (X, r), where X is a set, and r is a group homomorphism from G (resp. from the opposite group G^{op}) to the group $\mathfrak{S}(X)$ of permutations of X.*

Equivalently, a left G-set X is a set equipped with *a left G-action*, i.e. a map $G \times X \to X$ (denoted by $(g, x) \mapsto g \cdot x$, or $(g, x) \mapsto gx$) such that the following conditions hold:

1. If $g, h \in G$ and $x \in X$, then $g \cdot (h \cdot x) = (gh) \cdot x$.
2. If $x \in X$, then $1_G \cdot x = x$, where 1_G is the identity element of G.

2.1.3. The correspondence between the two points of view is the following: if (X, r) is a G-set in the sense of Definition 2.1.2, then set $g \cdot x = r(g)(x)$. Conversely, if the map $(g, x) \mapsto g \cdot x$ is given, define $r(g)$, for $g \in G$, as the map $x \mapsto g \cdot x$ from X to itself.

2.1.4. Similarly, a right G-set X is a set equipped with *a right G-action*, i.e. map $(x, g) \in X \times G \mapsto x \cdot g \in X$, such that

1. If $g, h \in G$ and $x \in X$, then $(x \cdot g) \cdot h = x \cdot (gh)$.
2. If $x \in X$, then $x \cdot 1_G = x$.

S. Bouc, *Biset Functors for Finite Groups*, Lecture Notes in Mathematics 1990, DOI 10.1007/978-3-642-11297-3_2, © Springer-Verlag Berlin Heidelberg 2010

2.1.5. If G is a group, and if X and Y are left G-sets, *a morphism of G-sets* from X to Y is a map $f : X \to Y$ such that $f(g \cdot x) = g \cdot f(x)$, for any $g \in G$ and $x \in X$. Such a map is also called *a G-equivariant map* from X to Y, and the set of such maps is denoted by $\mathrm{Hom}_G(X, Y)$. Morphisms of G-sets can be composed: if $f : X \to Y$ and $f' : Y \to Z$ are morphisms of G-sets, then the composition $f' \circ f : X \to Z$ is a morphism of G-sets. For any G-set X, the identity map Id_X is a morphism of G-sets from X to itself.

A morphism of G-sets $f : X \to Y$ is *an isomorphism* if there exists a morphism of G-sets $f' : Y \to X$ such that $f' \circ f = \mathrm{Id}_X$ and $f \circ f' = \mathrm{Id}_Y$. A morphism of G-sets is an isomorphism if and only if it is bijective.

2.1.6. Definition : *Let G be a group.* The category of G-sets, *denoted by G-Set, is the category whose objects are G-sets and morphisms are G-equivariant maps. The composition of morphisms is the usual composition of maps.*

The category G-set is the full subcategory of G-Set whose objects are finite *G-sets.*

2.1.7. Orbits and Fixed Points. Let G be a group and X be a G-set. If H is a subgroup of G, and $x \in X$, the set

$$H \cdot x = \{y \in X \mid \exists h \in H,\ h \cdot x = y\}$$

is called *the H-orbit* of x, or *the orbit of x under H*. It is also denoted by Hx. The set of H-orbits in X is denoted by $H\backslash X$. It is a partition of X. It has a natural structure of left $N_G(H)/H$-set, defined by

$$\forall nH \in N_G(H)/H,\ \forall x \in X,\ \ nH \cdot Hx = H \cdot nx\ .$$

The set of fixed points of H on X is the set

$$X^H = \{x \in X \mid \forall h \in H,\ h \cdot x = x\}\ .$$

It is also a left $N_G(H)/H$-set, for the action defined by

$$\forall nH \in N_G(H)/H,\ \forall x \in X,\ \ nH \cdot x = n \cdot x\ .$$

Similarly, if now X is a right G-set, one can define the orbit $x \cdot H$ of $x \in X$ under H. The set of these orbits is denoted by X/H, and the set of fixed points of H on X is again denoted by X^H. The sets X/H and X^H are right $N_G(H)/H$-sets.

2.1.8. Transitive G-Sets and Stabilizers. A (left or right) G-set X is called *transitive* if there is a single G-orbit on X. If X is a transitive left G-set, then for any element of X, the map

$$m_x : g \in G \mapsto g \cdot x \in X$$

is surjective. Moreover, two elements g and g' of G have the same image by m_x if and only if $g \in g'G_x$, where

$$G_x = \{g \in G \mid g \cdot x = x\}$$

is *the stabilizer* of x in G. It follows that m_x induces a bijection $\overline{m}_x$ from *the set G/G_x of left cosets of G_x in G* to X. Conversely, if H is any subgroup of G, the set G/H of left cosets of H in G is a G-set, for the action defined by

$$\forall g \in G,\ \forall \gamma H \in G/H,\ \ g \cdot \gamma H = g\gamma H\ ,$$

and the above bijection $\overline{m}_x : G/G_x \to X$ is an isomorphism of G-sets. This shows the first Assertion of the following lemma:

2.1.9. Lemma : *Let G be a group.*

1. *Any transitive left G-set is isomorphic to G/H, for some subgroup H of G.*
2. *If H and K are subgroups of G, then the G-sets G/H and G/K are isomorphic if and only if H and K are conjugate in G.*

Proof: If $K = {}^gH$, then the map $xK \in G/K \mapsto xKg = xgH \in G/H$ is an isomorphism of G-sets. Conversely, if $f : G/K \to G/H$ is an isomorphism of G-sets, then there exists $g \in G$ such that $f(K) = gH$, and then $f(xK) = xgH$ for any $x \in G$.

Since $kK = K$ for any $k \in K$, it follows that $kgH = gH$, hence $k^g \in H$. Thus $K^g \leq H$. Now if $h \in H$, then

$$f({}^ghK) = {}^ghgH = ghH = gH = f(K)\ .$$

It follows that ${}^gh \in K$, hence ${}^gH \leq K$, or $H \leq K^g$. Thus $H = K^g$, as was to be shown. □

2.2. Operations on G-Sets

2.2.1. Disjoint Union and Cartesian Product. If I is a set, and if $(X_i)_{i \in I}$ is a family of G-sets indexed by I, then the disjoint union $\bigsqcup_{i \in I} X_i$ is a G-set, and it is a coproduct of the family $(X_i)_{i \in I}$ in the category G-Set (and in the category G-set if I and all the X_i's are finite).

2.2.2. Lemma : *Let G be a group, and X be a left G-set. If $[G\backslash X]$ is a set of representatives of the G-orbits in X, then the map*

$$gG_x \in \bigsqcup_{x\in[G\backslash X]} G/G_x \mapsto g\cdot x \in X$$

is an isomorphism of G-sets.

Proof: This is obvious. □

In the same situation, the cartesian product $\prod\limits_{x\in I} X_i$ is also a G-set, for the diagonal G-action defined by $g\cdot(x_i)_{i\in I} = (g\cdot x_i)_{i\in I}$. It is a product of the family $(X_i)_{i\in I}$ in the category G-Set (and in the category G-set if I and all the X_i's are finite).

2.2.3. Maps of G-Sets. If X and Y are G-sets, the set $\mathrm{Map}(X,Y)$ of all maps from X to Y is a G-set for the following action:

$$\forall f\in \mathrm{Map}(X,Y),\ \forall g\in G,\ \forall x\in X,\ \ (g\cdot f)(x) = g\cdot f(g^{-1}\cdot x)\ .$$

With this notation, the set $\mathrm{Hom}_G(X,Y)$ is the set of fixed points under the action of G on $\mathrm{Map}(X,Y)$.

For a fixed G-set X, the correspondence $Z\mapsto \mathrm{Map}(X,Z)$ has an obvious structure of functor from G-Set to itself, and this functor is right adjoint to the functor $Y\mapsto Y\times X$, i.e. there are bijections

$$\mathrm{Hom}_G(Y\times X, Z) \cong \mathrm{Hom}_G\big(Y, \mathrm{Map}(X,Z)\big)\ ,$$

which are natural in Y and Z.

2.3. Bisets

2.3.1. Definition : *Let G and H be groups. Then* an (H,G)-biset *is a left $(H\times G^{op})$-set.*

So equivalently, an (H,G)-biset U is both a left H-set and a right G-set, such that the H-action and the G-action commute, i.e.

$$\forall h\in H,\ \forall u\in U,\ \forall g\in G,\ \ (h\cdot u)\cdot g = h\cdot(u\cdot g)\ .$$

This element will simply be denoted by $h\cdot u\cdot g$, or hug.

2.3.2. All notions which apply to G-sets also apply to (H,G)-bisets: one can consider disjoint union or products of bisets. If U and V are (H,G)-bisets, then a biset homomorphism f from U to V is a map $f : U \to V$ such that $f(h \cdot u \cdot g) = h \cdot f(u) \cdot g$, for any $h \in H$, $u \in U$, $g \in G$.

If U is an (H, G)-biset, then the set $(H \times G^{op})\backslash U$ is called *the set of* (H, G)-*orbits on* U, and it is denoted by $H\backslash U/G$. The biset U is called *transitive* if $H\backslash U/G$ has cardinality 1. Equivalently, for any elements $u, v \in U$, there exists $(h, g) \in H \times G$ such that $h \cdot u \cdot g = v$.

2.3.3. Example (identity bisets) : If G is a group, then the set G is a (G, G)-biset for the left and right actions of G on itself by multiplication. The biset is called *the identity* (G, G)*-biset*, and denoted by Id_G. More generally, if H is a subgroup of G, then the set G/H is a $\big(G, N_G(H)/H\big)$-biset, and $H\backslash G$ is a $\big(N_G(H)/H, G\big)$-biset.

2.3.4. Lemma : *Let G and H be groups.*

1. *If L is a subgroup of $H \times G$, then the set $(H \times G)/L$ is a transitive (H, G)-biset for the actions defined by*

$$\forall h \in H,\ \forall (b, a)L \in (H \times G)/L,\ \forall g \in G,\quad h \cdot (b, a)L \cdot g = (hb, g^{-1}a)L\ .$$

2. *If U is an (H, G)-biset, choose a set $[H\backslash U/G]$ of representatives of (H, G)-orbits on U. Then there is an isomorphism of (H, G)-bisets*

$$U \cong \bigsqcup_{u \in [H\backslash U/G]} (H \times G)/L_u\ ,$$

where $L_u = (H, G)_u$ is the stabilizer of u in $H \times G$, *i.e. the subgroup of $H \times G$ defined by*

$$(H, G)_u = \{(h, g) \in H \times G \mid h \cdot u = u \cdot g\}\ .$$

In particular, any transitive (H, G)-biset is isomorphic to $(H \times G)/L$, for some subgroup L of $H \times G$.

Proof: Assertion 1 is straightforward, and Assertion 2 is a reformulation of Lemma 2.2.2. □

2.3.5. Example : Let $f : G \to H$ be a group homomorphism. Then the set H has an (H, G)-biset structure given by

$$\forall h, k \in H,\ \forall g \in G,\quad h \cdot k \cdot g = hkf(g)\ .$$

This biset is isomorphic to $(H \times G)/\Delta_f(G)$, where $\Delta_f(G)$ is the graph of f

$$\Delta_f(G) = \{\big(f(g), g\big) \mid g \in G\}\ .$$

2.3.6. Definition : *Let G and H be groups. If U is an (H,G)-biset, then* the opposite biset U^{op} *is the (G,H)-biset equal to U as a set, with actions defined by*

$$\forall g \in G,\ \forall u \in U,\ \forall h \in H,\ \ g \cdot u \cdot h \text{ (in } U^{op}) = h^{-1}ug^{-1} \text{ (in } U) \ .$$

2.3.7. Example : If H is a subgroup of G, then the map $xH \mapsto Hx^{-1}$ is an isomorphism of $\big(G, N_G(H)/H\big)$-bisets from G/H to $(H\backslash G)^{op}$.

2.3.8. Example (opposite subgroup) : If G and H are groups, and L is a subgroup of $H \times G$, then the *opposite subgroup* $L^{\diamond}$ is the subgroup of $G \times H$ defined by

$$L^{\diamond} = \{(g,h) \in G \times H \mid (h,g) \in L\} \ .$$

With this notation, there is an isomorphism of (G,H)-bisets

$$\big((H \times G)/L\big)^{op} \cong (G \times H)/L^{\diamond} \ .$$

2.3.9. Elementary Bisets. Let G be a group. The following examples of bisets are fundamental:

- If H is a subgroup of G, then the set G is an (H,G)-biset for the actions given by left and right multiplication in G. It is denoted by Res_H^G (where Res means *restriction*).
- In the same situation, the set G is a (G,H)-biset for the actions given by left and right multiplication in G. It is denoted by Ind_H^G (where Ind means *induction*).
- If $N \trianglelefteq G$, and if $H = G/N$, the set H is a (G,H)-biset, for the right action of H by multiplication, and the left action of G by projection to H, and then left multiplication in H. It is denoted by Inf_H^G (where Inf means *inflation*).
- In the same situation, the set H is an (H,G)-biset, for the left action of H by multiplication, and the right action of G by projection to H, and then right multiplication in H. It is denoted by Def_H^G (where Def means *deflation*).
- If $f : G \to H$ is a group isomorphism, then the set H is an (H,G)-biset, for the left action of H by multiplication, and the right action of G given by taking image by f, and then multiplying on the right in H. It is denoted by $\mathrm{Iso}(f)$, or Iso_G^H if the isomorphism f is clear from the context.

2.3.10. Composition of Bisets.

2.3.11. Definition : *Let G, H, and K be groups. If U is an (H,G)-biset, and V is a (K,H)-biset,* the composition of V and U *is the set of H-orbits on the cartesian product $V \times U$, where the right action of H is defined by*

$$\forall (v,u) \in V \times U,\ \forall h \in H,\ \ (v,u)\cdot h = (v\cdot h, h^{-1}\cdot u)\ .$$

It is denoted by $V \times_H U$. The H-orbit of $(v,u) \in V \times U$ is denoted by $(v,_{_H} u)$. The set $V \times_H U$ is a (K,G)-biset for the actions defined by

$$\forall k \in K,\ \forall (v,_{_H} u) \in V \times_H U,\ \forall g \in G,\ \ k\cdot(v,_{_H} u)\cdot g = (k\cdot v,_{_H} u\cdot g)\ .$$

2.3.12. Definition : *Let G be a group.* A section (T,S) *of G is a pair of subgroups of G such that $S \trianglelefteq T$.* The associated subquotient *of G is the factor group T/S.*

2.3.13. Example ($\mathrm{Defres}^G_{T/S}$ **and** $\mathrm{Indinf}^G_{T/S}$) **:** Set G be a group, and let (T,S) be a section of G. Then there is an isomorphism of $(G,T/S)$-bisets

$$\mathrm{Ind}_T^G \times_T \mathrm{Inf}^T_{T/S} \xrightarrow{\cong} G/S$$

sending $(g,_{_T} tS)$ to gtS. For this reason, the $(G,T/S)$-biset G/S will be denoted by $\mathrm{Indinf}^G_{T/S}$. Similarly, there is an isomorphism of $(T/S,G)$-bisets

$$\mathrm{Def}^T_{T/S} \times_T \mathrm{Res}^G_T \xrightarrow{\cong} S\backslash G$$

sending $(tS,_{_T} g)$ to Stg. For this reason, the $(T/S,G)$-biset $S\backslash G$ will be denoted by $\mathrm{Defres}^G_{T/S}$.

2.3.14. Proposition : *Let G, H, K, and L be groups.*

1. *If U is an (H,G)-biset, if V is a (K,H)-biset, and if W is an (L,K)-biset, then there is a canonical isomorphism of (L,G)-bisets*

$$W \times_K (V \times_H U) \xrightarrow{\cong} (W \times_K V) \times_H U$$

given by $\big(w,_{_K} (v,_{_H} u)\big) \mapsto \big((w,_{_K} v),_{_H} u\big)$, for all $(w,v,u) \in W \times V \times U$.

2. *If U is an (H,G)-biset and V is a (K,H)-biset, then there is a canonical isomorphism of (G,K)-bisets*

$$(V \times_H U)^{op} \xrightarrow{\cong} U^{op} \times_H V^{op}$$

given by $(v,_{_H} u) \mapsto (u,_{_H} v)$.

3. *If* U *and* U' *are* (H,G)*-bisets and if* V *and* V' *are* (K,H)*-bisets, then there are canonical isomorphisms of* (K,G)*-bisets*

$$V \times_H (U \sqcup U') \cong (V \times_H U) \sqcup (V \times_H U')$$
$$(V \sqcup V') \times_H U \cong (V \times_H U) \sqcup (V' \times_H U)\ .$$

4. *If* U *is an* (H,G)*-biset, then there are canonical* (H,G)*-biset isomorphisms*

$$\mathrm{Id}_H \times_H U \xrightarrow{\cong} U \xleftarrow{\cong} U \times_G \mathrm{Id}_G$$

given by $(h,_{_H} u) \mapsto h \cdot u$ *and* $(u,_{_G} g) \mapsto u \cdot g$*, for all* $(h,u,g) \in H \times U \times G$.

Proof: This is straightforward. □

2.3.15. Remark : Assertion 1 allows for the notation $W \times_K V \times_H U$ instead of $(W \times_K V) \times_H U$ or $W \times_K (V \times_H U)$, and for the similar notation $(w,_{_K} v,_{_H} u)$ for the element $\big((w,_{_K} v),_{_H} u\big)$ of $(W \times_K V) \times_H U$.

2.3.16. Notation : *Let* G *and* H *be groups, and* U *be an* (H,G)*-biset.*

1. *If* L *is a subgroup of* H*, and* u *an element of* U*, then set*

$$L^u = \{g \in G \mid \exists l \in L,\ \ l \cdot u = u \cdot g\}\ .$$

Then L^u *is a subgroup of* G*. In particular* $\mathbf{1}^u$ *is the stabilizer of* u *in* G*.*

2. *If* K *is a subgroup of* G*, then set*

$${}^uK = \{h \in H \mid \exists k \in K,\ \ h \cdot u = u \cdot k\}\ .$$

Then ${}^uK$ *is a subgroup of* H*. In particular* ${}^u\mathbf{1}$ *is the stabilizer of* u *in* H*.*

2.3.17. Remark : If G is a group, if $U = G$ is the identity biset, and if $H \leq G$, then $H^u = u^{-1}Hu$, for $u \in G$, and ${}^uH = uHu^{-1}$. So Notation 2.3.16 is a generalization of the usual notation for conjugation of subgroups.

2.3.18. Proposition : *Let* G *and* H *be groups, and let* U *be an* (H,G)*-biset.*

1. *If* $u \in U$ *and* (T,S) *is a section of* H*, then* (T^u, S^u) *is a section of* G*. If* (Y,X) *is a section of* G*, then* $({}^uY, {}^uX)$ *is a section of* H*.*
2. *In particular, if* $u \in U$*, then* $\mathbf{1}^u \trianglelefteq H^u$ *and* ${}^u\mathbf{1} \trianglelefteq {}^uG$*, and there is a canonical group isomorphism*

$$\bar{c}_u : H^u/\mathbf{1}^u \xrightarrow{\cong} {}^uG/{}^u\mathbf{1} \ ,$$

defined by $\bar{c}_u(g\ \mathbf{1}^u) = h\ {}^u\mathbf{1}$, *where* $g \in H^u$, *and* $h \in H$ *is such that* $h \cdot u = u \cdot g$.

3. *The stabilizer* $(H,G)_u$ *of* u *in* $H \times G$ *is equal to the set of pairs* (h,g) *in* ${}^uG \times H^u$ *such that* $h\ {}^u\mathbf{1} = \bar{c}_u(g\ {}^u\mathbf{1})$.
4. *The group* ${}^u\mathbf{1} \times \mathbf{1}^u$ *is a normal subgroup of* $(H,G)_u$, *and there are canonical group isomorphisms*

$$ {}^uG/{}^u\mathbf{1} \xleftarrow{\cong} (H,G)_u/({}^u\mathbf{1} \times \mathbf{1}^u) \xrightarrow{\cong} H^u/\mathbf{1}^u \ ,$$

defined by $(h,g)({}^u\mathbf{1} \times \mathbf{1}^u) \mapsto h\ {}^u\mathbf{1}$ *and* $(h,g)({}^u\mathbf{1} \times \mathbf{1}^u) \mapsto g\ \mathbf{1}^u$.

Proof: Let $u \in U$ and (T,S) be a section of H. Then obviously $S^u \leq T^u$. If $g \in T^u$ and $g' \in S^u$, then there exist $t \in T$ and $s \in S$ such that $t \cdot u = u \cdot g$ and $s \cdot u = u \cdot g'$. Thus

$$u \cdot gg'g^{-1} = t \cdot u \cdot g'g^{-1} = ts \cdot u \cdot g^{-1} = tst^{-1} \cdot u \ ,$$

so $gg'g^{-1} \in S^u$, since $tst^{-1} \in S$. This shows that $S^u \trianglelefteq T^u$. The last part of Assertion 1 is similar.

In particular $\mathbf{1}^u \trianglelefteq H^u$. Now if $g \in H^u$ and $h \in H$ are such that $h \cdot u = u \cdot g$, then $h \in {}^uG$. If $h' \in H$ is another element such that $h' \cdot u = u \cdot g$, then $(h^{-1}h') \cdot u = u$, so $h' \in h\,{}^u\mathbf{1}$. It follows that the map

$$c_u : g \in H^u \mapsto h\ {}^u\mathbf{1} \in {}^uG/{}^u\mathbf{1} \ \text{ if } h \cdot u = u \cdot g \ ,$$

is well defined. It is straightforward to check that c_u is a group homomorphism. It is moreover surjective, since for any $h \in {}^uG$, there exists $g \in G$ with $h \cdot u = u \cdot g$, and such an element g is in H^u. Finally, the kernel of c_u is precisely equal to $\mathbf{1}^u$.

Assertion 3 is obvious, and it implies that $({}^u\mathbf{1} \times \mathbf{1}^u) \trianglelefteq (H,G)_u$. The last part of Assertion 4 is obvious. □

2.3.19. Notation : *Let* G, H *and* K *be groups. If* L *is a subgroup of* $H \times G$, *and if* M *is a subgroup of* $K \times H$, *set*

$$M * L = \{(k,g) \in K \times G \mid \exists h \in H,\ \ (k,h) \in M \ \text{ and } (h,g) \in L\} \ .$$

Then $M * L$ is obviously a subgroup of $K \times G$.

2.3.20. Lemma : *Let G, H, and K be groups, let U be an (H,G)-biset and V be a (K,H)-biset. Then if $u \in U$ and $v \in V$, the stabilizer of $(v,_H u)$ in $K \times G$ is equal to*

$$(K,G)_{(v,_H u)} = (K,H)_v * (H,G)_u \ .$$

Proof: Suppose that $(k,g) \in K \times G$ is such that $k(v,_H u) = (v,_H u)g$. Then $(kv,_H u) = (v,_H ug)$, so there exists $h \in H$ such that $kv = vh$ and $hu = ug$. Hence $(k,g) \in (K,H)_v * (H,G)_u$. Conversely, if $(k,g) \in (K,H)_v * (H,G)_u$, then there exists $h \in H$ such that $kv = vh$ and $hu = ug$. Thus

$$k(v,_H u) = (kv,_H u) = (vh,_H u) = (v,_H hu)(v,_H ug) = (v,_H u)g \ ,$$

hence $(k,g) \in (K,G)_{(v,_H u)}$. □

2.3.21. Notation : *If G and H are groups, and L is a subgroup of $H \times G$, then set*

$$\begin{aligned} p_1(L) &= \{h \in H \mid \exists g \in G, \ \ (h,g) \in L\} \\ p_2(L) &= \{g \in G \mid \exists h \in H, \ \ (h,g) \in L\} \\ k_1(L) &= \{h \in H \mid (h,1) \in L\} \\ k_2(L) &= \{g \in G \mid (1,g) \in L\} \\ q(L) &= L/\big(k_1(L) \times k_2(L)\big) \ . \end{aligned}$$

With this notation, the stabilizer in $H \times G$ of the element $u = (1,1)L$ of the biset $(H \times G)/L$ is the group L. The group H^u is equal to the projection $p_2(L)$ of L on G, and the group ${}^uG$ is equal to the projection $p_1(L)$ of L on H. The stabilizer ${}^u\mathbf{1}$ of u in H is the group $k_1(L)$ and the stabilizer $\mathbf{1}^u$ of u in G is the group $k_2(L)$. The isomorphism $\bar{c}_u$ of Proposition 2.3.18 is the map

$$gk_2(L) \in p_2(L)/k_2(L) \mapsto hk_1(L) \in p_1(L)/k_1(L), \ \ \text{for } (h,g) \in L \ .$$

Finally, Proposition 2.3.18 shows that $\big(k_1(L) \times k_2(L)\big) \trianglelefteq L$, and that there are canonical group isomorphisms

$$p_1(L)/k_1(L) \cong q(L) \cong p_2(L)/k_2(L) \ .$$

2.3.22. Lemma : *Let G, H and K be groups. Let L be a subgroup of $H\times G$, and M be a subgroup of $K\times H$.*

1. *There are exact sequences of groups*

$$\mathbf{1}\to k_1(M)\times k_2(L)\to M*L\to \big(p_2(M)\cap p_1(L)\big)/\big(k_2(M)\cap k_1(L)\big)\to \mathbf{1}\;,$$
$$\mathbf{1}\to k_1(M)\to k_1(M*L)\to \big(p_2(M)\cap k_1(L)\big)/\big(k_2(M)\cap k_1(L)\big)\to \mathbf{1}\;,$$
$$\mathbf{1}\to k_2(L)\to k_2(M*L)\to \big(k_2(M)\cap p_1(L)\big)/\big(k_2(M)\cap k_1(L)\big)\to \mathbf{1}\;.$$

2. *There are inclusions of subgroups*

$$k_1(M)\subseteq k_1(M*L)\subseteq p_1(M*L)\subseteq p_1(M)\;,$$
$$k_2(L)\subseteq k_2(M*L)\subseteq p_2(M*L)\subseteq p_2(L)\;.$$

In particular, the group $q(M*L)$ is isomorphic to a subquotient of $q(M)$ and $q(L)$.

Proof: Let $(k,g)\in M*L$. Then there exists $h\in H$ such that $(k,h)\in M$ and $(h,g)\in L$. In particular $h\in p_2(M)\cap p_1(L)$. If h' is another element of H such that $(k,h')\in M$ and $(h',g)\in L$, then $h^{-1}h'\in k_2(M)\cap k_1(L)$. So the map $\theta:(k,g)\mapsto h\big(k_2(M)\cap k_1(L)\big)$ is a well defined group homomorphism from $M*L$ to $\big(p_2(M)\cap p_1(L)\big)/\big(k_2(M)\cap k_1(L)\big)$.

This morphism is surjective: if $h\in p_2(M)\cap p_1(L)$, then there exists $k\in K$ such that $(k,h)\in M$, and there exists $g\in G$ such that $(h,g)\in L$. Thus $(k,g)\in M*L$, and $\theta\big((k,g)\big)=h\big(k_2(M)\cap k_1(L)\big)$.

Finally, if $k\in k_1(M)$ and $g\in k_2(L)$, then $(k,1)\in M$ and $(1,g)\in L$, thus $\theta\big((k,g)\big)=k_2(M)\cap k_1(L)$. In other words $k_1(M)\times k_2(L)\leq \mathrm{Ker}\;\theta$. Conversely, if $(k,g)\in\mathrm{Ker}\;\theta$, then there exist $h\in k_2(M)\cap k_1(L)$ such that $(k,h)\in M$ and $(h,g)\in L$. Thus $(1,h)\in M$, and $(k,1)=(k,h)(1,h)^{-1}\in M$. Similarly $(h,1)\in L$, and $(1,g)=(h,g)(h,1)^{-1}\in L$. It follows that $(k,g)\in k_1(M)\times k_2(L)$, thus $\mathrm{Ker}\;\theta=k_1(M)\times k_2(L)$. This shows the existence of the first exact sequence in Assertion 1.

Now an element k of K is in $k_1(M*L)$ if and only if there exists $h\in H$ such that $(k,h)\in M$ and $(h,1)\in L$. Thus $h\in p_2(M)\cap k_1(L)$. This shows that the image of the group $k_1(M*L)\times\mathbf{1}$ by the morphism θ is precisely equal to $\big(p_2(M)\cap k_1(L)\big)/\big(k_2(M)\cap k_1(L)\big)$. Moreover its intersection with the kernel $k_1(M)\times k_2(L)$ of θ is equal to $k_1(M)\times\mathbf{1}$. This yields the second exact sequence of Assertion 1. The third one is similar.

This completes the proof of Assertion 1. Assertion 2 is straightforward. □

2.3.23. Factorization of Transitive Bisets. If G and H are groups, then by Lemma 2.3.4, any (H,G)-biset is a disjoint union of transitive (H,G)-bisets, and a transitive (H,G)-biset is isomorphic to $(H\times G)/L$, for some subgroup L of $H\times G$.

2.3.24. Lemma (Mackey formula for bisets) : *Let G, H and K be groups. If L is a subgroup of $H \times G$, and if M is a subgroup of $K \times H$, then there is an isomorphism of (K,G)-bisets*

$$\big((K \times H)/M\big) \times_H \big((H \times G)/L\big) \cong \bigsqcup_{h \in [p_2(M)\backslash H/p_1(L)]} (K \times G)/(M * {}^{(h,1)}L) ,$$

where $[p_2(M)\backslash H/p_1(L)]$ is a set of representatives of double cosets.

Proof: Set $V = (K \times H)/M$ and $U = (H \times G)/L$. It is easy to check that the map

$$K\big((k,h)M,_{_H} (h',g)L\big)G \mapsto p_2(M)h^{-1}h'p_1(L)$$

is a bijection $K\backslash(V \times_H U)/G \to p_2(M)\backslash H/p_1(L)$, the inverse bijection being the map

$$p_2(M)hp_1(L) \mapsto K\big((1,1)M,_{_H} (h,1)L\big)G \; .$$

Moreover the stabilizer of $\big((1,1)M,_{_H} (h,1)L\big)$ in $K \times G$ is equal to $M * {}^{(h,1)}L$. □

Proposition 2.3.18 and the above remarks show the following, sometimes called the Goursat Lemma:

2.3.25. Lemma : *Let G and H be groups.*

1. *If (D,C) is a section of H and (B,A) is a section of G such that there exist a group isomorphism $f : B/A \to D/C$, then*

$$L_{(D,C),f,(B,A)} = \{(h,g) \in H \times G \mid h \in D,\; g \in B,\;\; hC = f(gA)\}$$

   is a subgroup of $H \times G$.
2. *Conversely, if L is a subgroup of $H \times G$, then there exists a unique section (D,C) of H, a unique section (B,A) of G, and a unique group isomorphism $f : B/A \to D/C$, such that $L = L_{(D,C),f,(B,A)}$.*

Proof: The only thing that remains to prove is the uniqueness part of Assertion 2. This is clear, since $p_1\big(L_{(D,C),f,(B,A)}\big) = D$, $k_1\big(L_{(D,C),f,(B,A)}\big) = C$, $p_2\big(L_{(D,C),f,(B,A)}\big) = B$, and $k_2\big(L_{(D,C),f,(B,A)}\big) = A$, and since the group isomorphism $f : B/A \to D/C$ is determined by $f(bA) = dC$ if $(d,b) \in L_{(D,C),f,(B,A)}$. □

The following result is elementary, but essential:

2.3.26. Lemma : *Let G and H be groups. If L is a subgroup of $H \times G$, let (D,C) and (B,A) be the sections of H and G respectively, and f be the group isomorphism $B/A \xrightarrow{\cong} D/C$ such that $L = L_{(D,C),f,(B,A)}$. Then there* *is an isomorphism of* (H,G)*-bisets*

$$(H\times G)/L \cong \mathrm{Ind}_D^H \times_D \mathrm{Inf}_{D/C}^D \times_{D/C} \mathrm{Iso}(f) \times_{B/A} \mathrm{Def}_{B/A}^B \times_B \mathrm{Res}_B^G .$$

Proof: Set $\Lambda = (H\times G)/L$, and let Γ denote the right hand side. Let φ be the map from Λ to Γ defined by

$$\varphi\big((h,g)L\big) = (h,_D C,_{D/C} C,_{B/A} A,_B g^{-1}) .$$

Let $\psi : \Gamma \to \Lambda$ be the map defined by

$$\psi\big((h,_D dC,_{D/C} d'C,_{B/A} bA,_B g)\big) = (hdd', g^{-1}b^{-1})L ,$$

for $h\in H$, $d,d'\in D$, $b\in B$ and $g\in G$.

First, the map φ is well defined: if $(h,g)\in H\times G$ and $(d,b)\in L$, then $d\in D$, $b\in B$, and $f(bA)=dC$. Thus

$$\begin{aligned}
\varphi\big((hd,gb)L\big) &= (hd,_D C,_{D/C} C,_{B/A} A,_B b^{-1}g^{-1})\\
&= (h,_D dC,_{D/C} C,_{B/A} Ab^{-1},_B g^{-1})\\
&= \big(h,_D C(dC),_{D/C} C,_{B/A} (b^{-1}A)A,_B g^{-1}\big)\\
&= \big(h,_D C,_{D/C} (dC)C(bA)^{-1},_{B/A} A,_B g^{-1}\big) ,
\end{aligned}$$

and this is equal to $(h,_D C,_{D/C} C,_{B/A} A,_B g^{-1})$ since $f(bA)=dC$.

The map ψ is also well defined: in other words, if $x\in D$, if $yC\in D/C$, if $zA\in B/A$, if $y'\in f(zA)$, and if $t\in B$, then the image by ψ of the element

$$E = (hx,_D x^{-1}dCy,_{D/C} y^{-1}d'Cy',_{B/A} z^{-1}bAt, t^{-1}g)$$

of Γ should be equal to $\psi(h,_D dC,_{D/C} d'C,_{B/A} bA,_B g)$. But:

$$\begin{aligned}
\psi(E) &= (hxx^{-1}dyy^{-1}d'y', g^{-1}tt^{-1}b^{-1}z)L\\
&= (hdd'y', g^{-1}b^{-1}z)L\\
&= (hdd', g^{-1}b^{-1})L ,
\end{aligned}$$

since $(y',z)\in L$.

Now it is easy to check that φ and ψ are biset homomorphisms. Moreover, it is clear that $\psi\circ\varphi = \mathrm{Id}_\Lambda$. Finally

$$(h,_D dC,_{D/C} d'C,_{B/A} bA,_B g) = (hdd',_D C,_{D/C} C,_{B/A} A,_B bg) ,$$

so φ is surjective. Since $\varphi\circ\psi\circ\varphi=\varphi$, it follows that $\varphi\circ\psi=\mathrm{Id}_\Gamma$, so φ and ψ are mutual inverse isomorphisms of bisets. □

2.4. Burnside Groups

Let G be a arbitrary group. At this level of generality, there are several possibilities for the definition of the Burnside group of G: it is always defined as the Grothendieck group of *some category of G-sets*, but this category depends on additional assumptions on G (the group G may be finite, compact, profinite,...). From now on, the group G will be supposed finite.

2.4.1. Definition : *Let G be a finite group.* The Burnside group $B(G)$ *of G is the Grothendieck group of the category G-*set*: it is defined as the quotient of the free abelian group on the set of isomorphism classes of finite G-sets, by the subgroup generated by the elements of the form*

$$[X \sqcup Y] - [X] - [Y]$$

where X and Y are finite G-sets, and $[X]$ denotes the isomorphism class of X.

2.4.2. Remark : The elements of $B(G)$ are sometimes called *virtual G-sets.*

2.4.3. Universal property : The Burnside group has the following universal property: if φ is a function defined on the class of finite G-sets, with values in an abelian group A, such that:

1. If X and Y are isomorphic finite G-sets, then $\varphi(X) = \varphi(Y)$.
2. For any finite G-sets X and Y

$$\varphi(X \sqcup Y) = \varphi(X) + \varphi(Y) .$$

Then there exists a unique group homomorphism $\tilde{\varphi} : B(G) \to A$ such that $\varphi(X) = \tilde{\varphi}([X])$, for any finite G-set X.

2.4.4. At this point, a natural question is to know, being given two finite G-sets X and Y, whether $[X] = [Y]$ in $B(G)$. The answer will be a consequence of the following fundamental result of Burnside ([24] Chap. XII Theorem I):

2.4.5. Theorem : [Burnside] *Let G be a finite group, and let X and Y be finite G-sets. The following conditions are equivalent:*

1. *The G-sets X and Y are isomorphic.*
2. *For any subgroup H of G, the sets of fixed points X^H and Y^H have the same cardinality.*

Proof: It is clear that (1) implies (2), since any G-set isomorphism $X \to Y$ induces a bijection $X^H \to Y^H$ on the sets of fixed points by any subgroup H of G.

To show the converse, observe that it follows from Lemma 2.2.2 that any finite G-set X can be written up to isomorphism as

$$X = \bigsqcup_{K \in [s_G]} a_K(X)\, G/K$$

where $[s_G]$ is a set of representatives of conjugacy classes of subgroups of G, for some $a_K(X) \in \mathbb{N}$. Here $a_K(X)\, G/K$ denotes the disjoint union of $a_K(X)$ copies of G/K.

Now if (2) holds, for any $H \in [s_G]$, there is an equation

$$\sum_{K \in [s_G]} \big(a_K(X) - a_K(Y)\big)|(G/K)^H| = 0 \,.$$

The matrix m of this system of equations is given by

$$m(H, K) = |(G/K)^H| = |\{x \in G/K \mid H^x \subseteq K\}|$$

for $K, H \in [s_G]$. In particular the entry $m(H, K)$ is non-zero if and only if some conjugate of H is contained in K.

If the set $[s_G]$ is given a total ordering $\preceq$ such that $H \preceq K$ implies $|H| \leq |K|$, then the matrix m is upper triangular, with non-zero diagonal coefficient $m(H, H) = |N_G(H) : H|$. In particular m is non-singular, and it follows that $a_K(X) = a_K(Y)$, for any $K \in [s_G]$, and the G-sets X and Y are isomorphic. □

2.4.6. Corollary : *Let X and Y be finite G-sets. Then $[X]$ and $[Y]$ have the same image in $B(G)$ if and only if X and Y are isomorphic.*

Proof: Indeed $[X]$ and $[Y]$ have the same image in $B(G)$ if and only if there exist positive integers $m \leq n$, finite G-sets Z_i and T_i, for $i = 1, \ldots n$, and an isomorphism of G-sets

$$X \sqcup \bigsqcup_{i=1}^{m}(Z_i \sqcup T_i) \sqcup \Big(\bigsqcup_{i=m+1}^{n} Z_i\Big) \sqcup \Big(\bigsqcup_{i=m+1}^{n} T_i\Big) \cong Y \sqcup \Big(\bigsqcup_{i=1}^{m} Z_i\Big) \sqcup \Big(\bigsqcup_{i=1}^{m} T_i\Big) \sqcup \bigsqcup_{i=m+1}^{n} (Z_i \sqcup T_i) \,.$$

Counting fixed points on each side by a subgroup H of G shows that $|X^H| = |Y^H|$. Since this holds for any H, the G-sets X and Y are isomorphic. □

2.4.7. Remark : Corollary 2.4.6 allows for an identification of the isomorphism class $[X]$ of a finite G-set X with its image in $B(G)$. This image will also be abusively simply denoted by X.

2.4.8. Remark : The Burnside group $B(G)$ has a natural ring structure, for the product defined by $[X][Y] = [X \times Y]$, for finite G-sets X and Y. The basic properties of this ring will be recalled in Sect. 2.5.

2.4.9. Definition : *Let G and H be finite groups.* The biset Burnside group *$B(H, G)$ is the Burnside group $B(H \times G^{op})$, i.e. the Grothendieck group of the category of finite (H, G)-bisets.*

The elements of $B(H, G)$ are sometimes called *virtual (H, G)-bisets.* Proposition 2.3.14 shows that virtual bisets can be composed. More precisely:

2.4.10. Notation : *Let G, H and K be finite groups.*

1. *There is a unique bilinear map*

$$\times_H : B(K, H) \times B(H, G) \to B(K, G)$$

such that $[V] \times_H [U] = [V \times_H U]$, whenever U is a finite (H, G)-biset and V is a finite (K, H)-biset.

2. *There is a unique linear map $u \in B(H, G) \mapsto u^{op} \in B(G, H)$ such that $[U]^{op} = [U^{op}]$ for any finite (H, G)-biset U.*

With this notation, Proposition 2.3.14 yields the following:

2.4.11. Proposition : *Let G, H, K and L be finite groups.*

1. *If $u \in B(H, G)$, if $v \in B(K, H)$, and if $w \in B(L, K)$, then*

$$w \times_K (v \times_H u) = (w \times_K v) \times_H u \quad \textit{in } B(L, G).$$

2. *If $u \in B(H, G)$ and $v \in B(K, H)$, then*

$$(v \times_H u)^{op} = u^{op} \times_H v^{op} \quad \textit{in } B(G, K).$$

3. *If $u, u' \in B(H, G)$ and $v, v' \in B(K, H)$, then*

$$\begin{aligned} v \times_H (u + u') &= (v \times_H u) + (v \times_H u') \\ (v + v') \times_H u &= (v \times_H u) + (v' \times_H u) \quad \textit{in } B(K, G). \end{aligned}$$

4. *If $u \in B(H, G)$, then*

$$[\mathrm{Id}_H] \times_H u = u = u \times_G [\mathrm{Id}_G] \quad \textit{in } B(H, G).$$

2.5. Burnside Rings

2.5.1. The Burnside Ring of a Finite Group. Let G be a finite group. The Burnside group $B(G)$ has a natural ring structure, for the product defined by $[X][Y] = [X \times Y]$, for finite G-sets X and Y. This ring is commutative, and the identity element is the class $[\bullet]$ of a G-set of cardinality 1.

If $H \leq G$, then there is a unique linear form $\phi_H : B(G) \to \mathbb{Z}$ such that $\phi_H([X]) = |X^H|$ for any finite G-set X. It is clear moreover that ϕ_H is a ring homomorphism, and Burnside's Theorem 2.4.5 implies that *the ghost map*

$$\phi = \prod_{H \in [s_G]} \phi_H : B(G) \to \prod_{H \in [s_G]} \mathbb{Z}$$

is injective. The ring $\prod\limits_{H \in [s_G]} \mathbb{Z}$ is called *the ghost ring*. It can be seen as *the ring of superclass functions* of G, i.e. the ring of functions from the set of all subgroups of G to $\mathbb{Z}$ which are constant on G-conjugacy classes.

The cokernel of the ghost map is finite, and has been explicitly described by Dress [33]. In particular, the ghost map $\mathbb{Q}\phi : \mathbb{Q}B(G) \to \prod_{H \in [s_G]} \mathbb{Q}$ is an algebra isomorphism, where $\mathbb{Q}B(G) = \mathbb{Q} \otimes_{\mathbb{Z}} B(G)$. This shows that $\mathbb{Q}B(G)$ is a split semi-simple commutative $\mathbb{Q}$-algebra, whose primitive idempotents are indexed by $[s_G]$. The following more precise theorem was proved by Gluck [35] and independently by Yoshida [60]:

2.5.2. Theorem : *Let G be a finite group. If H is a subgroup of G, denote by e_H^G the element of $\mathbb{Q}B(G)$ defined by*

$$e_H^G = \frac{1}{|N_G(H)|} \sum_{K \leq H} |K| \mu(K, H) [G/K] ,$$

where μ is the Möbius function of the poset of subgroups of G.

Then $e_H^G = e_K^G$ if the subgroups H and K are conjugate in G, and the elements e_H^G, for $H \in [s_G]$, are the primitive idempotents of the $\mathbb{Q}$-algebra $\mathbb{Q}B(G)$.

2.5.3. Remark : The idempotent e_H^G is the only (non zero) idempotent of $\mathbb{Q}B(G)$ such that $ue_H^G = |u^H| e_H^G$ for any $u \in \mathbb{Q}B(G)$, where $|u^H| = \mathbb{Q}\phi_H(u)$. It follows that

$$u = \sum_{H \in [s_G]} |u^H| e_H^G .$$

2.5.4. Remark : A lot more can be said about the ring structure of $B(G)$ (see, e.g. [8]). Some specific results will be used in Chaps. 5 and 11 of this book.

2.5.5. The Biset Burnside Ring of a Finite Group. Let G be a finite group. Then the biset Burnside group $B(G,G)$ has also a natural ring structure, for the product $\times_G$ defined by $[V]\times_G[U]=[V\times_G U]$, for finite (G,G)-bisets U and V. This ring is not commutative in general. The identity element is the class $[\mathrm{Id}_G]$ of the identity (G,G)-biset.

There is a natural ring homomorphism from $B(G)$ to $B(G,G)$, induced by the following construction, which makes sense for an arbitrary group G:

2.5.6. Notation : *Let G be a group, and X be a G-set. Let $\widetilde{X}$ denote the set $G\times X$, endowed with the (G,G)-biset structure defined by*

$$\forall a,b,g\in G,\ \forall x\in X,\ a\cdot(g,x)\cdot b=(agb,b^{-1}x)\ .$$

2.5.7. Remark : One can check that $\widetilde{X}$ is isomorphic to $\mathrm{Ind}_{\delta(G)}^{G\times G^{op}}\mathrm{Iso}(\delta)(X)$, where $\delta:G\to G\times G^{op}$ is the "twisted diagonal embedding" defined by $\delta(g)=(g,g^{-1})$, for $g\in G$.

2.5.8. Lemma : *Let G be a group.*

1. *For any G-sets X and Y, there is an isomorphism of (G,G)-bisets*

$$\widetilde{X\times Y}\cong\widetilde{X}\times_G\widetilde{Y}\ .$$

2. *If $X=\bullet$ is a G-set of cardinality 1, then $\widetilde{X}$ is isomorphic to the identity (G,G)-biset Id_G.*
3. *If X is a G-set, if H is a group, and U is a (G,H)-biset, then $\widetilde{X}\times_G U$ is isomorphic to $X\times U$, with (G,H)-biset structure given by*

$$\forall g\in G,\ \forall h\in H,\ \forall x\in X,\ \forall u\in U,\ \ g(x,u)h=(gx,guh)\ .$$

Similarly, if V is an (H,G)-biset, then $V\times_G\widetilde{X}$ is isomorphic to $V\times X$, with (H,G)-biset structure given by

$$\forall h\in H,\ \forall g\in G,\ \forall v\in V,\ \forall x\in X,\ \ h(v,x)g=(hvg,g^{-1}x)\ .$$

4. *For any G-sets X and Y, there is an isomorphism of G-sets*

$$\widetilde{X}\times_G Y\cong X\times Y\ .$$

5. *If G is finite, then the correspondence sending a finite G-set X to the finite (G,G)-biset $\widetilde{X}$ induces a ring homomorphism from $B(G)$ to $B(G,G)$, which preserves identity elements.*

Proof: For Assertion 1, the map

$$\alpha : \widetilde{X \times Y} \to \widetilde{X} \times_G \widetilde{Y}$$

defined by $\alpha\big((g,x,y)\big) = \big((g,x) ,_G (1,y)\big)$ is a map of (G,G)-bisets: indeed if $a, b \in G$, then

$$\begin{aligned}\alpha\big(a \cdot (g,x,y) \cdot b\big) &= \alpha\big((agb, b^{-1}x, b^{-1}y)\big) = \big((agb, b^{-1}x) ,_G (1, b^{-1}y)\big)\\ &= \big((ag,x)b ,_G (1, b^{-1}y)\big) = \big((ag,x) ,_G b(1, b^{-1}y)\big)\\ &= \big((ag,x) ,_G (b, b^{-1}y)\big) = \big((ag,x) ,_G (1,y)b\big)\\ &= a\big((g,x) ,_G (1,y)\big)b = a\alpha\big((g,x,y)\big)b \ .\end{aligned}$$

Conversely, the map

$$\beta : \widetilde{X} \times_G \widetilde{Y} \to \widetilde{X \times Y}$$

sending $\big((g,x) ,_G (h,y)\big)$ to $(gh, h^{-1}x, y)$, for $g, h \in G$ and $(x,y) \in X \times Y$, is well defined: indeed if $a \in G$, then

$$\begin{aligned}\beta\Big(\big((g,x)a ,_G a^{-1}(h,y)\big)\Big) &= \beta\Big(\big((ga, a^{-1}x) ,_G (a^{-1}h, y)\big)\Big)\\ &= (gaa^{-1}h, h^{-1}aa^{-1}x, y) = (gh, h^{-1}x, y)\\ &= \beta\Big(\big((g,x) ,_G (h,y)\big)\Big) \ .\end{aligned}$$

Moreover $\beta \circ \alpha$ is clearly the identity map, and conversely

$$\begin{aligned}\alpha \circ \beta\Big(\big((g,x) ,_G (h,y)\big)\Big) &= \alpha\big((gh, h^{-1}x, y)\big) = \big((gh, h^{-1}x) ,_G (1,y)\big)\\ &= \big((g,x)h ,_G (1,y)\big) = \big((g,x) ,_G h(1,y)\big)\\ &= \big((g,x) ,_G (h,y)\big) \ .\end{aligned}$$

so $\alpha \circ \beta$ is also equal to the identity. It follows that α and β are mutual inverse bijections, hence mutual inverse isomorphisms of (G,G)-bisets, since α is a morphism of bisets.

For Assertion 2, if $X = \bullet$, then $\widetilde{X} = G \times \bullet \cong G$, and with this identification, the biset structure is given by left and right multiplication in G.

Now for Assertion 3, observe that the following map

$$a : \big((g,x) ,_G u\big) \in \widetilde{X} \times_G U \mapsto (gx, gu) \in X \times U$$

is well defined. Let b denote the map

$$b : (x,u) \in X \times U \mapsto \big((1,x) ,_G u\big) \in \widetilde{X} \times_G U \ .$$

Then clearly $a \circ b$ is equal to the identity, and

$$(b \circ a)\big((g, x),_G u\big) = \big((1, gx),_G gu\big) = \big((1, gx)g,_G u\big) = \big((g, x),_G u\big) ,$$

so a and b are mutually inverse bijections. They yield the following (G, H)-biset structure on $X \times U$:

$$g(x, u)h = a\big(gb(x, u)h\big) = a\Big(g\big((1, x),_G u\big)h\Big) = a\Big(\big((g, x),_G uh\big)\Big) = (gx, guh) .$$

The proof of the second part of Assertion 3 is similar.

Assertion 4 is a special case of Assertion 3, when $H = \mathbf{1}$.

Now for any G-sets X and Y, the (G, G)-bisets $\widetilde{X \sqcup Y}$ and $\widetilde{X} \sqcup \widetilde{Y}$ are clearly isomorphic. Thus when G is finite, the correspondence $X \mapsto \widetilde{X}$ induces a group homomorphism $B(G) \to B(G, G)$, which is a ring homomorphism by Assertion 1, and maps the identity element of $B(G)$ to the identity element of $B(G, G)$ by Assertion 2. This completes the proof of Lemma 2.5.8. □

2.5.9. Notation : *If G is a finite group, and R is a commutative ring, the map of R-algebras $RB(G) \to RB(G, G)$ defined as the R-linear extension of the assignment $[X] \mapsto [\widetilde{X}]$, for a finite G-set X, will be denoted by $u \mapsto \widetilde{u}$.*

2.5.10. Proposition : *Let G and H be groups, and let U be an (H, G)-biset.*

1. *If X is an H-set, then the map $\alpha_{U,X} : \widetilde{X} \times_H U \to U \times_G (\widetilde{U^{op} \times_H X})$ given by*
$$\alpha_{U,X} : \big((h, x),_H u\big) \mapsto \Big(hu,_G \big(1, (u,_H x)\big)\Big)$$
for $h \in H$, $x \in X$, and $u \in U$, is a well defined morphism of (H, G)-bisets.
2. *If Y is a G-set, then the map $\beta_{U,Y} : \widetilde{U \times_G Y} \to U \times_G \widetilde{Y} \times_G U^{op}$ given by*
$$\beta_{U,Y} : \big(h, (u,_G y)\big) \mapsto \big(hu,_G (1, y),_G u\big) ,$$
for $h \in H$, $u \in U$, and $y \in Y$, is a well defined morphism of (H, H)-bisets.
3. *If U is left-free, then $\alpha_{U,X}$ is injective, for any X, and $\beta_{U,Y}$ is injective, for any Y.*
4. *If U is left-transitive, then $\alpha_{U,X}$ is surjective, for any X, and $\beta_{U,Y}$ is surjective, for any Y.*

Proof: If $u \in U$, denote by u^{op} the element u, viewed in the (G, H)-biset U^{op}. With this notation, if $g \in G$ and $h \in H$, then $(hug)^{op} = g^{-1}u^{op}h^{-1}$.

- For Assertion 1, with this notation, the map $\alpha_{U,X}$ is given by

$$\alpha_{U,X}\Big(\big((h,x),_{H} u\big)\Big) = \Big(hu,_{G} \big(1,(u^{op},_{H} x)\big)\Big) ,$$

for $h \in H$, $x \in X$, and $u \in U$. This map is well defined, for if $k \in H$, then the element

$$\big((h,x)k,_{H} k^{-1}u\big) = \big((hk,k^{-1}x),_{H} k^{-1}u\big) ,$$

for $h,k \in H$, $x \in X$ and $u \in U$, is mapped to

$$\begin{aligned}\Big(hkk^{-1}u,_{G} \big(1,((k^{-1}u)^{op},_{H} k^{-1}x)\big)\Big) &= \Big(hu,_{G} \big(1,(u^{op}k,_{H} k^{-1}x)\big)\Big)\\ &= \Big(hu,_{G} \big(1,(u,_{H} x)\big)\Big) .\end{aligned}$$

Moreover $\alpha_{U,X}$ is a morphism of (H,G)-bisets, for if $b \in H$ and $a \in G$, then

$$\begin{aligned}\alpha_{U,X}\Big(b\big((h,x),_{H} u\big)a\Big) &= \alpha_{U,X}\Big(\big((bh,x),_{H} ua\big)\Big)\\ &= \Big(bhua,_{G} \big(1,((ua)^{op},_{H} x)\big)\Big)\\ &= b\Big(hu,_{G} a\big(1,(a^{-1}u^{op},_{H} x)\big)\Big)\\ &= b\Big(hu,_{G} \big(a,a^{-1}(u^{op},_{H} x)\big)\Big)\\ &= b\Big(hu,_{G} \big(1,(u^{op},_{H} x)\big)a\Big)\\ &= b\Big(hu,_{G} \big(1,(u^{op},_{H} x)\big)\Big)a .\end{aligned}$$

- For Assertion 2, the map $\beta_{U,Y}$ is given by

$$\beta_{U,Y}\Big(\big(h,(u,_{G} y)\big)\Big) = \big(hu,_{G} (1,y),_{G} u^{op}\big) ,$$

for $h \in H$, $u \in U$, and $y \in Y$. It is well defined, for if $k \in G$, then

$$\begin{aligned}\beta_{U,Y}\Big(\big(h,(uk,_{G} k^{-1}y)\big)\Big) &= \big(huk,_{G} (1,k^{-1}y),_{G} (uk)^{op}\big)\\ &= \big(hu,_{G} k(1,k^{-1}y),_{G} k^{-1}u^{op}\big)\\ &= \big(hu,_{G} (k,k^{-1}y)k^{-1},_{G} u^{op}\big)\\ &= \big(hu,_{G} (1,y),_{G} u^{op}\big) .\end{aligned}$$

Moreover, it is a morphism of (H,H)-bisets: if $a,b \in H$, then

$$\begin{aligned}\beta_{U,Y}\Big(a\big(h,(u,_G y)\big)b\Big) &= \beta_{U,Y}\Big(\big(ahb,(b^{-1}u,_G y)\big)\Big)\\ &= \big(ahbb^{-1}u,_G (1,y),_G (b^{-1}u)^{op}\big)\\ &= \big(ahu,_G (1,y),_G u^{op}b\big)\\ &= a\big(hu,_G (1,y),_G u^{op}\big)b\ .\end{aligned}$$

• For Assertion 3, suppose that $\alpha_{U,X}\Big(\big((h,x),_H u\big)\Big) = \alpha_{U,X}\Big(\big((h',x'),_H u'\big)\Big)$, for some $h,h'\in H$, $x,x'\in X$, and $u,u'\in U$. This means that

$$\text{(2.5.11)}\qquad \Big(hu,_G \big(1,(u^{op},_H x)\big)\Big) = \Big(h'u',_G \big(1,(u'^{op},_H x')\big)\Big)\ .$$

This is equivalent to the existence of $a\in G$ such that

$$h'u' = hua^{-1},\qquad \big(1,(u'^{op},_H x')\big) = a\big(1,(u^{op},_H x)\big) = \big(a,(u^{op},_H x)\big)\ .$$

The last equality implies $a=1$, so Equation 2.5.11 is equivalent to $h'u'=hu$ and $(u'^{op},_H x') = (u^{op},_H x)$, which in turn is equivalent to the existence of $b\in H$ such that $u'^{op} = u^{op}b$ and $x' = b^{-1}x$, i.e. finally

$$h'u' = hu\ ,\qquad u' = b^{-1}u\ ,\qquad x' = b^{-1}x\ .$$

If U is left-free, the first two equalities imply $h'^{-1}h = b^{-1}$, and then the third one gives $x' = h'^{-1}hx$. Thus

$$\begin{aligned}\big((h',x'),_H u'\big) &= \big((h',h'^{-1}hx),_H h'^{-1}hu\big)\\ &= \big((1,hx)h',_H h'^{-1}hu\big)\\ &= \big((1,hx)h,_H u\big)\\ &= \big((h,x),_H u\big)\ ,\end{aligned}$$

so $\alpha_{U,Y}$ is injective.

Suppose now that $\beta_{U,Y}\Big(\big(h,(u,_G y)\big)\Big) = \beta_{U,Y}\Big(\big(h',(u',_G y')\big)\Big)$, for some $h,h'\in H$, $u,u'\in U$, and $y,y'\in Y$. This means that

$$\big(hu,_G (1,y),_G u^{op}\big) = \big(h'u',_G (1,y'),_G u'^{op}\big)\ .$$

Equivalently, there exist $a,b\in G$ such that

$$h'u' = hua,\qquad (1,y') = a^{-1}(1,y)b = (a^{-1}b, b^{-1}y)\qquad u'^{op} = b^{-1}u^{op}\ .$$

The center equalities imply $b = a$ and $y' = a^{-1}y$, and then the last one gives $u' = ua$. Then the first equality implies $h'u' = hu'$, hence $h = h'$ if U is left-free. Now

$$\big(h', (u',_G y')\big) = \big(h, (ua,_G a^{-1}y)\big) = \big(h, (u,_G y)\big) ,$$

so $\beta_{U,Y}$ is injective.

• For Assertion 4, assuming U left-transitive, let $\Big(u,_G \big(g, (u'^{op},_H x)\big)\Big)$ be any element of $U \times_G (\widetilde{U^{op} \times_H X})$, where $u, u' \in U$, $g \in G$, and $x \in X$. Choose an element $h \in H$ such that $hu' = ug$. Then

$$\begin{aligned}
\alpha_{U,X}\Big(\big((h,x),_G u'\big)\Big) &= \Big(hu',_G \big(1, (u'^{op},_H x)\big)\Big) \\
&= \Big(ug,_G \big(1, (u'^{op},_H x)\big)\Big) \\
&= \Big(u,_G g\big(1, (u'^{op},_H x)\big)\Big) \\
&= \Big(u,_G \big(g, (u'^{op},_H x)\big)\Big) ,
\end{aligned}$$

so $\alpha_{U,X}$ is surjective.

Similarly, if $\big(u,_G (g,y),_G u'^{op}\big)$ is any element of $U \times_G \widetilde{Y} \times_G U^{op}$, where $u, u' \in U$, $g \in G$, and $y \in Y$, choose an element $h \in H$ such that $hu' = ug$. Then

$$\begin{aligned}
\beta_{U,Y}\Big(\big(h, (u',_G y)\big)\Big) &= \big(hu',_G (1,y),_G, u'^{op}\big) \\
&= \big(ug,_G (1,y),_G, u'^{op}\big) \\
&= \big(u,_G g(1,y),_G, u'^{op}\big) \\
&= \big(u,_G (g,y),_G, u'^{op}\big) ,
\end{aligned}$$

so $\beta_{U,Y}$ is surjective, and this completes the proof of the proposition. □

2.5.12. Corollary : *Let G be a group.*

1. *Let H be a subgroup of G, and X be a G-set. Then there is an isomorphism of (G,H)-bisets*

$$\widetilde{X} \times_G \mathrm{Ind}_H^G \cong \mathrm{Ind}_H^G \times_H \widetilde{\mathrm{Res}_H^G X} ,$$

and an isomorphism of (H,G)-bisets

$$\mathrm{Res}_H^G \times_G \widetilde{X} \cong \widetilde{\mathrm{Res}_H^G X} \times_H \mathrm{Res}_H^G .$$

2. *Let H be a subgroup of G, and Y be an H-set. Then there is an isomorphism of (G, G)-bisets*

$$\mathrm{Ind}_H^G \times_H \widetilde{Y} \times_H \mathrm{Res}_H^G \cong \widetilde{\mathrm{Ind}_H^G Y} \ .$$

3. *Let N be a normal subgroup of G, and X be a (G/N)-set. Then there is an isomorphism of $(G/N, G)$-bisets*

$$\widetilde{X} \times_{G/N} \mathrm{Def}_{G/N}^G \cong \mathrm{Def}_{G/N}^G \times_G \widetilde{\mathrm{Inf}_{G/N}^G X} \ ,$$

and an isomorphism of $(G, G/N)$-bisets

$$\mathrm{Inf}_{G/N}^G \times_{G/N} \widetilde{X} \cong \widetilde{\mathrm{Inf}_{G/N}^G X} \times_G \mathrm{Inf}_{G/N}^G \ .$$

4. *Let N be a normal subgroup of G, and Y be a G-set. Then there is an isomorphism of $(G/N, G/N)$-bisets*

$$\mathrm{Def}_{G/N}^G \times_G \widetilde{Y} \times_G \mathrm{Inf}_{G/N}^G \cong \widetilde{\mathrm{Def}_{G/N}^G Y} \ .$$

Proof: Assertion 2 and the first isomorphism of Assertion 1 follow from Proposition 2.5.10, by exchanging the positions of G and H, and taking for U the (G, H)-biset Ind_H^G. Recall that this is the set G itself, with biset structure given by left multiplication in G and right multiplication by elements of H. So U is both left-free and left-transitive, so the maps $\alpha_{U,X}$ and $\beta_{U,Y}$ are isomorphisms. The second isomorphism of Assertion 1 follows from the first one and from Assertion 2 of Proposition 2.3.14, by considering the opposite bisets, and observing that the (G, G)-bisets $\widetilde{X}^{op}$ and $\widetilde{X}$ are isomorphic.

Similarly, Assertion 4 and the first isomorphism of Assertion 3 follow from the case $H = G/N$, and $U = \mathrm{Def}_{G/N}^G$. Recall that this is the set G/N, with (H, G)-biset structure given by left multiplication in G/N, and projection $G \to G/N$ followed by right multiplication. Then U is also left-free and left transitive, so $\alpha_{U,X}$ and $\beta_{U,Y}$ are isomorphisms. The second isomorphism of Assertion 3 follows from the first one, by considering opposite bisets. □

2.5.13. Groups of Coprime Orders.

2.5.14. Proposition : *Let G and H be finite groups.*

1. *If X is a G-set and Y is an H-set, then $X \times Y$ is a $(G \times H)$-set with the following action*

$$(g, h) \cdot (x, y) = (gx, hy) \ ,$$

for $g \in G$, $h \in H$, $x \in X$, and $y \in Y$.

The correspondence $(X, Y) \mapsto X \times Y$ *induces a bilinear map from* $B(G) \times B(H)$ *to* $B(G \times H)$*, hence an homomorphism*

$$\pi : B(G) \otimes_{\mathbb{Z}} B(H) \to B(G \times H) ,$$

which is an injective ring homomorphism, preserving identity elements. If G *and* H *have coprime orders, this map is an isomorphism.*

2. *If* U *is a* (G, G)*-biset, and* V *is an* (H, H)*-biset, then* $U \times V$ *is a* $(G \times H, G \times H)$*-biset for the structure given by*

$$(g, h) \cdot (u, v) \cdot (g', h') = (gug', huh') ,$$

for $g, g' \in G$*,* $h, h' \in H$*,* $u \in U$*,* $v \in V$*.*
The correspondence $(U, V) \mapsto U \times V$ *induces a bilinear map from* $B(G, G) \times B(H, H)$ *to* $B(G \times H, G \times H)$*, hence a linear map*

$$\pi_2 : B(G, G) \otimes_{\mathbb{Z}} B(H, H) \to B(G \times H, G \times H) ,$$

which is an injective ring homomorphism, preserving identity elements. If G *and* H *have coprime orders, this map is an isomorphism.*

Proof: In Assertion 1, the correspondence $(X, Y) \mapsto X \times Y$ induces an obviously bilinear map from $B(G) \times B(H)$ to $B(G \times H)$, hence a linear map π from $B(G) \otimes_{\mathbb{Z}} B(H)$ to $B(G \times H)$. Now if X and X' are G-sets, and if Y and Y' are H-sets, the map

$$\big((x, x'), (y, y')\big) \in (X \times X') \times (Y \times Y') \mapsto \big((x, y), (x', y') \in (X \times Y) \times (X' \times Y')$$

is an isomorphism of $(G \times H)$-sets. This shows that π is a ring homomorphism. Moreover if X is a G-set of cardinality 1 and Y is an H-set of cardinality 1, then $X \times Y$ is a $(G \times H)$-set of cardinality 1, so π preserves identity elements.

Let $[s_G]$ and $[s_H]$ denote sets of representatives of conjugacy classes of subgroups of G and H, respectively. Then $B(G)$ is a free abelian group with basis $\{[G/S] \mid S \in [s_G]\}$, and $B(H)$ is a free abelian group with basis $\{[H/T] \mid T \in [s_H]\}$. Hence $B(G) \otimes_{\mathbb{Z}} B(H)$ is a free abelian group with basis $\mathcal{B} = \{[G/S] \otimes [H/T] \mid (S, T) \in [s_G] \times [s_H]\}$.

Now obviously $\pi([G/S] \otimes [H/T]) = [(G \times H)/(S \times T)]$. Since the subgroups $S \times T$, for $(S, T) \in [s_G] \times [s_H]$, lie in different conjugacy classes of subgroups of $G \times H$, it follows that the image by π of the basis $\mathcal{B}$ is a subset of a $\mathbb{Z}$-basis of $B(G \times H)$. So π is injective.

If G and H have coprime orders, and L is a subgroup of $G \times H$, then $q(L) = \mathbf{1}$, because $q(L)$ is isomorphic to a subquotient of both G and H. It follows that $L = p_1(L) \times p_2(L)$, so any subgroup of $G \times H$ is equal to $S \times T$, for some $S \leq G$ and some $T \leq H$. This shows that π is surjective, hence it is an isomorphism.

For Assertion 2, set $G_2 = G \times G^{op}$ and $H_2 = H \times H^{op}$. Then Assertion 1 yields a linear map

$$\pi : B(G_2) \otimes B(H_2) \to B(G_2 \times H_2) \ ,$$

and the group $G_2 \times H_2$ is isomorphic to $(G \times H)_2 = (G \times H) \times (G \times H)^{op}$. This gives a map $B(G_2) \otimes B(H_2) \to B\big((G \times H)_2\big)$, which clearly identifies with the map

$$\pi_2 : B(G,G) \otimes_{\mathbb{Z}} B(H,H) \to B(G \times H, G \times H) \ ,$$

after identification of $B(G_2)$ with $B(G,G)$, of $B(H_2)$ with $B(H,H)$, and $B\big((G \times H)_2\big)$ with $B(G \times H, G \times H)$. Assertion 1 shows that π_2 is injective, and even an isomorphism if the orders of G and H are coprime.

It remains to see that π_2 is a ring homomorphism. This follows from the fact that, if U and U' are (G,G)-bisets, and V and V' are (H,H)-bisets, then there is an isomorphism of $(G \times H, G \times H)$-bisets

$$(U \times_G U') \times (V \times_H V') \cong (U \times V) \times_{G \times H} (U' \times V') \ ,$$

given by $\big((u,_{_G} u'), (v,_{_H} v')\big) \mapsto \big((u,v),_{_{G \times H}} (u',v')\big)$.

Finally, if $U = G$ is the identity biset Id_G and $V = H$ is the identity biset Id_H, then $U \times V$ is obviously isomorphic to the identity biset $\mathrm{Id}_{G \times H}$, so π_2 preserves identity elements. □

Chapter 3
Biset Functors

3.1. The Biset Category of Finite Groups

3.1.1. Definition : The biset category $\mathcal{C}$ *of finite groups is the category defined as follows:*

- *The objects of* $\mathcal{C}$ *are finite groups.*
- *If* G *and* H *are finite groups, then* $\mathrm{Hom}_{\mathcal{C}}(G,H) = B(H,G)$.
- *If* G, H, *and* K *are finite groups, then the composition* $v \circ u$ *of the morphism* $u \in \mathrm{Hom}_{\mathcal{C}}(G,H)$ *and the morphism* $v \in \mathrm{Hom}_{\mathcal{C}}(H,K)$ *is equal to* $v \times_H u$.
- *For any finite group* G, *the identity morphism of* G *in* $\mathcal{C}$ *is equal to* $[\mathrm{Id}_G]$.

3.1.2. Remark (presentation of the biset category) : It follows from this definition that the category $\mathcal{C}$ is *a preadditive category*, in the sense of Mac Lane ([39] I Sect 8): the sets of morphisms in $\mathcal{C}$ are abelian groups, and the composition of morphisms is bilinear.

If G and H are finite groups, then any morphism from G to H in $\mathcal{C}$ is a linear combination with integral coefficients of morphisms of the form $[(H \times G)/L]$, where L is some subgroup of $H \times G$. By Lemma 2.3.26, any such morphism factors in $\mathcal{C}$ as the composition

$$G \xrightarrow{\mathrm{Res}_B^G} B \xrightarrow{\mathrm{Def}_{B/A}^B} B/A \xrightarrow{\mathrm{Iso}(f)} D/C \xrightarrow{\mathrm{Inf}_{D/C}^D} D \xrightarrow{\mathrm{Ind}_D^H} H\ ,$$

for suitable sections (B,A) and (D,C) of G and H respectively, and a group isomorphism $f : B/A \to D/C$. In other words, the category $\mathcal{C}$ is *generated* as a preadditive category by the five types of morphisms above, associated to elementary bisets of Sect. 2.3.9.

Conversely, let ${}^*\mathcal{C}$ denote the preadditive category whose objects are finite groups, and morphisms are generated by elementary morphisms

S. Bouc, *Biset Functors for Finite Groups*, Lecture Notes in Mathematics 1990, DOI 10.1007/978-3-642-11297-3_3, © Springer-Verlag Berlin Heidelberg 2010

$$^*\mathrm{Res}_H^G : G \to H \text{ and } ^*\mathrm{Ind}_H^G : H \to G ,$$

$$^*\mathrm{Inf}_{G/N}^G : G/N \to G \text{ and } ^*\mathrm{Def}_{G/N}^G : G \to G/N ,$$

$$^*\mathrm{Iso}(\varphi) : G \to G' ,$$

where G is a finite group, H is a subgroup of G, N is a normal subgroup of G, and φ is a group isomorphism $G \to G'$. These morphisms are subject to the complete list of relations obtained from Sect. 1.1.3 by replacing each elementary operation by the elementary morphism with the same name (so Ind_H^G becomes $^*\mathrm{Ind}_H^G$, etc.).

The correspondence $\Theta : {}^*\mathcal{C} \to \mathcal{C}$ obtained by sending each group to itself, and removing *'s on the elementary morphisms, is easily seen to be a functor: this amounts to check that the elementary bisets satisfy all the relations listed in Sect. 1.1.3.

Conversely, there is a unique correspondence $\Psi : \mathcal{C} \to {}^*\mathcal{C}$ which is the identity on objects, and sends the morphism $G \to H$ defined by a transitive biset $(H \times G)/L$ to the morphism

$$\textbf{(3.1.3)} \qquad ^*\mathrm{Ind}_D^H \circ {}^*\mathrm{Inf}_{D/C}^S \circ {}^*\mathrm{Iso}(\varphi) \circ {}^*\mathrm{Def}_{B/A}^B \circ {}^*\mathrm{Res}_B^G ,$$

where $D = p_1(L)$, $C = k_1(L)$, $B = p_2(L)$, $A = k_2(L)$, and φ is the canonical isomorphism $B/A \to D/C$. Using the triviality relations and the commutation relations of Sect. 1.1.3, one can show that this morphism does not depend on L up to conjugation in $H \times G$, so Ψ is well defined.

It is a tedious, but straightforward task to show that Ψ is a functor: this is essentially equivalent to checking, using the complete list of relations of Sect. 1.1.3, that any composition of elementary morphisms in $^*\mathcal{C}$ is a sum of morphisms of the form 3.1.3.

Then it is clear that Θ and Ψ are mutual inverse equivalences of categories. In other words, the elementary morphisms, together with the relations of Sect. 1.1.3, form *a presentation of the biset category* $\mathcal{C}$.

3.1.4. Remark : Lemma 2.4.11 shows that there is a functor from the biset category to the opposite category, which maps any object to itself, and any morphism $u \in \mathrm{Hom}_{\mathcal{C}}(G,H) = B(H,G)$ to $u^{op} \in B(G,H) = \mathrm{Hom}_{\mathcal{C}^{op}}(G,H)$. This functor is obviously an equivalence of categories.

3.1.5. Extension of Coefficients. In some applications, it is natural to consider other coefficient rings instead of integers:

3.1.6. Definition : *Let R be a commutative ring with identity element. The category $R\mathcal{C}$ is defined as follows:*

- *The objects of $R\mathcal{C}$ are finite groups.*
- *If G and H are finite groups, then*

$$\mathrm{Hom}_{R\mathcal{C}}(G,H) = R \otimes_{\mathbb{Z}} B(H,G) .$$

- *The composition of morphisms in $R\mathcal{C}$ is the R-linear extension of the composition in $\mathcal{C}$.*
- *For any finite group G, the identity morphism of G in $R\mathcal{C}$ is equal to $R \otimes_{\mathbb{Z}} \mathrm{Id}_G$.*

The category $R\mathcal{C}$ is *an R-linear category*, i.e. the sets of morphisms in $R\mathcal{C}$ are R-modules, and the composition in $R\mathcal{C}$ is R-bilinear.

3.2. Biset Functors

Recall that a functor $F : \mathcal{A} \to \mathcal{B}$ between preadditive categories is called *additive* if for any objects X and Y of $\mathcal{A}$, the map $f \mapsto F(f)$ from $\mathrm{Hom}_{\mathcal{A}}(X,Y)$ to $\mathrm{Hom}_{\mathcal{B}}\big(F(X), F(Y)\big)$ is a group homomorphism. A *preadditive subcategory* of $\mathcal{B}$ is a subcategory $\mathcal{A}$ such that the inclusion functor $\mathcal{A} \to \mathcal{B}$ is additive.

Similarly, when R is a commutative ring with identity, a functor between R-linear categories is called *R-linear* if the maps it induces between sets of morphisms is R-linear. An *R-linear subcategory* of an R-linear category is a subcategory such that the inclusion functor is R-linear.

3.2.1. Example : Let $\mathcal{D}$ be a preadditive subcategory of the biset category $\mathcal{C}$. Then $R\mathcal{D}$ can be viewed as an R-linear subcategory of $R\mathcal{C}$.

The universal property of Burnside groups (see Remark 2.4.3) shows that the conditions (B1)–(B4) of Sect. 1.3 can be formulated equivalently as follows:

3.2.2. Definition : *Let R be a commutative ring with identity, and let $\mathcal{D}$ be a preadditive subcategory of $\mathcal{C}$. A* biset functor defined on $\mathcal{D}$ (or over $\mathcal{D}$), with values in R-Mod *is an R-linear functor from $R\mathcal{D}$ to R-*Mod.

*Biset functors over $\mathcal{D}$, with values in R-*Mod *are the objects of a category, denoted by $\mathcal{F}_{\mathcal{D},R}$, where morphisms are natural transformation of functors, and composition of morphisms is composition of natural transformations.*

If F is an object of $\mathcal{F}_{\mathcal{D},R}$, then a minimal group for F *is an object H of $\mathcal{D}$ such that $F(H) \neq \{0\}$, but $F(K) = \{0\}$ for any object K of $\mathcal{D}$ with $|K| < |H|$. The class of minimal groups for F is denoted by* $\mathrm{Min}(F)$.

3.2.3. Remark : In particular $\mathrm{Min}(F)$ is empty if and only if F is the zero object of $\mathcal{F}_{\mathcal{D},R}$, i.e. if $F(H) = \{0\}$ for any object H of $\mathcal{D}$.

3.2.4. Example (p-biset functors) : Let p be a prime number. The full subcategory of the biset category $\mathcal{C}$ whose objects are finite p-groups is denoted by $\mathcal{C}_p$. A biset functor over $\mathcal{C}_p$ (with values in $\mathbb{Z}$-Mod) is called *a p-biset functor*, and the category of p-biset functors is simply denoted by $\mathcal{F}_p$.

3.2.5. Example (inflation functors) : Let $\mathcal{I}$ denote the following subcategory of $\mathcal{C}$: the objects of $\mathcal{I}$ are all finite groups (so $\mathcal{I}$ has the same objects as $\mathcal{C}$). If G and H are finite groups, then $\mathrm{Hom}_{\mathcal{I}}(G,H)$ is the subgroup of $\mathrm{Hom}_{\mathcal{C}}(G,H) = B(H,G)$ generated by the classes of *finite right free* (H,G)-*bisets*. Then the biset functors on $\mathcal{I}$ are the inflation functors introduced in Sect. 1.2.4.

3.2.6. Example (global Mackey functors) : Let $\mathcal{M}$ denote the following subcategory of $\mathcal{C}$: the objects of $\mathcal{M}$ are all finite groups (so $\mathcal{M}$ has the same objects as $\mathcal{C}$). If G and H are finite groups, then $\mathrm{Hom}_{\mathcal{M}}(G,H)$ is the subgroup of $\mathrm{Hom}_{\mathcal{C}}(G,H) = B(H,G)$ generated by the classes of *finite left and right free* (H,G)-*bisets*. Then the biset functors on $\mathcal{M}$ are the global Mackey functors introduced in Sect. 1.2.5.

3.2.7. Example (dual of a biset functor) : Let $\mathcal{D}$ be a preadditive subcategory of $\mathcal{C}$. Then the opposite category $\mathcal{D}^{op}$ is isomorphic to the following preadditive subcategory $\mathcal{D}^{\diamond}$ of $\mathcal{C}$: the objects of $\mathcal{D}^{\diamond}$ are the objects of $\mathcal{D}$, and morphisms are defined by

$$\mathrm{Hom}_{\mathcal{D}^{\diamond}}(G,H) = \{\alpha \in \mathrm{Hom}_{\mathcal{C}}(G,H) \mid \alpha^{op} \in \mathrm{Hom}_{\mathcal{D}}(H,G)\} .$$

Fix an R-module M. If $F \in \mathcal{F}_{\mathcal{D},R}$, *the* M*-dual of* F is the object $\mathrm{Hom}_R(F,M)$ of $\mathcal{F}_{\mathcal{D}^{\diamond},R}$ defined by

$$\mathrm{Hom}_R(F,M)(G) = \mathrm{Hom}_R\big(F(G),M\big) ,$$

for any object G of $\mathcal{D}$, and by

$$\mathrm{Hom}_R(F,M)(\alpha) = {}^tF(\alpha^{op}) ,$$

if $\alpha : G \to H$ is a morphism in $\mathcal{D}^{\diamond}$, where

$${}^tF(\alpha^{op}) : \mathrm{Hom}_R\big(F(G),M\big) \to \mathrm{Hom}_R\big(F(H),M\big)$$

is the transposed map of $F(\alpha^{op}) : F(H) \to F(G)$.

3.2.8. Proposition : *Let R be a commutative ring with identity, and let $\mathcal{D}$ be a preadditive subcategory of $\mathcal{C}$.*

1. *The category $\mathcal{F}_{\mathcal{D},R}$ is an R-linear abelian category: if $f : F \to F'$ is a morphism of biset functors, then for any object G of $\mathcal{D}$*

$$(\mathrm{Ker}\, f)(G) = \mathrm{Ker}\, f_G \qquad\qquad (\mathrm{Coker}\, f)(G) = \mathrm{Coker}\, f_G .$$

where $f_G : F(G) \to F'(G)$ is the evaluation of f at G.

2. *A sequence* $0 \longrightarrow F \xrightarrow{f} F' \xrightarrow{f'} F" \longrightarrow 0$ *is an exact sequence in* $\mathcal{F}_{\mathcal{D},R}$ *if and only if for any object* G *of* $\mathcal{D}$*, the sequence*

$$0 \longrightarrow F(G) \xrightarrow{f_G} F'(G) \xrightarrow{f'_G} F"(G) \longrightarrow 0$$

is an exact sequence of R*-modules.*

3. *If* I *is a set, and* $(F_i)_{i\in I}$ *is a family of objects of* $\mathcal{F}_{\mathcal{D},R}$*, then the direct sum* $\bigoplus_{i\in I} F_i$ *and the direct product* $\prod_{i\in I} F_i$ *exist: for any object* G *of* $\mathcal{D}$

$$(\bigoplus_{i\in I} F_i)(G) = \bigoplus_{i\in I} F_i(G) \qquad (\prod_{i\in I} F_i)(G) = \prod_{i\in I} F_i(G)\,.$$

Proof: This is straightforward. □

3.2.9. Remark (subfunctor generated by a family of elements) : Let F be an object of $\mathcal{F}_{\mathcal{D},R}$. If $(F_i)_{i\in I}$ is a set of subfunctors of F, then the intersection $\bigcap_{i\in I} F_i$ is the subfunctor of F whose evaluation at the object G of $\mathcal{D}$ is equal to

$$(\bigcap_{i\in I} F_i)(G) = \bigcap_{i\in I} F_i(G)\,.$$

In particular, let $\mathcal{G}$ be a set of objects of $\mathcal{C}$, and for each group G in $\mathcal{G}$, let Γ_G be a subset of $F(G)$. The subfunctor $F_{\mathcal{G},\Gamma}$ of F *generated by the data* $(\mathcal{G},\Gamma)$ is by definition the intersection of all subfunctors F' of F such that $F'(G) \supseteq \Gamma_G$ for all G in $\mathcal{G}$. If H is an object of $\mathcal{D}$, it is easy to see that

$$F_{\mathcal{G},\Gamma}(H) = \sum_{\substack{G\in\mathcal{G}\\ \gamma\in\Gamma_G}} \mathrm{Hom}_{R\mathcal{D}}(G,H)(\gamma)\,.$$

3.2.10. Definition : *A subcategory* $\mathcal{D}$ *of* $\mathcal{C}$ *is said to* contain group isomorphisms *if whenever* $f : G \to H$ *is an isomorphism of finite groups, and* G *is an object of* $\mathcal{D}$*, the group* H *is also an object of* $\mathcal{D}$*, and* $\mathrm{Iso}(f)$ *is a morphism from* G *to* H *in* $\mathcal{D}$*.*

3.2.11. Remark : If $\mathcal{D}$ contains group isomorphisms, then in particular the equivalence classes of objects of $\mathcal{D}$ modulo group isomorphism form a subset of the set of isomorphism classes of finite groups.

3.2.12. Corollary : *If* $\mathcal{D}$ *contains group isomorphisms, then the category* $\mathcal{F}_{\mathcal{D},R}$ *has enough projective objects.*

Proof: For any object D of $\mathcal{D}$, the Yoneda functor

$$Y_D : G \mapsto \mathrm{Hom}_{R\mathcal{D}}(D, G)$$

is a projective object of $\mathcal{F}_{\mathcal{D},R}$: indeed, evaluation at D preserves exact sequences, and by Yoneda's lemma, for any object F of $\mathcal{F}_{\mathcal{D},R}$, the map

$$f \in F(D) \mapsto \phi_{D,f} \in \mathrm{Hom}_{\mathcal{F}_{\mathcal{D},R}}(Y_D, F)$$

is an isomorphism, where the evaluation of $\phi_{D,f}$ at an object G of $\mathcal{D}$ is defined by

$$\phi_{D,f,G} : u \in Y_D(G) = \mathrm{Hom}_{R\mathcal{D}}(D, G) \mapsto F(u)(f) \in F(G) .$$

In particular, the map $\phi_{D,f}$ maps the identity morphism of D in $R\mathcal{D}$ to the element f of $F(D)$. So if $\mathcal{S}$ is the set of isomorphism classes of finite groups which are objects of $\mathcal{D}$, if $[\mathcal{S}]$ is a set of representatives of $\mathcal{S}$, and if for $S \in [\mathcal{S}]$, the set Γ_S is a set of generators of the R-module $F(S)$, then the evaluation of the map

$$\Phi = \bigoplus_{\substack{S\in[\mathcal{S}]\\ f\in\Gamma_S}} \phi_{S,f} : \bigoplus_{\substack{S\in[\mathcal{S}]\\ f\in\Gamma_S}} Y_S \to F$$

at $S \in [\mathcal{S}]$ is obviously surjective. Since $\mathcal{D}$ contains groups isomorphisms, it follows that $(\mathrm{Coker}\ f)(G) \cong (\mathrm{Coker}\ f)(S) = \{0\}$, for any object G of $\mathcal{D}$ isomorphic to S as a group. Hence Φ is surjective. Moreover the functor $\bigoplus_{\substack{S\in[\mathcal{S}]\\ f\in\Gamma_S}} Y_S$ is projective, since it is a direct sum of projective objects. □

3.2.13. Corollary : *If $\mathcal{D}$ contains group isomorphisms, then $\mathcal{F}_{\mathcal{D},R}$ also has enough injective objects.*

Proof: There are several steps:

• The $\mathbb{Z}$-module $\mathbb{Q}/\mathbb{Z}$ is an injective cogenerator for the category $\mathbb{Z}$-Mod. It follows that the module $J = \mathrm{Hom}_{\mathbb{Z}}(R, \mathbb{Q}/\mathbb{Z})$ is an injective cogenerator for R-Mod, since

$$\mathrm{Hom}_R\big(M, \mathrm{Hom}_{\mathbb{Z}}(R, \mathbb{Q}/\mathbb{Z})\big) \cong \mathrm{Hom}_{\mathbb{Z}}(M, \mathbb{Q}/\mathbb{Z}) .$$

• If M is an R-module, denote by M^o the R-module $\mathrm{Hom}_R(M, J)$. It follows that the canonical map $\eta : M \to M^{oo}$ is injective: indeed, if $m \in M$, then $\eta(m)$ is the map $M^o \to J$ defined by

$$\eta(m)(\phi) = \phi(m) ,$$

so $\mathrm{Ker}\,\eta = \bigcap_{\phi\in M^o} \mathrm{Ker}\,\phi$. Thus if $\alpha \in \mathrm{Hom}_R(\mathrm{Ker}\,\eta, J)$, the diagram

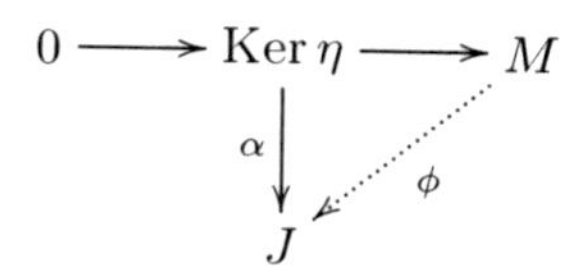

can be completed to a commutative diagram by a map $\phi : M \to J$, i.e. an element of M^o. Since the restriction of ϕ to $\mathrm{Ker}\,\eta$ is equal to 0, it follows that $\alpha = 0$, thus $\mathrm{Hom}_R(\mathrm{Ker}\,\eta, J) = \{0\}$, and $\mathrm{Ker}\,\eta = \{0\}$ since J is a cogenerator for R-Mod.

- If F is a biset functor over $\mathcal{D}$, denote by F^o the J-dual of F, defined in Example 3.2.7. Recall that it is the functor from $R\mathcal{D}^\diamond$ to R-Mod defined by $F^o(G) = \big(F(G)\big)^o$, for an object G of $\mathcal{D}$, and by $F^o(f) = {}^tF(f^{op})$ for a morphism in $R\mathcal{D}$, where ${}^tF(f^{op})$ is the transposed map. A similar argument as in the proof of Corollary 3.2.12 shows that the category $\mathcal{F}_{\mathcal{D}^\diamond,R}$ has enough projective objects. In particular, there is a surjective morphism $P \to F^o$ in this category.
- It follows that the morphism $F^{oo} \to P^o$ is injective. By composition, this yields an injective morphism $F \to P^o$. Finally, it is easy to prove that P^o is an injective object of $\mathcal{F}_{\mathcal{D},R}$, since for any object F of $\mathcal{F}_{\mathcal{D},R}$

$$\mathrm{Hom}_{\mathcal{F}_{\mathcal{D},R}}(F, P^o) \cong \mathrm{Hom}_{\mathcal{F}_{\mathcal{D}^\diamond,R}}(P, F^o)\ ,$$

and since the functor $F \mapsto F^o$ from $\mathcal{F}_{\mathcal{D},R}$ to $\mathcal{F}_{\mathcal{D}^\diamond,R}$ is exact, as J is an injective R-module. □

3.2.14. Remark : Another proof of Corollaries 3.2.12 and 3.2.13 will be given in Sect. 3.3.5.

3.3. Restriction to Subcategories

3.3.1. Let $\mathcal{D}' \subseteq \mathcal{D}$ be subcategories of the biset category $\mathcal{C}$, both containing group isomorphisms, and let R be a commutative ring with identity. There is an obvious restriction functor

$$\mathcal{R}es_{R\mathcal{D}'}^{R\mathcal{D}} : \mathcal{F}_{\mathcal{D},R} \to \mathcal{F}_{\mathcal{D}',R}\ ,$$

that will be denoted abusively by $\mathcal{R}es_{\mathcal{D}'}^{\mathcal{D}}$ for simplicity. This functor is R-linear, and exact, by Proposition 3.2.8. By standard arguments from category theory, it has left and right adjoints, called *left and right induction*, whose description requires the following notation (see Chap. IX of [39] for details on ends and coends):

3.3.2. Notation : *Let $\mathcal{D}' \subseteq \mathcal{D}$, and R as in 3.3.1. If F is an object of $\mathcal{F}_{\mathcal{D}',R}$, denote by ${}^l\mathcal{I}nd_{\mathcal{D}'}^{\mathcal{D}}$ the coend*

$$ {}^l\mathcal{I}nd_{\mathcal{D}'}^{\mathcal{D}}(F) = \int^{d'} \mathrm{Hom}_{R\mathcal{D}}(d', -) \otimes_R F(d') \; , $$

and by ${}^r\mathcal{I}nd_{\mathcal{D}'}^{\mathcal{D}}$ the end

$$ {}^r\mathcal{I}nd_{\mathcal{D}'}^{\mathcal{D}}(F) = \int_{d'} \mathrm{Hom}_R\big(\mathrm{Hom}_{R\mathcal{D}}(-, d'), F(d')\big) \; . $$

Recall that for an object d of $\mathcal{D}$, the evaluation at d of ${}^l\mathcal{I}nd_{\mathcal{D}'}^{\mathcal{D}}(F)$ can be computed by

$$ {}^l\mathcal{I}nd_{\mathcal{D}'}^{\mathcal{D}}(F)(d) = \Big(\bigoplus_{d' \in \mathcal{S}'} \mathrm{Hom}_{R\mathcal{D}}(d', d) \otimes_R F(d') \Big) / \mathcal{I} \; , $$

where $\mathcal{S}'$ is a set of representatives of objects of $\mathcal{D}'$ up to group isomorphisms (see Remark 3.2.11), and $\mathcal{I}$ is the R-submodule generated by the elements

$$ (u \circ \alpha) \otimes_R f - u \otimes_R F(\alpha)(f) \; , $$

for any elements d'_1 and d'_2 of $\mathcal{S}'$, any morphism $\alpha : d'_1 \to d'_2$ in $R\mathcal{D}'$, any f in $F(d'_1)$ and any u in $\mathrm{Hom}_{R\mathcal{D}}(d'_2, d)$.

If $v : d \to e$ is a morphism in $R\mathcal{D}$, the map

$$ {}^l\mathcal{I}nd_{\mathcal{D}'}^{\mathcal{D}}(F)(v) : {}^l\mathcal{I}nd_{\mathcal{D}'}^{\mathcal{D}}(F)(d) \to {}^l\mathcal{I}nd_{\mathcal{D}'}^{\mathcal{D}}(F)(e) $$

is the R-linear map induced by composition on the left in $R\mathcal{D}$, i.e.

$$ {}^l\mathcal{I}nd_{\mathcal{D}'}^{\mathcal{D}}(F)(v)(u \otimes f) = (v \circ u) \otimes f \; , $$

for any $d' \in \mathcal{S}'$, any $u \in \mathrm{Hom}_{R\mathcal{D}}(d', d)$ and any $f \in F(d')$.

Similarly, the evaluation of ${}^r\mathcal{I}nd_{\mathcal{D}'}^{\mathcal{D}}(F)$ at d is the set of natural transformations from the functor $\mathrm{Hom}_{R\mathcal{D}}(d, -)$ to the functor F, i.e. the R-submodule

$$ {}^r\mathcal{I}nd_{\mathcal{D}'}^{\mathcal{D}}(F)(d) \subseteq \prod_{d' \in \mathcal{S}'} \mathrm{Hom}_R\big(\mathrm{Hom}_{R\mathcal{D}}(d, d'), F(d')\big) $$

consisting of sequences $(\rho_{d'})_{d' \in \mathcal{S}'}$ such that all the diagrams

$$ \begin{array}{ccc} \mathrm{Hom}_{R\mathcal{D}}(d, d'_1) & \xrightarrow{\rho_{d'_1}} & F(d'_1) \\ {\scriptstyle \mathrm{Hom}_{R\mathcal{D}}(d,\alpha)} \downarrow & & \downarrow {\scriptstyle F(\alpha)} \\ \mathrm{Hom}_{R\mathcal{D}}(d, d'_1) & \xrightarrow{\rho_{d'_2}} & F(d'_2) \end{array} $$

are commutative. If $v : d \to e$ is a morphism in $R\mathcal{D}$, the image of the sequence $(\rho_{d'})_{d'\in\mathcal{S}'}$ by the map

$$ {}^r\mathcal{I}nd_{\mathcal{D}'}^{\mathcal{D}}(F)(v) : {}^r\mathcal{I}nd_{\mathcal{D}'}^{\mathcal{D}}(F)(d) \to {}^r\mathcal{I}nd_{\mathcal{D}'}^{\mathcal{D}}(F)(e) $$

is the sequence $(\sigma_{d'})_{d'\in\mathcal{S}'}$ defined by

$$ \sigma_{d'} = \rho_{d'} \circ \mathrm{Hom}_{R\mathcal{D}}(v, d') \ . $$

In other words $\sigma_{d'}$ is the map from $\mathrm{Hom}_{R\mathcal{D}}(e, d')$ to $F(d')$ defined by

$$ \forall u \in \mathrm{Hom}_{R\mathcal{D}}(e, d'), \ \ \sigma_{d'}(u) = \rho_{d'}(u \circ v) \ . $$

Clearly, the correspondences $F \mapsto {}^l\mathcal{I}nd_{\mathcal{D}'}^{\mathcal{D}}(F)$ and $F \mapsto {}^r\mathcal{I}nd_{\mathcal{D}'}^{\mathcal{D}}(F)$ are R-linear functors from $\mathcal{F}_{\mathcal{D}',R}$ to $\mathcal{F}_{\mathcal{D},R}$.

3.3.3. Proposition : *Let $\mathcal{D}' \subseteq \mathcal{D}$ be subcategories of $\mathcal{C}$, both containing group isomorphisms, and let R be a commutative ring. Then the functor ${}^l\mathcal{I}nd_{\mathcal{D}'}^{\mathcal{D}}$ (resp. the functor ${}^r\mathcal{I}nd_{\mathcal{D}'}^{\mathcal{D}}$) is left adjoint (resp. right adjoint) to the functor $\mathcal{R}es_{\mathcal{D}'}^{\mathcal{D}}$.*

3.3.4. Proposition : *Let $\mathcal{D}$ be a subcategory of $\mathcal{C}$, containing group isomorphisms, and let $\mathcal{D}'$ be a* full *subcategory of $\mathcal{D}$. Then the functors*

$$ \mathcal{R}es_{\mathcal{D}'}^{\mathcal{D}} \circ {}^l\mathcal{I}nd_{\mathcal{D}'}^{\mathcal{D}} \qquad \text{and} \qquad \mathcal{R}es_{\mathcal{D}'}^{\mathcal{D}} \circ {}^r\mathcal{I}nd_{\mathcal{D}'}^{\mathcal{D}} $$

are isomorphic to the identity functor of $\mathcal{F}_{\mathcal{D}',R}$.

Proof: Since $\mathcal{R}es_{\mathcal{D}'}^{\mathcal{D}} \circ {}^l\mathcal{I}nd_{\mathcal{D}'}^{\mathcal{D}}$ is left adjoint to $\mathcal{R}es_{\mathcal{D}'}^{\mathcal{D}} \circ {}^r\mathcal{I}nd_{\mathcal{D}'}^{\mathcal{D}}$, it is enough to prove result for the latter. But since $\mathcal{D}'$ is a full subcategory of $\mathcal{D}$, for any object d' of $\mathcal{D}'$, the functor $\mathrm{Hom}_{R\mathcal{D}}(d', -)$, from $R\mathcal{D}'$ to R-Mod, is equal to the functor $\mathrm{Hom}_{R\mathcal{D}'}(d', -)$, i.e. to the Yoneda functor $Y_{d'}$ on the category $R\mathcal{D}'$. By the Yoneda Lemma, for any object F of $\mathcal{F}_{\mathcal{D}',R}$, the set of natural transformations from $Y_{d'}$ to F is isomorphic to $F(d')$, and this isomorphism is functorial in d'. It follows that the functor $\mathcal{R}es_{\mathcal{D}'}^{\mathcal{D}} \circ {}^r\mathcal{I}nd_{\mathcal{D}'}^{\mathcal{D}}$ is isomorphic to the identity functor of $\mathcal{F}_{\mathcal{D}',R}$, as was to be shown. □

3.3.5. Example : the evaluation functors. Let $\mathcal{D}$ be a subcategory of the biset category $\mathcal{C}$, containing group isomorphisms. If G is an object of $\mathcal{D}$, and if $\mathcal{D}'$ is the full subcategory of $\mathcal{D}$ consisting of the single object G, then an object F of $\mathcal{F}_{\mathcal{D}',R}$ is entirely determined by its value $F(G)$, together with the maps $F(f) : F(G) \to F(G)$, for $f \in \mathrm{End}_{R\mathcal{D}}(G)$. In other words F is determined by the $\mathrm{End}_{R\mathcal{D}}(G)$-module structure of $F(G)$. More precisely, the functor $F \mapsto F(G)$ is an equivalence of categories from $\mathcal{F}_{\mathcal{D}',R}$ to $R\mathrm{End}_{\mathcal{D}}(G)$-Mod.

It follows that in this case, the restriction functor $\mathcal{R}es^{\mathcal{D}}_{\mathcal{D}'}$ is a functor

$$Ev_G : \mathcal{F}_{\mathcal{D},R} \to \mathrm{End}_{R\mathcal{D}}(G)\text{-}\mathsf{Mod}\ .$$

The left adjoint ${}^l\mathcal{I}nd^{\mathcal{D}}_{\mathcal{D}'}$ of this functor maps the $\mathrm{End}_{R\mathcal{D}}(G)$-module V to the biset functor $L_{G,V}$ on $\mathcal{D}$ defined by

$$L_{G,V}(H) = \mathrm{Hom}_{R\mathcal{D}}(G,H) \otimes_{\mathrm{End}_{R\mathcal{D}}(G)} V$$

for an object H of $\mathcal{D}$, where $\mathrm{Hom}_{R\mathcal{D}}(G,H)$ is a right $\mathrm{End}_{R\mathcal{D}}(G)$-module by composition of morphisms. If $\varphi : H \to H'$ is a morphism in $R\mathcal{D}$, then the map

$$L_{G,V}(\varphi) : L_{G,V}(H) \to L_{G,V}(H')$$

is given by composition on the left with φ, i.e. it is defined by

$$L_{G,V}(\varphi)(f \otimes v) = (\varphi f) \otimes v\ ,$$

for $f \in \mathrm{Hom}_{R\mathcal{D}}(G,H)$ and $v \in V$.

Similarly, the right adjoint ${}^r\mathcal{I}nd^{\mathcal{D}}_{\mathcal{D}'}$ of the functor Ev_G maps the $\mathrm{End}_{R\mathcal{D}}(G)$-module V to the biset functor $L^o_{G,V}$ on $\mathcal{D}$ defined by

$$L^o_{G,V}(H) = \mathrm{Hom}_{\mathrm{End}_{R\mathcal{D}}(G)}\big(\mathrm{Hom}_{R\mathcal{D}}(H,G), V\big)\ ,$$

where $\mathrm{Hom}_{R\mathcal{D}}(H,G)$ is a left $\mathrm{End}_{R\mathcal{D}}(G)$-module by composition. When $\varphi : H \to H'$ is a morphism in $R\mathcal{D}$, the map

$$L^o_{G,V}(\varphi) : L^o_{G,V}(H) \to L^o_{G,V}(H')$$

is defined by

$$L^o_{G,V}(\varphi)(\theta)(f) = \theta(f\varphi)\ ,$$

for $\theta \in L^o_{G,V}(H)$ and $f \in \mathrm{Hom}_{R\mathcal{D}'}(H',G)$.

Note that in accordance with Proposition 3.3.4, there are isomorphisms of $\mathrm{End}_{R\mathcal{D}}(G)$-modules $V \cong L_{G,V}(G)$ and $V \cong L^o_{G,V}(G)$. Let $\mathcal{S}$ denote a set of representatives of equivalence classes of objects of $\mathcal{D}$ modulo group isomorphism. For each G in $\mathcal{S}$, choose a projective $\mathrm{End}_{R\mathcal{D}}(G)$-module P_G, such that there exists a surjective map of $\mathrm{End}_{R\mathcal{D}}(G)$-modules $P_G \to F(G)$. By adjunction, this gives a map $\Lambda_G : L_{G,P_G} \to F$, which is surjective when evaluated at G. Since the evaluation functor Ev_G is exact, its left adjoint maps projective objects to projective objects, so L_{G,P_G} is a projective object of $\mathcal{F}_{\mathcal{D},R}$. Clearly, the map

$$\Lambda = \bigoplus_{G\in\mathcal{S}} \Lambda_G : \bigoplus_{G\in\mathcal{S}} L_{G,P_G} \to F$$

is surjective: the map $Ev_G(\Lambda)$ is surjective by construction if $G \in \mathcal{S}$, hence for any G, since $\mathcal{D}$ contains group isomorphisms. This gives another proof of Corollary 3.2.12: the category $\mathcal{F}_{\mathcal{D},R}$ has enough projective objects.

Similarly, for each $G \in \mathcal{S}$, choose an injective $\mathrm{End}_{R\mathcal{D}}(G)$-module I_G such that there is an injective map of $\mathrm{End}_{R\mathcal{D}}(G)$-modules $F(G) \to I_G$. The dual version of the previous argument shows that the functor L^o_{G,I_G} is an injective object of $\mathcal{F}_{\mathcal{D},R}$, and that there is an injective morphism of biset functors

$$F \to \prod_{G \in \mathcal{S}} L^o_{G,I_G} ,$$

so the category $\mathcal{F}_{\mathcal{D},R}$ has enough injective objects (Corollary 3.2.13).

Chapter 4
Simple Functors

This chapter examines the structure of the simple objects of $\mathcal{F}_{\mathcal{D},R}$, under various assumptions on the subcategory $\mathcal{D}$ of the biset category. In particular, a rather complete description of the evaluations of simple functors can be stated when $\mathcal{D}$ is *an admissible subcategory*, as defined in the next section.

4.1. Admissible Subcategories

4.1.1. The examples of Chap. 1 show that it is natural for applications to consider some subcategories of the biset category $\mathcal{C}$, such as the categories obtained by restricting morphisms to right free bisets. These subcategories need not be full subcategories in general, but the first property that seems reasonable to expect from them is that they contain group isomorphisms (see Definition 3.2.10).

4.1.2. The following requirements ensure moreover that if $\mathcal{D}$ is an admissible subcategory of $\mathcal{C}$, then a weak form of the factorization of Remark 3.1.2 holds in $\mathcal{D}$. More precisely:

4.1.3. Definition : *A subcategory $\mathcal{D}$ of $\mathcal{C}$ is called* admissible *if it contains group isomorphisms, and if the following conditions are fulfilled:*

A1. If G and H are objects of $\mathcal{D}$, then there is a subset $S(H,G)$ of the set of subgroups of $H \times G$, invariant under $(H \times G)$-conjugation, such that $\mathrm{Hom}_{\mathcal{D}}(G,H)$ *is the subgroup of* $\mathrm{Hom}_{\mathcal{C}}(G,H)$ *generated by the elements $[(H \times G)/L]$, for $L \in S(H,G)$.*

A2. If G and H are objects of $\mathcal{D}$, and if $L \in S(H,G)$, then $q(L)$ is an object of $\mathcal{D}$. Moreover $\mathrm{Defres}^G_{p_2(L)/k_2(L)}$ *and* $\mathrm{Indinf}^H_{p_1(L)/k_1(L)}$ *are morphisms in $\mathcal{D}$, in other words*

$$\{(xk_2(L),x) \mid x \in p_2(L)\} \in S\big(p_2(L)/k_2(L),G\big)$$
$$\{(x,xk_1(L)) \mid x \in p_1(L)\} \in S\big(H,p_1(L)/k_1(L)\big)\ .$$

S. Bouc, *Biset Functors for Finite Groups*, Lecture Notes in Mathematics 1990, DOI 10.1007/978-3-642-11297-3_4, © Springer-Verlag Berlin Heidelberg 2010

4.1.4. Remark : Since $\mathcal{D}$ contains group isomorphisms, if G is an object of $\mathcal{D}$, and $f : G \to H$ is a group isomorphism, then H is an object of $\mathcal{D}$, and the group $\Delta_f(G) = \{(f(x), x) \mid x \in G\}$ is an element of $S(H, G)$: indeed $\mathrm{Iso}(f) \cong (H \times G)/\Delta_f(G)$.

4.1.5. Factorization Property. An admissible subcategory $\mathcal{D}$ of $\mathcal{C}$ is preadditive. Moreover, if G and H are objects of $\mathcal{D}$, then any morphism from G to H in $\mathcal{D}$ is a linear combination of transitive ones, of the form $(H \times G)/L$, for $L \in S(H, G)$, and any such morphism factors in $\mathcal{D}$ as the composition

$$G \xrightarrow{\mathrm{Defres}^G_{B/A}} B/A \xrightarrow{\mathrm{Iso}(f)} D/C \xrightarrow{\mathrm{Indinf}^H_{D/C}} H\ ,$$

for suitable sections (B, A) and (D, C) of G and H respectively, and some group isomorphism $f : B/A \to D/C$. Note that in this case, the groups B/A and D/C are both isomorphic to $q(L)$, hence they are objects of $\mathcal{D}$.

4.1.6. Remark : The sets $S(H, G)$ appearing in the definition of an admissible subcategory of $\mathcal{C}$ have the following properties:

S1. If G and H are objects of $\mathcal{D}$, then the set $S(H, G)$ is invariant under $(H \times G)$-conjugation.
S2. If G and H are objects of $\mathcal{D}$, and if $L \in S(H, G)$, then $q(L)$ is an object of $\mathcal{D}$. Moreover

$$\begin{aligned}&\{(xk_2(L), x) \mid x \in p_2(L)\} \in S\big(p_2(L)/k_2(L), G\big)\\ &\{(x, xk_1(L)) \mid x \in p_1(L)\} \in S\big(H, p_1(L)/k_1(L)\big)\ .\end{aligned}$$

S3. If $f : G \to H$ is a group isomorphism, and if G is an object of $\mathcal{D}$, then H is an object of $\mathcal{D}$, and $\Delta_f(G) \in S(H, G)$.
S4. If G, H, and K are finite groups, if $M \in S(K, H)$ and $L \in S(H, G)$, then $M * L \in S(K, G)$.

Conditions S1 to S3 follow from the definition. Condition S4 follows from the Mackey formula (Lemma 2.3.24), and from Condition A1 above.

Conversely, suppose that a class $\mathcal{D}$ of finite groups is given, together with a set $S(H, G)$ of subgroups of $H \times G$, for any elements G and H of $\mathcal{D}$, such that Conditions S1 to S4 are fulfilled.

Then for G and H in $\mathcal{D}$, *define* $\mathrm{Hom}_{\mathcal{D}}(G, H)$ as the subgroup of $B(H, G)$ generated by the elements $(H \times G)/L$, for $L \in S(H, G)$. Then $\mathcal{D}$ is a subcategory of $\mathcal{C}$: Condition S1 and S4, and the Mackey formula, imply that the composition of two morphisms in $\mathcal{D}$ is again a morphism in $\mathcal{D}$, and Condition S3 implies that if $G \in \mathcal{D}$, then Id_G is a morphism in $\mathcal{D}$, since

$$\mathrm{Id}_G = [(G \times G)/\{(g, g) \mid g \in G\}]$$

for any finite group G. Obviously then $\mathcal{D}$ is an admissible subcategory of $\mathcal{C}$.

4.1.7. Definition (replete subcategories) : *A class $\mathcal{D}$ of finite groups is said to be* closed under taking subquotients *if any group isomorphic to a subquotient of an element of $\mathcal{D}$ is in $\mathcal{D}$.*

A subcategory $\mathcal{D}$ of $\mathcal{C}$ is called replete *if it is a full subcategory whose class of objects is closed under taking subquotients.*

4.1.8. Example : If $\mathcal{D}$ is a replete subcategory of $\mathcal{C}$, then $\mathcal{D}$ is admissible: in this case, the set $S(H,G)$ is the set of all subgroups of $H \times G$. Condition A2 holds since for a subgroup L of $H \times G$, the group $q(L)$ is isomorphic to a subquotient of both G and H (see Notation 2.3.21).

4.1.9. Example ((P, Q)-free bisets) : A class $\mathcal{P}$ of finite groups is said to be *closed under taking subquotients and extensions*, if the following conditions are fulfilled, where G, H and N are finite groups:

1. If $G \in \mathcal{P}$, and if $H \leq G$, then $H \in \mathcal{P}$.
2. If $N \trianglelefteq G$, then $G \in \mathcal{P}$ if and only if $N \in \mathcal{P}$ and $G/N \in \mathcal{P}$.

Let $\mathcal{P}$ and $\mathcal{Q}$ be classes of finite groups with these two properties. If G and H are finite groups, *a $(\mathcal{P},\mathcal{Q})$-free (H,G)-biset* U is an (H,G)-biset such that for any $u \in U$, the stabilizer of u in H belongs to $\mathcal{P}$ and the stabilizer of u in G belongs to $\mathcal{Q}$. Let $\mathcal{D}$ be the subcategory of $\mathcal{C}$ whose objects are finite groups, but with less morphisms: if G and H are finite groups, then $\mathrm{Hom}_{\mathcal{D}}(G,H)$ is the Grothendieck group of the category of finite $(\mathcal{P},\mathcal{Q})$-free (H,G)-bisets. Then $\mathcal{D}$ is an admissible subcategory of $\mathcal{C}$, corresponding to the case where $S(H,G)$ is the set of subgroups L of $H \times G$ such that $k_1(L) \in \mathcal{P}$, and $k_2(L) \in \mathcal{Q}$. Conditions S1, S2, and S3 are obviously satisfied. Condition S4 follows from Lemma 2.3.22: with the notation of S4, there is an exact sequence of groups

$$\mathbf{1} \mapsto k_1(M) \to k_1(M * L) \to \big(p_2(M) \cap k_1(L)\big)/\big(k_2(M) \cap k_1(L)\big) \to \mathbf{1}\ .$$

Now the group $\big(p_2(M) \cap k_1(L)\big)/\big(k_2(M) \cap k_1(L)\big)$ is a subquotient of $k_1(L)$, and $k_1(M * L)$ is an extension of this group by $k_1(M)$. Thus $k_1(M * L) \in \mathcal{P}$, and similarly $k_2(M * L) \in \mathcal{Q}$.

4.1.10. Remark : The previous example shows that the subcategories $\mathcal{I}$ and $\mathcal{M}$ introduced in Sects. 3.2.5 and 3.2.6 are admissible subcategories of $\mathcal{C}$: the category $\mathcal{I}$, corresponding to inflation functors, is the case where $\mathcal{P}$ consists of all finite groups, and $\mathcal{Q}$ consists of only the trivial group. The category $\mathcal{M}$, corresponding to global Mackey functors, is the case where both $\mathcal{P}$ and $\mathcal{Q}$ consist of only the trivial group.

4.2. Restriction and Induction

4.2.1. Let $\mathcal{D}$ be a subcategory of the biset category $\mathcal{C}$, and let R be a commutative ring with identity element. *A simple biset functor* on $\mathcal{D}$, with values in R-Mod, is by definition a simple object of $\mathcal{F}_{\mathcal{D},R}$: it is a non zero functor F, whose only subfunctors are F itself and the zero functor. It will be shown in Corollary 4.2.4 that such simple biset functors always exist, as long as $\mathcal{D}$ contains group isomorphisms. The first step to show this is to consider restriction of simple functors to subcategories of $\mathcal{D}$:

4.2.2. Proposition : *Let R be a commutative ring with identity element, let $\mathcal{D}$ be a subcategory of the biset category $\mathcal{C}$, containing group isomorphisms, and let $\mathcal{D}'$ be a full subcategory of $\mathcal{D}$.*

If F is a simple object of $\mathcal{F}_{\mathcal{D},R}$, and if $\mathcal{R}es^{\mathcal{D}}_{\mathcal{D}'}F \neq \{0\}$, then $\mathcal{R}es^{\mathcal{D}}_{\mathcal{D}'}F$ is a simple object of $\mathcal{F}_{\mathcal{D}',R}$.

Proof: Let M be a non zero subfunctor of $\mathcal{R}es^{\mathcal{D}}_{\mathcal{D}'}F$. By adjunction, the (non zero) inclusion morphism $M \to \mathcal{R}es^{\mathcal{D}}_{\mathcal{D}'}F$ yields a non zero morphism

$$s : {}^{l}\mathcal{I}nd^{\mathcal{D}}_{\mathcal{D}'}M \to F \ .$$

The image of s is a non zero subfunctor of F, so it is equal to F, and s is surjective. Hence $\mathcal{R}es^{\mathcal{D}}_{\mathcal{D}'}s$ is surjective, since the functor $\mathcal{R}es^{\mathcal{D}}_{\mathcal{D}'}$ is exact. So the map

$$\mathcal{R}es^{\mathcal{D}}_{\mathcal{D}'}s : \mathcal{R}es^{\mathcal{D}}_{\mathcal{D}'} \circ {}^{l}\mathcal{I}nd^{\mathcal{D}}_{\mathcal{D}'}M \to \mathcal{R}es^{\mathcal{D}}_{\mathcal{D}'}F$$

is surjective. But by Proposition 3.3.4, the functor $\mathcal{R}es^{\mathcal{D}}_{\mathcal{D}'} \circ {}^{l}\mathcal{I}nd^{\mathcal{D}}_{\mathcal{D}'}$ is isomorphic to the identity functor. It follows that the map $\mathcal{R}es^{\mathcal{D}}_{\mathcal{D}'}s$ is isomorphic to the inclusion map

$$M \to \mathcal{R}es^{\mathcal{D}}_{\mathcal{D}'}F \ ,$$

so this map is surjective, and $M = \mathcal{R}es^{\mathcal{D}}_{\mathcal{D}'}F$. □

Conversely, in the same situation, one can try to start with a simple functor on $\mathcal{D}'$, and take its image by the functor ${}^{l}\mathcal{I}nd^{\mathcal{D}}_{\mathcal{D}'}$:

4.2.3. Proposition : *Let R be a commutative ring with identity element, let $\mathcal{D}$ be a subcategory of the biset category $\mathcal{C}$, containing group isomorphisms, and let $\mathcal{D}'$ be a full subcategory of $\mathcal{D}$.*

If F is a simple object of $\mathcal{F}_{\mathcal{D}',R}$, then the functor $I = {}^{l}\mathcal{I}nd^{\mathcal{D}}_{\mathcal{D}'}F$ has a unique proper maximal subfunctor J, and the quotient $S = I/J$ is a simple object of $\mathcal{F}_{\mathcal{D},R}$, such that $\mathcal{R}es^{\mathcal{D}}_{\mathcal{D}'}S \cong F$. Moreover, the functor $\mathcal{R}es^{\mathcal{D}}_{\mathcal{D}'}$ induces an R-linear isomorphism of skew fields

$$\mathrm{End}_{\mathcal{F}_{\mathcal{D},R}}(S) \cong \mathrm{End}_{\mathcal{F}_{\mathcal{D}',R}}(F) \ .$$

Proof: Let M be a subfunctor of $I = {}^l\mathcal{I}nd_{\mathcal{D}'}^{\mathcal{D}}F$. Then $\mathcal{R}es_{\mathcal{D}'}^{\mathcal{D}}M$ is a subfunctor of $\mathcal{R}es_{\mathcal{D}'}^{\mathcal{D}}I$, which is isomorphic to F by Proposition 3.3.4. Thus $\mathcal{R}es_{\mathcal{D}'}^{\mathcal{D}}M$ is equal to $\{0\}$ or F.

If $\mathcal{R}es_{\mathcal{D}'}^{\mathcal{D}}M = F$, then ${}^l\mathcal{I}nd_{\mathcal{D}'}^{\mathcal{D}} \circ \mathcal{R}es_{\mathcal{D}'}^{\mathcal{D}}M = \mathcal{I}nd_{\mathcal{D}'}^{\mathcal{D}}F$, and there is a factorization

$$\begin{array}{ccc} {}^l\mathcal{I}nd_{\mathcal{D}'}^{\mathcal{D}} \circ \mathcal{R}es_{\mathcal{D}'}^{\mathcal{D}}M & \xrightarrow{\cong} & \mathcal{I}nd_{\mathcal{D}'}^{\mathcal{D}}F = I \\ & {\scriptstyle\eta}\searrow \quad \nearrow{\scriptstyle i} & \\ & M & \end{array}$$

where η is the counit of the adjunction, and i is the inclusion map. In particular i is surjective, hence $M = I$ in this case.

Thus if M is a proper subfunctor of I, then $\mathcal{R}es_{\mathcal{D}'}^{\mathcal{D}}M = \{0\}$, and for any object G of $\mathcal{D}$, the R-module $M(G)$ is contained in the submodule $J(G)$ of $I(G)$ defined by

$$J(G) = \bigcap_{f:G\to H\in Ob(\mathcal{D}')} \mathrm{Ker}\, I(f) ,$$

where the intersection runs through all morphisms f in $R\mathcal{D}$ from G to an object H of $\mathcal{D}'$.

Now if $\varphi : G \to G'$ is a morphism in $R\mathcal{D}$, it is clear that

$$I(f)\big(J(G)\big) \subseteq J(G') ,$$

so J is actually a subfunctor of I. Since $\mathcal{R}es_{\mathcal{D}'}^{\mathcal{D}}I \cong F \neq \{0\}$, the functor J is a proper subfunctor of I, and it is the largest such subfunctor. In particular, the quotient functor I/J is a simple object of $\mathcal{F}_{\mathcal{D},R}$, and $\mathcal{R}es_{\mathcal{D}'}^{\mathcal{D}}S \cong F$, since the restriction of J to $\mathcal{D}'$ is equal to $\{0\}$.

Finally, by adjunction, the functor $\mathcal{R}es_{\mathcal{D}'}^{\mathcal{D}}$ induces an isomorphism of R-modules

$$\begin{aligned} \mathrm{Hom}_{\mathcal{F}_{\mathcal{D},R}}({}^l\mathcal{I}nd_{\mathcal{D}'}^{\mathcal{D}}F, S) &\cong \mathrm{Hom}_{\mathcal{F}_{\mathcal{D}',R}}(F, \mathcal{R}es_{\mathcal{D}'}^{\mathcal{D}}S) \\ &\cong \mathrm{End}_{\mathcal{F}_{\mathcal{D}',R}}(F) . \end{aligned}$$

Let f be any morphism from $I = {}^l\mathcal{I}nd_{\mathcal{D}'}^{\mathcal{D}}F$ to S. If $f(J) \neq \{0\}$, then $f(J) = S = f(I)$, since S is simple, so $\mathrm{Ker}\, f + J = I$. Since J is the largest proper subfunctor of I, it follows that $\mathrm{Ker}\, f = I$, i.e. $f = 0$, so $f(J) = \{0\}$. This contradiction shows that any morphism $I \to S$ factors through the projection morphism $I \to I/J = S$. In other words $\mathrm{Hom}_{\mathcal{F}_{\mathcal{D},R}}(I, S) \cong \mathrm{End}_{\mathcal{F}_{\mathcal{D},R}}(S)$, so the functor $\mathcal{R}es_{\mathcal{D}'}^{\mathcal{D}}$ induces is an isomorphism of R-modules

$$\mathrm{End}_{\mathcal{F}_{\mathcal{D},R}}(S) \cong \mathrm{End}_{\mathcal{F}_{\mathcal{D}',R}}(F) ,$$

which is easily seen to be an isomorphism of R-algebras. Both algebras are moreover skew fields, by the Schur Lemma. □

4.2.4. Corollary : *Let R be a commutative ring with identity element, let $\mathcal{D}$ be a subcategory of the biset category $\mathcal{C}$, containing group isomorphisms.*

1. *If F is a simple object of $\mathcal{F}_{\mathcal{D},R}$, and G is an object of $\mathcal{D}$ such that $F(G) \neq \{0\}$, then $F(G)$ is a simple $\mathrm{End}_{R\mathcal{D}}(G)$-module, and evaluation at G induces an R-linear isomorphism of skew fields*

$$\mathrm{End}_{\mathrm{End}_{R\mathcal{D}}(G)}\big(F(G)\big) \cong \mathrm{End}_{\mathcal{F}_{\mathcal{D},R}}(S) \ .$$

2. *If G is an object of $\mathcal{D}$, and V is a simple $\mathrm{End}_{R\mathcal{D}}(G)$-module, then the functor $L_{G,V}$ has a unique proper maximal subfunctor $J_{G,V}$, and the quotient $S_{G,V} = L_{G,V}/J_{G,V}$ is a simple object of $\mathcal{F}_{\mathcal{D},R}$, such that $S_{G,V}(G) \cong V$.*

Proof: It is the special case where $\mathcal{D}'$ consists of the single object G. □

4.2.5. Remark : The existence of simple $\mathrm{End}_{R\mathcal{D}}(G)$-modules follows from Zorn's Lemma: there exists a maximal proper left ideal in any ring with identity element. It follows that there exist simple biset functors on $\mathcal{D}$, as long as $\mathcal{D}$ contains group isomorphisms.

4.2.6. Remark : It is worth recording the structure of $J_{G,V}$ in this case: if H is any object of $\mathcal{D}$, then $J_{G,V}(H)$ is equal to the set of finite sums $\sum_{i=1}^{n} \varphi_i \otimes v_i$ in $L_{G,V}(H)$, where $\varphi_i \in \mathrm{Hom}_{R\mathcal{D}}(G,H)$ and $v_i \in V$, such that $\sum_{i=1}^{n} (\psi\varphi_i) \cdot v_i = 0$ for any $\psi \in \mathrm{Hom}_{R\mathcal{D}}(H,G)$, where $(\psi\varphi_i) \cdot v_i$ denote the image of the element v_i of V under the action of the endomorphism $\psi\varphi_i$ of G.

4.3. The Case of an Admissible Subcategory

4.3.1. Endomorphism Algebras. When $\mathcal{D}$ is an admissible subcategory of $\mathcal{C}$ (in the sense of Definition 4.1.3), the description of simple objects of $\mathcal{F}_{\mathcal{D},R}$ can be made much more precise.

4.3.2. Proposition : *Let R be a commutative ring with identity, and let $\mathcal{D}$ be an admissible subcategory of the biset category $\mathcal{C}$. If G is an object of $\mathcal{D}$, denote by I_G the R-submodule of $\mathrm{End}_{R\mathcal{D}}(G)$ generated by all endomorphisms of G which can be factored through some object H of $\mathcal{D}$ with $|H| < |G|$. Then I_G is a two sided ideal of $\mathrm{End}_{R\mathcal{D}}(G)$, and there is a decomposition*

$$\mathrm{End}_{R\mathcal{D}}(G) = A_G \oplus I_G$$

where A_G is an R-subalgebra, isomorphic to the group algebra $R\mathrm{Out}(G)$ of the group of outer automorphisms of G.

Proof: First I_G is obviously a two sided ideal of $\mathrm{End}_{R\mathcal{D}}(G)$. Now since $\mathcal{D}$ is admissible, the R-module $\mathrm{End}_{R\mathcal{D}}(G)$ is a free R-module on the set of elements $[(G\times G)/L]$, for $L \in [S(G,G)]$, by Condition A1 of Definition 4.1.3, where $[S(G,G)]$ is some set of representatives of $(G\times G)$-conjugacy classes of elements of $S(G,G)$. Moreover, by Condition A2 of this definition, for each of these subgroups L, the group $q(L)$ is an object of $\mathcal{D}$, and the endomorphism $[(G\times G)/L]$ factors through $q(L)$, by the Factorization Property 4.1.5.

If $|q(L)| < |G|$, this shows that $[(G\times G)/L] \in I_G$. Denote by σ_G the subset of $|S(G,G)|$ consisting of such subgroups, and set $[\sigma_G] = \sigma_G \cap [S(G,G)]$. Denote by J_G the free R-submodule of $\mathrm{End}_{R\mathcal{D}}(G)$ generated by the elements $[(G\times G)/L]$, for $L \in [\sigma_G]$. Thus $J_G \subseteq I_G$.

On the other hand $|q(L)| = |G|$ if and only if $p_1(L)$ and $p_2(L)$ are both equal to G, and if $k_1(L)$ and $k_2(L)$ are both trivial, or equivalently if

$$L = \Delta_f(G) = \{\big(f(x),x\big) \mid x \in G\}\ ,$$

where f is some automorphism of G.

Conversely, if f is an automorphism of G, then $\Delta_f(G) \in S(G,G)$, by Remark 4.1.4, and the (G,G)-biset $(G\times G)/\Delta_f(G)$ is isomorphic to $\mathrm{Iso}(f)$. It is easy to see that if f and f' are automorphisms of G, then the subgroups $\Delta_f(G)$ and $\Delta'_{f'}G)$ are conjugate in $G\times G$ (or equivalently the (G,G)-bisets $\mathrm{Iso}(f)$ and $\mathrm{Iso}(f')$ are isomorphic) if and only if $f^{-1}f'$ is an inner automorphism of G.

Denote by $\Sigma(G)$ the set of subgroups of $G\times G$ of the form $\Delta_f(G)$, for some automorphism f of G, and set $[\Sigma(G)] = \Sigma(G)\cap[S(G,G)]$. Then $[\Sigma(G)]$ is in one to one correspondence with $\mathrm{Out}(G)$. Let A_G denote the free R-submodule of $\mathrm{End}_{R\mathcal{D}}(G)$ generated by the elements $[(G\times G)/L]$, for $L \in [\Sigma(G)]$. Since $\mathrm{Iso}(f)\times_G \mathrm{Iso}(f') = \mathrm{Iso}(ff')$ for any automorphisms f and f' of G, it follows that A_G is actually an R-subalgebra of $\mathrm{End}_{R\mathcal{D}}(G)$, and A_G is isomorphic to the group algebra $R\mathrm{Out}(G)$.

Now clearly $[S(G,G)] = [\Sigma(G)] \sqcup [\sigma_G]$, and this gives a decomposition

$$\mathrm{End}_{R\mathcal{D}}(G) = A_G \oplus J_G\ .$$

Since $J_G \subseteq I_G$, it just remains to show that $I_G \subseteq J_G$. An element α of I_G can be written as a sum

$$\alpha = \sum_{i=1}^{n} \psi_i \circ \phi_i\ ,$$

where $\phi_i : G \to H_i$ and $\psi_i : H_i \to G$ are morphism in $R\mathcal{D}$, and H_i is an object of $\mathcal{D}$ with $|H_i| < |G|$, for $i \in \{1\ldots n\}$. Now each ϕ_i is a linear combination of morphisms $[(H_i \times G)/L_{i,j}]$, for $j \in \{1\ldots n_i\}$, and each ψ_i is a linear combination of morphisms $[(G\times H_i)/M_{i,j}]$, for $j \in \{1\ldots m_i\}$. The Mackey formula 2.3.24 implies that β is a linear combination of morphisms of the form

$$[(G \times G)/M_{i,j} * L'_{i,k}] ,$$

for some integers j and k, where $L'_{i,k}$ is some conjugate of $L_{i,k}$ in $H_i \times G$. By Lemma 2.3.22, the group $q(M_{i,j} * L'_{i,k})$ is isomorphic to a subquotient of $q(M_{i,j})$ and $q(L'_{i,k})$. In particular, it is isomorphic to a subquotient of H_i, hence $|q(M_{i,j} * L'_{i,k})| < |G|$, thus $I_G \subseteq J_G$, as was to be shown. □

4.3.3. Remark : It will be useful to record that in this situation, the ideal I_G of $\mathrm{End}_{R\mathcal{D}}(G)$ is a free R-module on the set of elements $[(G \times G)/L]$, for $L \in [S(G,G)]$ with $|q(L)| < |G|$.

4.3.4. Notation : *If G is an object of $\mathcal{D}$, let π_G denote the projection map $\mathrm{Aut}(G) \to \mathrm{Out}(G)$, and ϖ_G denote the projection map $\mathrm{End}_{R\mathcal{D}}(G) \to R\mathrm{Out}(G)$ defined by*

$$\varpi([(G \times G)/L]) = \begin{cases} 0 & \text{if } |q(L)| < |G| \\ \pi_G(\theta) & \text{if } \theta \in \mathrm{Aut}(G),\ L = \Delta_\theta(G) . \end{cases}$$

4.3.5. Parametrization of Simple Functors. Let R be a commutative ring with identity element, and $\mathcal{D}$ be an admissible subcategory of the biset category $\mathcal{C}$. It means in particular that $\mathcal{D}$ contains group isomorphisms, thus by Corollary 4.2.4, there exist simple objects in $\mathcal{F}_{\mathcal{D},R}$.

Let S be such a simple functor. Thus S is non zero, and there exists a minimal group G for S (see Definition 3.2.2). If f is an endomorphism of G in $R\mathcal{D}$ which factors through an object H of $\mathcal{D}$ with $|H| < |G|$, then $S(f) = 0$, since $S(f)$ factors through $S(H) = \{0\}$. It follows that I_G acts by 0 on $S(G)$, i.e. that $S(G)$ is a module for the quotient algebra

$$\mathrm{End}_{R\mathcal{D}}(G)/I_G = (A_G \oplus I_G)/I_G \cong A_G \cong R\mathrm{Out}(G) .$$

By Proposition 4.2.4, this module $S(G)$ is a simple $R\mathrm{Out}(G)$-module.

Conversely, suppose that G is an object of $\mathcal{D}$, and V is a simple $R\mathrm{Out}(G)$-module. Then V becomes a (simple) $\mathrm{End}_{R\mathcal{D}}(G)$-module via the projection algebra homomorphism $\mathrm{End}_{R\mathcal{D}}(G) \to A_G$. This module is still denoted abusively by V. By Proposition 4.2.4, the functor $L_{G,V}$ has a unique simple quotient $S_{G,V}$, and $S_{G,V}(G) \cong V$.

4.3.6. Definition : *Let R be a commutative ring with identity element, and $\mathcal{D}$ be an admissible subcategory of $\mathcal{C}$. A pair (G,V), where G is an object of $\mathcal{D}$, and V is a simple $R\mathrm{Out}(G)$-module, is called* a seed *of $R\mathcal{D}$.*

If (G,V) and (G',V') are seeds of $R\mathcal{D}$, then an isomorphism *from (G,V) to (G',V') is a pair (φ,ψ), where $\varphi : G \to G'$ is a group isomorphism, and $\psi : V \to V'$ is an R-module isomorphism such that*

$$\forall v \in V,\ \forall a \in \mathrm{Out}(G),\ \psi(a \cdot v) = (\varphi a \varphi^{-1}) \cdot \psi(v)\ .$$

Two seeds (G,V) and (G',V') of $R\mathcal{D}$ are said to be isomorphic *if there exists an isomorphism from (G,V) to (G',V').*

If (G,V) is a seed of $R\mathcal{D}$, the associated simple functor *is the unique simple quotient $S_{G,V}$ of $L_{G,V}$.*

4.3.7. Remark : Isomorphism is clearly an equivalence relation on the class of seeds of $R\mathcal{D}$. Moreover, if (G,V) and (G',V') are isomorphic seeds of $R\mathcal{D}$, then $L_{G,V} \cong L_{G',V'}$, and $S_{G,V} \cong S_{G',V'}$.

4.3.8. Remark : In the case $G = G'$, an isomorphism (φ,ψ) from (G,V) to (G,V') consists of an automorphism φ of G, and an isomorphism $\psi : V \to V'$ such that the map $v \mapsto \varphi^{-1} \cdot \psi(v)$ commutes with the action of $\mathrm{Out}(G)$. This shows that (G,V) and (G,V') are isomorphic if and only if the $R\mathrm{Out}(G)$-modules V and V' are isomorphic.

4.3.9. Lemma : *Let R be a commutative ring with identity element, and let D be an admissible subcategory of the biset category. Let $S_{G,V}$ denote the simple functor associated to the seed (G,V) of $R\mathcal{D}$.*

If H is an object of $\mathcal{D}$ such that $S_{G,V}(H) \neq \{0\}$, then G is isomorphic to a subquotient of H.

Proof: Indeed $S_{G,V}(H) \neq \{0\}$ if and only if $L_{G,V}(H) \neq J_{G,V}(H)$. By Remark 4.2.6, it means that there exists elements $\varphi \in \mathrm{Hom}_{R\mathcal{D}}(G,H)$ and $\psi \in \mathrm{Hom}_{R\mathcal{D}}(H,G)$ such that $\psi\varphi$ has a non zero action on V. In particular $\psi\varphi \notin I_G$. It follows that there exist a subgroup L of $H \times G$ and a subgroup M of $G \times H$ such that the product

$$(G \times H)/M \times_H (H \times G)/L = \sum_{x \in p_2(M)\backslash H/p_1(L)} (G \times G)/(M *^{(x,1)} L)$$

is not in I_G.

So there exists $x \in H$ such that the group $Q = q(M *^{(x,1)} L)$ does not have order smaller than $|G|$. Since Q is always a subquotient of G, it follows that $Q \cong G$. Now Q is also a subquotient of $q(M)$, by Proposition 2.3.22, and $q(M)$ is a subquotient of G. It follows that $q(M) \cong G$, so that M is equal to

$$M = \{\big(f(x), x\big) \mid x \in K\}\ ,$$

where K is some subgroup of H and f is a surjective homomorphism from K to G. Thus G is isomorphic to a quotient of K, i.e. to a subquotient of H. □

4.3.10. Theorem : *Let R be a commutative ring with identity element, and let D be an admissible subcategory of the biset category. There is a one to one correspondence between the set of isomorphism classes of simple objects of $\mathcal{F}_{\mathcal{D},R}$ and the set of isomorphisms classes of seeds of $R\mathcal{D}$, sending the class of the simple functor S to the isomorphism class of a pair $\big(G,S(G)\big)$, where G is any minimal group for S. The inverse correspondence maps the class of the seed (G,V) to the class of the functor $S_{G,V}$.*

Proof: Let S be a simple object in $\mathcal{F}_{\mathcal{D},R}$. Let G be a minimal group for S, and set $V=S(G)$. Then (G,V) is a seed of $R\mathcal{D}$. Since $V=S(G)$, it follows by adjunction that there is a non zero morphism $L_{G,V}\to S$, and this morphism is surjective since S is simple. But $L_{G,V}$ has a unique simple quotient $S_{G,V}$, thus $S\cong S_{G,V}$.

If G' is another minimal group for S, and if $V'=S(G')$, then $S\cong S_{G',V'}$ also. Since $S_{G,V}(G')\neq\{0\}$, it follows from Lemma 4.3.9 that G is isomorphic to a subquotient of G'. By symmetry of G and G', the group G' is also a subquotient of G, so there exists a group isomorphism $\varphi: G'\to G$. Then obviously $S_{G',V'}\cong S_{G,{}^{\varphi}V'}$, where ${}^{\varphi}V'$ is the $R\mathrm{Out}(G)$-module equal to V' as an R-module, with $\mathrm{Out}(G)$ action defined by

$$\forall a\in \mathrm{Out}(G),\ \forall v'\in V',\ \ a.v'\ (\text{in } {}^{\varphi}V')=(\varphi^{-1}a\varphi)v'\ (\text{in } V)\ .$$

Now the functors $S_{G,V}$ and $S_{G,{}^{\varphi}V'}$ are isomorphic, so there exists an isomorphism ψ of $R\mathrm{Out}(G)$-modules from ${}^{\varphi}V'$ to V. Clearly, the pair (φ,ψ) is an isomorphism from the seed (G',V') to the seed (G,V).

This shows that there is a well defined map s from the set $\mathcal{A}$ of isomorphism classes of simple objects in $\mathcal{F}_{\mathcal{D},R}$ to the set $\mathcal{B}$ of isomorphisms classes of seeds of $R\mathcal{D}$, sending the class of S to the class of the seed (G,V) defined above.

Conversely, there a map $t:\mathcal{B}\to\mathcal{A}$, sending the class of (G,V) to the class of $S_{G,V}$. Now by Lemma 4.3.9, and since $S_{G,V}(G)\cong V$, the group G is minimal for $S_{G,V}$, and $s\circ t=\mathrm{Id}_{\mathcal{B}}$. Conversely, if S is a simple object of $\mathcal{F}_{\mathcal{D},R}$, if G is a minimal group for S, and if $V=S(G)$, then $S\cong S_{G,V}$ by the above argument, thus $t\circ s=\mathrm{Id}_{\mathcal{A}}$. It follows that s and t are mutual inverse bijections between $\mathcal{A}$ and $\mathcal{B}$. □

4.3.11. Linked Sections and Isomorphisms. Let R be a commutative ring with identity element, and $\mathcal{D}$ be an admissible subcategory of the biset category $\mathcal{C}$. Theorem 4.3.10 shows that the simple objects of $\mathcal{F}_{\mathcal{D},R}$ are parametrized up to isomorphism by pairs (G,V), where G is an object of $\mathcal{D}$ and V is a simple $R\mathrm{Out}(G)$-module. In this situation, the evaluations of the simple functor $S_{G,V}$ can be computed from combinatorial information on the relative positions of sections of G:

4.3.12. Definition : *Let G be a group. Two sections (D,C) and (B,A) of G are said to be* linked *(notation $(D,C) — (B,A)$) if*

$$D \cap A = C \cap B \qquad (D \cap B)C = D \qquad (D \cap B)A = B\ .$$

They are said to be linked modulo G *(notation $(D,C) —_G (B,A)$) if there exists $g \in G$ such that $(D,C) — ({}^gB, {}^gA)$.*

4.3.13. Remark : The sections (D,C) and (B,A) of G are linked if and only if $D \cap A = C \cap B$ and $DA = CB$ (note that the sets DA and CB need not be subgroups of G).

4.3.14. Remark : When (B,A) and (D,C) are arbitrary sections of G, the Zassenhaus's butterfly diagram is the following picture:

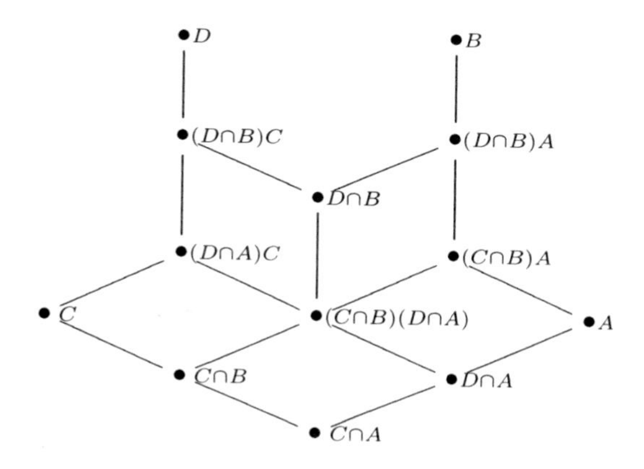

In this situation, the set $U = C\backslash DB/A$ is a transitive $(D/C, B/A)$-biset. The stabiliser H of the point CA of this biset is equal to

$$H = \{(yC, xA) \mid x \in B,\ y \in\ D, yx^{-1} \in CA\}\ .$$

Thus $p_1(H) = (D \cap B)C$, $k_1(H) = (D \cap A)C$, $p_2(H) = (D \cap B)A$, and $k_2(H) = (C \cap B)A$. So there is an isomorphism of $(D/C, B/A)$-bisets

$$U \cong \mathrm{Indinf}_{(D\cap B)C/(D\cap A)C}^{D/C} \circ \mathrm{Iso}_{(D\cap B)A/(C\cap B)A}^{(D\cap B)C/(D\cap A)C} \circ \mathrm{Defres}_{(D\cap B)A/(C\cap B)A}^{B/A}\ ,$$

where $\mathrm{Iso}_{(D\cap B)A/(C\cap B)A}^{(D\cap B)C/(D\cap A)C}$ is the biset associated to the group isomorphism

$$(D \cap B)A/(C \cap B)A \to (D \cap B)C/(D \cap A)C$$

sending xA to xC, for $x \in D \cap B$.

Then (B,A) and (D,C) are linked if and only if this diagram reduces to

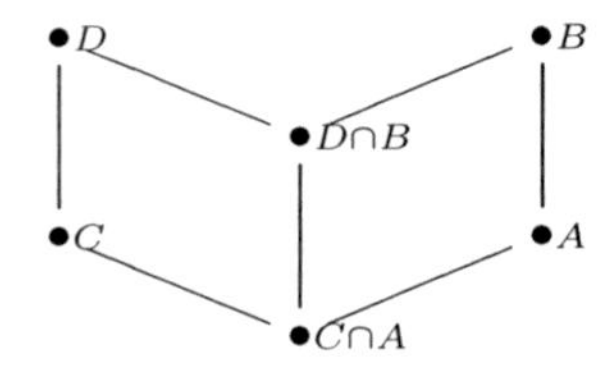

This situation yields a natural group isomorphism $\varphi_{(B,A)}^{(D,C)} : B/A \to D/C$, sending xA to xC, for $x \in D \cap B$.

More generally, when $g \in G$ is such that $(D^g, C^g) — (B,A)$, the composition of $\varphi_{(B,A)}^{(D^g,C^g)} : B/A \to D^g/C^g$ with the isomorphism $D^g/C^g \to D/C$ induced by the map $x \mapsto {}^g x$ from D^g to D, yields a group isomorphism $\theta_g : B/A \to D/C$.

4.3.15. Lemma : *Let G be a group, let (B,A) and (D,C) be two sections of G, and let g be an element of G. The following conditions are equivalent:*

1. *$(D^g, C^g) — (B,A)$.*
2. *There exists a group isomorphism $\varphi : B/A \to D/C$ such that the $(D/C, B/A)$-biset $C\backslash DgB/A$ is isomorphic to* $\mathrm{Iso}(\varphi)$.

Moreover, if these conditions hold, then φ is equal to the isomorphism θ_g defined above, up to some inner automorphism of B/A.

Proof: Suppose first that $(D^g, C^g) — (B,A)$. Then the group isomorphism $\theta_g : B/A \to D/C$ yields a $(D/C, B/A)$-biset $\mathrm{Iso}(\theta_g)$, equal to D/C as a set. Define a map

$$f : \mathrm{Iso}(\theta_g) \to C\backslash DgB/A$$

by $f(dC) = dCgA = CdgA$, for $d \in D$. Then f in injective: if $f(d'C) = f(dC)$, for $d, d' \in D$, then $d'g \in CdgA$, i.e.

$$d' \in Cd({}^g A) \cap D = Cd({}^g A \cap D) = Cd({}^g B \cap C) = Cd = dC \; .$$

The map f is also surjective: if $d \in D$ and $b \in B$, then there exists $e \in B \cap D$ and $a \in A$ such that ${}^g b = e({}^g a)$. Hence

$$CdgbA = Cd({}^g b)gA = Cde({}^g a)gA = CdegA = f(deC) \; .$$

Thus f is a bijection. It is an isomorphism of $(D/C, B/A)$-bisets: if $d, e \in D$ and $b \in B$, then the $(D/C, B/A)$-biset structure on $\mathrm{Iso}(\theta_g) = D/C$ is defined by

$$dC \cdot eC \cdot bA = dC\theta_g(bA) \; ,$$

where the right hand side is the product of dC and $\theta_g(bA)$ in the group D/C. If $i \in D^g \cap B$ is such that $i^{-1}b \in A$, then $\theta_g(bA) = {}^g iC$, and

$$\begin{aligned} f(dC \cdot eC \cdot bA) &= f\big(de({}^g i)C\big) = Cde({}^g i)gA = CdegiA \\ &= CdebA = dC \cdot CegA \cdot bA = dCf(eC)bA \ . \end{aligned}$$

Conversely, suppose that Condition 2 holds: then in particular the stabilizer in D/C of the element CgA of $C\backslash DgB/A \cong \mathrm{Iso}(\varphi)$ is trivial. Since

$$(D \cap {}^g A) \cdot CgA = C(D \cap {}^g A)gA = Cg(D^g \cap A)A = CgA \ ,$$

it follows that $D \cap {}^g A \leq C$, hence $D \cap {}^g A \leq C \cap {}^g B$. But the stabilizer of CgA in B/A is also trivial, and a similar argument shows that $C^g \cap B \leq D^g \cap A$. Thus $C^g \cap B = D^g \cap A$.

Now the group D/C acts transitively on the set $C\backslash DgB/A \cong \mathrm{Iso}(\varphi)$. It follows that $D \cdot CgA = DgB$, hence $DgB = DgA$, and $B \leq D^g A$, i.e. $B = (D^g \cap B)A$. Similarly B/A acts transitively on $C\backslash DgB/A$, thus $CgB = DgB$, and $D^g = (D^g \cap B)C^g$. Thus $(D^g, C^g) — (B, A)$, as was to be shown.

Now it follows from the first part of the proof that the $(D/C, B/A)$-bisets $\mathrm{Iso}(\theta_g)$ and $\mathrm{Iso}(\varphi)$ are isomorphic. The following lemma shows that this happens if and only if $\varphi^{-1}\theta_g$ is an inner automorphism of B/A, and this completes the proof. □

4.3.16. Lemma : *Let G and H be groups, and let φ and ψ be groups isomorphisms from G to H. Then the bisets $\mathrm{Iso}(\varphi)$ and $\mathrm{Iso}(\psi)$ are isomorphic if and only if $\psi^{-1}\varphi$ is an inner automorphism of G.*

Proof: Since $\mathrm{Iso}(\psi^{-1}\varphi) \cong \mathrm{Iso}(\psi)^{-1} \circ \mathrm{Iso}(\varphi)$, this lemma is equivalent to the special case where $G = H$ and $\psi = \mathrm{Id}$. So let φ be an automorphism of G, and f be an isomorphism of (G, G)-bisets from $\mathrm{Iso}(\mathrm{Id})$ to $\mathrm{Iso}(\varphi)$. Thus $f : G \to G$ is a map such that

$$f(xyz) = xf(y)\varphi(z) \ ,$$

for all $x, y, z \in G$. Setting $x = y = 1$ gives $f(z) = f(1)\varphi(z)$ for any $z \in G$, and it follows that $f(1)\varphi(xyz) = xf(1)\varphi(y)\varphi(z) = f(1)\varphi(x)\varphi(y)\varphi(z)$. Thus $\varphi(x) = x^{f(1)}$ for any $x \in G$, and φ is inner.

Conversely, if there exists $a \in G$ such that $\varphi(x) = x^a$, for any $x \in G$, then the map $x \mapsto xa$ is an isomorphism of (G, G)-bisets from $\mathrm{Iso}(\mathrm{Id})$ to $\mathrm{Iso}(\varphi)$, and this completes the proof. □

4.3.17. Computing the Evaluations of Simple Functors. Let R be a commutative ring with identity element, let $\mathcal{D}$ be an admissible subcategory of the biset category $\mathcal{C}$, and let (G, V) be a seed of $R\mathcal{D}$. Recall from Remark 4.2.6 that for any object H of $\mathcal{D}$, the evaluation $S_{G,V}(H)$ is equal to the quotient of

$$L_{G,V}(H) = \mathrm{Hom}_{R\mathcal{D}}(G,H) \otimes_{\mathrm{End}_{R\mathcal{D}}(G)} V$$

by the submodule

$$J_{G,V}(H) = \{\sum_{i=1}^{n} \varphi_i \otimes v_i \mid \sum_{i=1}^{n} (\psi\varphi_i)\cdot v_i = 0,\ \forall \psi \in \mathrm{Hom}_{R\mathcal{D}}(H,G)\} .$$

4.3.18. Notation : *If H is an object of $\mathcal{D}$, let $\sigma_G(H)$ be a set of representatives of H-conjugacy classes of sections (T,S) of H such that $T/S \cong G$ and $\mathrm{Defres}^H_{T/S}$ is a morphism in $\mathcal{D}$, i.e. such that*

$$\{(tS,t) \mid t \in T\} \in \mathcal{S}(T/S,H) .$$

For each $(T,S) \in \sigma_G(H)$, let $s_{T,S} : T \to G$ denote a surjective group homomorphism with kernel S, and by $\bar{s}_{T,S} : T/S \to G$ the associated group isomorphism.

Similarly, let $\tau_G(H)$ be a set of representatives of H-conjugacy classes of sections (T,S) of H such that $T/S \cong G$ and $\mathrm{Indinf}^H_{T/S}$ is a morphism in $\mathcal{D}$, i.e. such that

$$\{(t,tS) \mid t \in T\} \in \mathcal{S}(H,T/S) .$$

For each $(T,S) \in \tau_G(H)$, let $t_{T,S} : T \to G$ denote a surjective group homomorphism with kernel S, and by $\bar{t}_{T,S} : T/S \to G$ the associated group isomorphism.

4.3.19. Notation : *If $(T,S) \in \sigma_G(H)$ and $(B,A) \in \tau_G(H)$, denote by $\langle (B,A) \parallel (T,S) \rangle_G$ the element of $R\mathrm{Out}(G)$ defined by*

$$\langle (B,A) \parallel (T,S) \rangle_G = \sum_{\substack{h \in [B\backslash H/T] \\ (B,A) \,—\, ({}^hT,{}^hS)}} \pi_G\big(\bar{t}_{B,A} \circ \theta_h \circ (\bar{s}_{T,S})^{-1}\big) ,$$

where $\theta_h : T/S \to B/A$ is the group isomorphism defined by $\theta_h(tS) = bA$, if $t \in T$ and $b \in B$ are such that $b^{-1}\cdot{}^ht \in A$.

4.3.20. Theorem : *Let R be a commutative ring with identity element, and let $\mathcal{D}$ be an admissible subcategory of the biset category $\mathcal{C}$. Let (G,V) be a seed of $R\mathcal{D}$, and let H be an object of $\mathcal{D}$.*

Then $S_{G,V}(H)$ is isomorphic to the image of the map

$$\Psi : \bigoplus_{(T,S)\in\sigma_G(H)} V \to \bigoplus_{(B,A)\in\tau_G(H)} V$$

whose block from the component indexed by $(T,S) \in \sigma_G(H)$ to the component indexed by $(B,A) \in \tau_G(H)$ is the map defined by

$$\Psi_{(B,A),(T,S)}(v) = \langle (B,A) \parallel (T,S) \rangle_G \cdot v \ .$$

Proof: Recall that $\mathrm{Hom}_{R\mathcal{D}}(G,H)$ is a free R-module with basis the set of elements $[(H \times G)/K]$, for $K \in [S(H,G)]$. Now for $v \in V$, the element $\gamma = [(H \times G)/K] \otimes v$ is in $J_{G,V}(H)$ if and only if for any $L \in [S(G,H)]$, the element

$$w_{L,K}(v) = ([(G \times H)/L] \times_H [(H \times G)/K]) \cdot v$$

is equal to 0 in V. By Lemma 2.3.24

$$\textbf{(4.3.21)} \qquad w_{L,K}(v) = \sum_{x \in [p_2(L) \backslash H / p_1(K)]} [(G \times G)/(L * {}^{(x,1)}K)] \cdot v \ .$$

The only possibly non zero terms in this summation are those for which $q(L*^{(x,1)}K) \cong G$, for otherwise $[(G \times G)/(L*^{(x,1)}K)] \in I_G$, and $I_G \cdot V = \{0\}$ since V is an $R\mathrm{Out}(G)$-module. By Lemma 2.3.22, this implies $q(L) \cong G \cong q({}^{(x,1)}K)$, hence $q(K) \cong G$, and this has two consequences:

(C1). If $\gamma \notin J_{G,V}(H)$, then there is a subgroup M of H and a surjective homomorphism $s : M \to G$ such that $K = \{(m, s(m) \mid m \in M\}$. Hence, there exists an element $h \in H$ and a section $(T,S) \in \sigma_G(H)$ such that $(M, \mathrm{Ker}\ s) = ({}^hT, {}^hS)$. Then $K = \{({}^hx, s({}^hx)) \mid x \in T\}$ is conjugate to $\{(x, s({}^hx)) \mid x \in T\}$ in $H \times G$.
Moreover, since $\mathrm{Ker}\ s = {}^hS = {}^h\mathrm{Ker}\ s_{T,S}$, there is an automorphism ξ of G such that $s({}^hx) = \xi \circ s_{T,S}(x)$ for any $x \in T$. Now in $L_{G,V}(H)$, the element $\gamma = [(H \times G)/K] \otimes v$ is equal to

$$\begin{aligned} \gamma &= [(H \times G)/\{(t, \xi s_{T,S}(t) \mid t \in T\}] \otimes v \\ &= [(H \times G)/\{(t, s_{T,S}(t) \mid t \in T\}] \times_G [(G \times G)/\Delta_{\xi^{-1}}(G)] \otimes v \\ &= [(H \times G)/K_{T,S}] \otimes \pi_G(\xi^{-1}) \cdot v \ , \end{aligned}$$

where $K_{T,S} = \{(t, s_{T,S}(t) \mid t \in T\}$. Thus $S_{G,V}(H)$ is generated by the images of the elements

$$\gamma_{T,S}(v) = [(H \times G)/K_{T,S}] \otimes v \ ,$$

for $(T,S) \in \sigma_G(H)$ and $v \in V$.

(C2). If there exists $K \leq H \times G$ and $v \in V$ such that $w_{L,K}(v) \neq 0$, then there exists a subgroup N of H and a surjective homomorphism $t : N \to G$ such that $L = \{(t(n), n) \mid n \in N\}$. Hence there exists an element $l \in H$ and a section $(B,A) \in \tau_G(H)$ such that $(N, \mathrm{Ker}\ t) = ({}^lB, {}^lA)$.

It follows as above that there exists an automorphism ψ of G such that $t({}^l y) = \psi \circ t_{B,A}(y)$ for $y \in B$, and then

$$\begin{aligned}[(G \times H)/L] &= [(G \times G)/\Delta_\psi(G)] \times_G [(G \times H)/\{\big(t_{B,A}(y), y\big) \mid y \in B\}] \\ &= \mathrm{Iso}(\psi) \circ [(G \times H)/L_{B,A}] \ ,\end{aligned}$$

where $L_{B,A} = \{\big(t_{B,A}(y), y\big) \mid y \in B\}$. In particular

$$w_{L,K}(v) = \pi_G(\psi) \cdot \big(w_{L_{B,A},K}(v)\big) \ .$$

Now if $K_i \leq G \times H$ and $v_i \in V$, for $i = 1 \ldots n$, the sum

$$\Sigma = \sum_{i=1}^{n} w_{L,K_i}(v_i)$$

is equal to the image by $\pi_G(\psi)$ of the sum $\Sigma' = \sum_{i=1}^{n} w_{L_{B,A},K_i}(v_i)$. In particular $\Sigma = 0$ if and only if $\Sigma' = 0$.

So for each $(T, S) \in \sigma_G(H)$, let $v_{T,S} \in V$. The sum $\sum_{(T,S)\in\sigma_G(H)} \gamma_{T,S}(v_{T,S})$ is in $J_{G,V}(H)$ if and only if, for any $(B, A) \in \tau_G(H)$, the sum

$$\mathcal{S}_{B,A} = \sum_{(T,S)\in\sigma_G(H)} \big([(G \times H)/L_{B,A}] \times_H [(H \times G)/K_{T,S}]\big) \cdot v_{T,S}$$

is equal to 0 in V. By Lemma 2.3.24, this is equal to

$$\mathcal{S}_{B,A} = \sum_{\substack{(T,S)\in\sigma_G(H)\\ h\in[B\backslash H/T]}} [(G \times G)/D_h] \cdot v_{T,S} \ ,$$

where $D_h = \{\big(t_{B,A}(b), s_{T,S}(b^h)\big) \mid b \in B \cap {}^hT\}$. Now $q(D_h) \cong G$ if and only $k_1(D_h) = k_2(D_h) = \mathbf{1}$ and $p_1(D_h) = p_2(D_h) = G$, i.e. if

$$B \cap {}^h\mathrm{Ker}\, s_{T,S} = \mathrm{Ker}\, t_{B,A} \cap {}^hT \ ,$$

$$\mathrm{Ker}\, t_{B,A}(B \cap {}^hT) = B \ , \qquad {}^h\mathrm{Ker}\, s_{T,S}(B \cap {}^hT) = {}^hT \ ,$$

in other words if $(B, A) — ({}^hT, {}^hS)$. In this case moreover

$$[(G \times G)/D_h] = \mathrm{Iso}(\varphi_h) \ ,$$

where φ_h is the automorphism of G such that

$$\varphi_h s_{T,S}(b^h) = t_{B,A}(b) \ ,$$

for any $b \in B \cap {}^hT$. In other words $\varphi_h = \bar{t}_{B,A} \circ \theta_h \circ (\bar{s}_{T,S})^{-1}$, where θ_h is the group isomorphism from T/S to B/A defined by $\theta_h(xS) = yB$ if $x \in T$, $y \in B$, and $y^{-1}({}^hx) \in A$. This gives finally

$$\mathcal{S}_{B,A} = \sum_{(T,S)\in\sigma_G(H)} \langle (B,A) \parallel (T,S) \rangle_G \cdot v_{T,S} \ .$$

It follows that the map $\Psi' : \bigoplus_{(T,S)\in\sigma_G(H)} V \to S_{G,V}(H)$ sending $\sum_{(T,S)} v_{T,S}$ to the image of $\sum_{(T,S)} \gamma_{T,S}(v_{T,S})$ is surjective, and its kernel is equal to the kernel of the map Ψ defined in the statement of Theorem 4.3.20. Since Ψ and Ψ' have the same domain and the same kernel, the image $S_{G,V}(H)$ of Ψ' is isomorphic to the image of Ψ, as was to be shown. □

4.4. Examples

4.4.1. Simple $R\mathrm{Out}(G)$-Modules. Let R be a commutative ring with identity element. The next two results are devoted to showing that if Ω is a finite group, e.g. if $\Omega = \mathrm{Out}(G)$ for a finite group G, then the simple $R\Omega$-modules are actually simple $K\Omega$-modules, where $K = R/I$ is a quotient of R by some maximal ideal I.

4.4.2. Lemma : *Let R be a commutative ring with identity element. If there exists a non zero finitely generated R-module V such that for any $r \in R-\{0\}$, the map $v \mapsto rv$ is a bijection from V to V, then R is a field.*

Proof: First R has no zero divisors, since for $r, s \in R - \{0\}$, the map $v \mapsto srv$ from V to V is a bijection, being the composition of the two bijections $v \mapsto rv$ and $v \mapsto sv$. So $rs \neq 0$.

Consider finite subsets $v_1, \ldots, v_k$ of V with the following property (P): for any $v \in V$, there exist $r \in R - \{0\}$ and $r_1, \ldots, r_k \in R$ such that

$$rv = r_1v_1 + \ldots + r_kv_k \ .$$

For example the finite generating subsets of V as an R-module have the property (P).

Let $W = \{w_1, \ldots, w_l\}$ be a minimal such subset of V. Then W is linearly independent: if there is a non trivial linear relation

$$s_1w_1 + \ldots + s_lw_l = 0$$

with, e.g. $s_l \neq 0$, then $s_lw_l = -s_1w_1 - \ldots - s_{l-1}w_{l-1}$. Now for any $v \in V$, there exists $r \in R - \{0\}$ and $r_1, \ldots, r_l \in R$ with

$$rv = r_1w_1 + \ldots + r_lw_l \ .$$

It follows that

$$\begin{aligned} s_l r v &= s_l r_1 w_1 + \ldots + s_l r_{l-1} w_{l-1} - r_l (s_1 w_1 + \ldots + s_{l-1} w_{l-1}) \\ &= (s_l r_1 - r_l s_1) w_1 + \ldots + (s_l r_{l-1} - r_l s_{l-1}) w_{l-1} \end{aligned}$$

Moreover $s_l r \neq 0$ since $s_l \neq 0$ and $r \neq 0$. It follows that the subset $\{w_1, \ldots, w_{l-1}\}$ has the property (P), and this contradicts the minimality of W.

Hence W generates a free R-submodule $M \cong R^l$ of V. Moreover, for any $v \in V$, there exists $r \in R - \{0\}$ such that $rv \in M$. Let $u_1, \ldots, u_m$ be a generating set of V as an R-module. For each $i = 1, \ldots n$, let $r_i \in R - \{0\}$ such that $r_i u_i \in M$, and set $r = r_1 \ldots r_n$. It follows that $r \neq 0$, and $rV \subseteq M$, since the u_i's generate V. Thus $rV = V = M$. Now for any $r \in R - \{0\}$, the multiplication by r in R^l is a bijection. So the multiplication by r in R is a bijection, and R is a field. □

4.4.3. Corollary : *Let Ω be a finite group, and V be a simple $R\Omega$-module. Then the annihilator of V in R is a maximal ideal of R.*

Proof: Indeed V is non zero by definition of a simple module, it is finitely generated over R by the (finite) Ω-orbit of any non zero vector of V. Moreover, the kernel and image of the multiplication by $r \in R$ on V are $R\Omega$-submodules of V. So the multiplication by r is either 0 or a bijection. If I is the annihilator of V in R, Lemma 4.4.2 applies to R/I, so R/I is a field, and I is a maximal ideal of R. □

4.4.4. Remark : Let R be a commutative ring with identity element, and $\mathcal{D}$ be an admissible subcategory of $\mathcal{C}$. Let (G, V) be a seed of $R\mathcal{D}$, and let I be the annihilator of V in R. Then $k = R/I$ is a field, and V is a finite dimensional k-vector space. Hence the map Ψ of Theorem 4.3.20 is a map of k-vector spaces. It follows that $S_{G,V}(H)$ is also a k-vector space, whose dimension is equal to the rank of the matrix of Ψ. In particular, this dimension can be computed by the usual methods of linear algebra.

4.4.5. The Case of a Trivial Module. Let I be a maximal ideal of R. Then $k = R/I$ is a field, that can be viewed as a simple $R\mathrm{Out}(G)$-module with trivial $\mathrm{Out}(G)$-action. If $(B, A) \in \tau_G(H)$ and $(T, S) \in \sigma_G(H)$, then the action of $\langle (B, A) \parallel (T, S) \rangle_G$ on k is given by

$$\begin{aligned} \langle (B, A) \parallel (T, S) \rangle_G \cdot 1_k &= \sum_{\substack{h \in [B \backslash H / T] \\ (B,A) \,\text{---}\, ({}^hT, {}^hS)}} \pi_G\big(\bar{t}_{B,A} \circ \theta_h \circ (\bar{s}_{T,S})^{-1}\big) \cdot 1_k \\ &= \big| \{ h \in [B \backslash H / T] \mid (B, A) \,\text{---}\, ({}^hT, {}^hS) \} \big|_k \,, \end{aligned}$$

where $n_k = n \cdot 1_k$ denotes the image of the integer n in k.

The map Ψ of Theorem 4.3.20 is a map $k^{\sigma_G(H)} \to k^{\tau_G(H)}$. Its matrix m in the canonical bases is a rectangular matrix with coefficients in k, whose coefficient in line $(B,A) \in \tau_G(H)$ and column $(T,S) \in \sigma_G(H)$ is equal to

$$m_{(B,A),(T,S)} = \left|\{h \in [B\backslash H/T] \mid (B,A) — ({}^hT, {}^hS)\}\right|_k .$$

The module $S_{G,V}(H)$ is isomorphic to the image of this map. In particular, it is a k-vector space of dimension equal to the rank of the matrix m.

In the case of Example 4.1.8, this gives ([6] Proposition 16, or [12] Proposition 7.1):

4.4.6. Proposition : *Let $\mathcal{D}$ be a replete subcategory of $\mathcal{C}$. Let R be a commutative ring with identity element, let I be a maximal ideal of R, and let $k = R/I$. For any objects G and H of $\mathcal{D}$, let $kB_G(H)$ be the k-vector space with basis the set of conjugacy classes of sections (T,S) of H such that $T/S \cong G$.*

Then the module $S_{G,k}(H)$ is isomorphic to the quotient of $kB_G(H)$ by the radical of the k-valued symmetric bilinear form on $kB_G(H)$ defined by

$$\langle (B,A) \mid (T,S) \rangle_G = \left|\{h \in B\backslash H/T \mid (B,A) — ({}^hT, {}^hS)\}\right|_k .$$

Proof: Indeed since $\mathcal{D}$ is a full subcategory, one can take $\tau_G(H) = \sigma_G(H)$, and $k^{\sigma(G)}$ identifies with $B_H(G)$. The matrix of Ψ is a square matrix, and it is moreover symmetric, since there is a bijection

$$\{h \in B\backslash H/T \mid (B,A) — ({}^hT, {}^hS)\} \to \{h \in T\backslash H/B \mid (T,S) — ({}^hB, {}^hA)\}$$

given by $BhT \mapsto (BhT)^{-1} = Th^{-1}B$. □

4.4.7. Remark : In particular, the k-dimension of $S_{G,k}$ is equal to the rank of the above bilinear form. However, it is very difficult in general to express this rank in terms of other combinatorial data on the group H. The following is an example where such a result is known:

4.4.8. Proposition : *Let k be a field of characteristic 0, and $\mathcal{D}$ be a replete subcategory of $\mathcal{C}$. Then the restriction of the functor $k\mathcal{R}_{\mathbb{Q}}$ to $k\mathcal{D}$ is isomorphic to the simple functor $S_{1,k}$. In particular, for any object H of $\mathcal{D}$, the k-dimension of $S_{1,k}(H)$ is equal to the number of conjugacy classes of cyclic subgroups of H.*

Proof: Let G be a finite group. It follows from Artin's induction Theorem that the natural morphism from $kB(G)$ to the group $kR_{\mathbb{Q}}(G)$ (see Remark 1.2.3) is surjective. So the morphism from kB (restricted to $k\mathcal{D}$) to $kR_{\mathbb{Q}}$ is surjective. But clearly, the functor kB is isomorphic to the functor $L_{1,k}$, which has a unique simple quotient $S_{1,k}$.

Hence $S_{\mathbf{1},k}$ is the only simple quotient of $kR_{\mathbb{Q}}$, so it is enough to show that the restriction of $kR_{\mathbb{Q}}$ to $\mathcal{D}$ is simple: indeed, the group $R_{\mathbb{Q}}(H)$ is a free abelian group of with basis a set $[\mathrm{Irr}_{\mathbb{Q}}(H)]$ of representatives of isomorphism classes of rational irreducible representations of H (cf. [45] Chapitre 13 Théorème 29 Corollaire 1), and the number of such representations is equal to the number $c(H)$ of conjugacy classes of cyclic subgroups of H. Thus $\dim_k kR_{\mathbb{Q}}(H) = c(H)$.

To show that $kR_{\mathbb{Q}}$ is simple, observe that there is a unique k-valued k-bilinear form $\langle\ ,\ \rangle_H$ on $kR_{\mathbb{Q}}(H)$ such that

$$\langle V, W\rangle_H = \big(\dim_{\mathbb{Q}} \mathrm{Hom}_{\mathbb{Q}G}(V, W)\big)_k$$

for any finite dimensional $\mathbb{Q}H$-modules V and W. The basis $[\mathrm{Irr}_{\mathbb{Q}}(H)]$ is orthogonal for this bilinear form, and $\langle V, V\rangle_H \neq 0_k$ for $V \in [\mathrm{Irr}_{\mathbb{Q}}(H)]$, so $\langle\ ,\ \rangle_H$ is non degenerate. Moreover for any subgroup T of H and any $\mathbb{Q}H$-module V, the element $\mathrm{Defres}^H_{T/T} V$ of $R_{\mathbb{Q}}(\mathbf{1}) \cong \mathbb{Z}$ is equal to the dimension of the space V_T of coinvariants of T on V, and this is also equal to $\langle \mathbb{Q}H/T, V\rangle_H$.

Now if F is a proper subfunctor of $kR_{\mathbb{Q}}$, then F is contained in the unique proper maximal subfunctor J of $kR_{\mathbb{Q}}$, and for any object H of $\mathcal{D}$, the module $J(H)$ is equal to the intersection of the kernels of the maps $kR_{\mathbb{Q}}(\varphi)$, for all morphisms $\varphi \in \mathrm{Hom}_{k\mathcal{D}}(H, \mathbf{1})$. In particular $F(H)$ is contained in the kernel of all the maps $\mathrm{Defres}^H_{T/T}$. It follows that

$$\langle \mathbb{Q}H/T, F(H)\rangle_H = \{0\} ,$$

for any $T \leq H$. But the elements $\mathbb{Q}H/T$ generate the whole of $kR_{\mathbb{Q}}(H)$, since the linearization morphism $kB(H) \to kR_{\mathbb{Q}}(H)$ is surjective. Hence $F(H) = \{0\}$ since the bilinear form $\langle\ ,\ \rangle_H$ is non degenerate. It follows that $kR_{\mathbb{Q}}$ is simple, hence isomorphic to $S_{\mathbf{1},k}$. □

Part II
Biset Functors on Replete Subcategories

Chapter 5
The Burnside Functor

For the remainder of this book, the framework will be the following:

Hypothesis : *The ring R is a commutative ring with identity. The subcategory $\mathcal{D}$ is a replete subcategory of the biset category $\mathcal{C}$ (see Definition 4.1.7). A biset functor is an R-linear functor from $R\mathcal{D}$ to R-*Mod.

5.1. The Burnside Functor

5.1.1. Definition : *The Burnside functor on $R\mathcal{D}$ is the restriction to $R\mathcal{D}$ of the Burnside functor RB on $R\mathcal{C}$, and will be also denoted by RB. It can be defined as*

$$RB = \mathrm{Hom}_{R\mathcal{D}}(\mathbf{1}, -) .$$

In other words RB is the Yoneda functor corresponding to the trivial group, which is an object of $\mathcal{D}$ since the class of objects of $\mathcal{D}$ is closed under taking quotients. Thus for an object G of $\mathcal{D}$, the group $RB(G)$ is equal to $R \otimes_{\mathbb{Z}} B(G)$. If H is another object of $\mathcal{D}$, and if U is a finite (H, G)-biset, then the map $RB(U) : RB(G) \to RB(H)$ is induced by the correspondence sending a finite G-set X to the H-set $U \times_G X$.

5.1.2. Proposition :

1. *The Burnside functor RB is a projective object of $\mathcal{F}_{\mathcal{D},R}$. More precisely, if F is an object of $\mathcal{F}_{\mathcal{D},R}$, then*

$$\mathrm{Hom}_{\mathcal{F}_{\mathcal{D},R}}(RB, F) \cong F(\mathbf{1}) .$$

2. *In particular there is an isomorphism of R-algebras*

$$\mathrm{End}_{\mathcal{F}_{\mathcal{D},R}}(RB) \cong R .$$

3. *The simple quotients of RB are the functors $S_{\mathbf{1},R/I}$, where I is a maximal ideal of R.*

S. Bouc, *Biset Functors for Finite Groups*, Lecture Notes in Mathematics 1990, DOI 10.1007/978-3-642-11297-3_5, © Springer-Verlag Berlin Heidelberg 2010

Proof: Assertion 1 is exactly the Yoneda Lemma. Recall that the isomorphism

$$\mathrm{Hom}_{\mathcal{F}_{\mathcal{D},R}}(RB, F) \cong F(\mathbf{1})$$

sends $\theta \in \mathrm{Hom}_{\mathcal{F}_{\mathcal{D},R}}(RB, F)$ to $\theta_{\mathbf{1}}(\bullet) \in F(\mathbf{1})$, where $\bullet$ is the identity element of $RB(\mathbf{1}) \cong R$. If $F = RB$, then $\theta_{\mathbf{1}}$ is just multiplication by a scalar $r_\theta \in R$, and the map $\theta \mapsto r_\theta$ is an algebra isomorphism.

For Assertion 3, consider a simple functor $S_{H,V}$, where (H, V) is a seed of $R\mathcal{D}$. Assertion 1 implies that $S_{H,V}$ is a quotient of RB if and only if $S_{H,V}(\mathbf{1}) \neq \{0\}$. By Lemma 4.3.9, it implies that H is isomorphic to a subquotient of $\mathbf{1}$, i.e. that $H = \mathbf{1}$. Conversely $S_{\mathbf{1},V}(\mathbf{1}) \cong V \neq \{0\}$. So the simple quotients of RB are exactly the functors $S_{\mathbf{1},V}$, where V is a simple $R\mathrm{Out}(\mathbf{1})$-module, i.e. a simple R-module. Thus $V = R/I$, for some maximal ideal I of R. □

5.1.3. Remark : If R is a local ring with maximal ideal M, it follows that RB has a unique maximal subfunctor, equal to $J_{\mathbf{1},R/M}$, with the notation of Remark 4.2.6: if G is an object of $\mathcal{D}$, then $J_{\mathbf{1},R/M}(G)$ is the set of elements $u \in RB(G)$ such that for any morphism $\varphi : G \to \mathbf{1}$ in $\mathcal{D}$, the element $\varphi(u)$ lies in M. This is equivalent to saying that $|H\backslash u|_R \in M$ for any subgroup H of G, where the map $u \in RB(G) \mapsto |H\backslash u|_R \in R$ is defined as the R-linear extension of the map $X \mapsto |H\backslash X|$ from $B(G)$ to $\mathbb{Z}$.

In other words $J_{\mathbf{1},R/M}(G)$ is equal to the radical of the R-bilinear form $\langle\ ,\ \rangle_G$ on $RB(G)$, with values in R/M, defined by

$$\langle [G/H], [G/K] \rangle_G = |H\backslash G/K|_{R/M} \ .$$

In this case in particular, the functor RB is a projective cover of $S_{\mathbf{1},R/M}$ in the category $\mathcal{F}_{\mathcal{D},R}$.

5.2. Effect of Biset Operations on Idempotents

5.2.1. Lemma : *Let F be a subfunctor of the biset functor RB. Then for any object G of $\mathcal{D}$, the R-module $F(G)$ is an ideal of the algebra $RB(G)$.*

Proof: This follows from Assertion 4 of Lemma 2.5.8, by R-linearity. □

Lemma 5.2.1 shows that the structure of the lattice of subfunctors of the Burnside functor RB has some connection with the Burnside ring structure. So it is a natural question to look at the effect of elementary biset operations on idempotents of the Burnside ring, in the case $R = \mathbb{Q}$. The case of deflation will require the following notation:

5.2.2. Notation : *If G is a finite group, and N is a normal subgroup of G, denote by $m_{G,N}$ the rational number defined by*

$$m_{G,N} = \frac{1}{|G|} \sum_{XN=G} |X|\mu(X,G) ,$$

where μ is the Möbius function of the poset of subgroups of G.

5.2.3. Example : If N is contained in the Frattini subgroup $\Phi(G)$ of G, then $m_{G,N} = 1$: indeed if $X \leq G$ and $XN = G$, then $X\Phi(G) = G$, thus $X = G$. In particular $m_{G,\mathbf{1}} = 1$. The same argument shows more generally that, if $N \trianglelefteq G$, then $m_{G,N} = m_{G,N\Phi(G)}$. Since moreover $\mu(X,G) = 0$ if $X \not\geq \Phi(G)$, it follows that $m_{G,N} = m_{G/\Phi(G),N\Phi(G)/\Phi(G)}$.

5.2.4. Theorem : *Let G be a finite group.*

1. *Let H and K be subgroups of G. Then*

$$\mathrm{Res}^G_K e^G_H = \sum_{x \in [N_G(H)\backslash T_G(H,K)/K]} e^K_{H^x} ,$$

where x runs through a set of representatives of $(N_G(H),K)$-orbits on the set

$$T_G(H,K) = \{g \in G \mid H^g \subseteq K\} .$$

In other words $\mathrm{Res}^G_K e^G_H$ is equal to the sum of idempotents e^K_L corresponding to subgroups L of K which are conjugate to H in G, up to K-conjugation.

2. *Let $K \leq H$ be subgroups of G. Then*

$$\mathrm{Ind}^G_H e^H_K = \frac{|N_G(K)|}{|N_H(K)|} e^G_K .$$

3. *Let $N \trianglelefteq G$. Then for any subgroup H of G containing N*

$$\mathrm{Inf}^G_{G/N} e^{G/N}_{H/N} = \sum_{\substack{KN =_G H \\ K \mathrm{mod}. G}} e^G_K .$$

4. *Let $N \trianglelefteq G$. Then*

$$\mathrm{Def}^G_{G/N} e^G_G = m_{G,N} e^{G/N}_{G/N} .$$

More generally, if H is a subgroup of G, then

$$\mathrm{Def}^G_{G/N} e^G_H = \frac{|N_G(HN)/HN|}{|N_G(H)/H|} m_{H,H\cap N} e^{G/N}_{HN/N} .$$

5. *If* $\varphi : G \to G'$ *is a group isomorphism, and* $H \leq G$*, then*

$$\mathrm{Iso}(\varphi)(e_H^G) = e_{\varphi(H)}^{G'} \cdot$$

Proof: Recall from Remark 2.5.3 that for $H \leq G$, the idempotent e_H^G has the property that $e_H^G u = |u^H| e_H^G$ for any $u \in \mathbb{Q}B(G)$, where $u \mapsto |u^H|$ is the $\mathbb{Q}$-linear extension of the assignment sending the isomorphism class of a finite G-set X to the cardinality $|X^H|$ of the set of fixed points of H on X. In particular $|(e_H^G)^H| = 1$, and $|(e_K^G)^H| = 0$ if K is a subgroup of G which is not conjugate to H in G.

For Assertion 1, observe that $\mathrm{Res}_K^G : \mathbb{Q}B(G) \to \mathbb{Q}B(K)$ is a ring homomorphism. It follows that $\mathrm{Res}_K^G e_H^G$ is an idempotent of $\mathbb{Q}B(K)$, hence a sum of idempotents e_L^K, for *some* subgroups L of K. The idempotent e_L^K appears in this sum if and only if

$$0 \neq e_L^K \mathrm{Res}_K^G e_H^G = |(\mathrm{Res}_K^G e_H^G)^L| e_L^K = |(e_H^G)^L| e_L^K \ ,$$

i.e. if and only if L is conjugate to H in G.

Assertion 2 is a straightforward consequence of Theorem 2.5.2, since $\mathrm{Ind}_H^G [H/L] = [G/L]$ for any subgroup L of H.

For Assertion 3, observe that $\mathrm{Inf}_{G/N}^G : \mathbb{Q}B(G/N) \to \mathbb{Q}B(G)$ is also a ring homomorphism. It follows that $\mathrm{Inf}_{G/N}^G e_{H/N}^{G/N}$ is an idempotent of $\mathbb{Q}B(G)$, hence a sum of idempotents e_X^G, for *some* subgroups X of G. The idempotent e_X^G appears in this sum if and only if

$$0 \neq e_X^G \mathrm{Inf}_{G/N}^G e_{H/N}^{G/N} = |(\mathrm{Inf}_{G/N}^G e_{H/N}^{G/N})^X| e_X^G = |(e_{H/N}^{G/N})^{XN/N}| e_X^G \ ,$$

i.e. if and only if XN is conjugate to H in G.

For Assertion 4, observe that if X is a G-set, then there is an isomorphism of G/N-sets

$$(G/N) \times_G X \cong N \backslash X \ .$$

So the map $\mathrm{Def}_{G/N}^G : B(G) \to B(G/N)$ is such that $\mathrm{Def}_{G/N}^G [X] = [N\backslash X]$, for any finite G-set X.

Now if Y is a G/N-set, and X is a G-set, there is an isomorphism of G/N-sets

$$Y \times (N\backslash X) \cong N\backslash\big((\mathrm{Inf}_{G/N}^G Y) \times X\big) \ ,$$

and it follows that

$$v \mathrm{Def}_{G/N}^G u \cong \mathrm{Def}_{G/N}^G \big((\mathrm{Inf}_{G/N}^G v) u\big) \ ,$$

for any $v \in \mathbb{Q}B(G/N)$ and $u \in \mathbb{Q}B(G)$. For $u = e_H^G$, this gives

$$v \mathrm{Def}_{G/N}^G e_H^G = |(\mathrm{Inf}_{G/N}^G v)^H| \mathrm{Def}_{G/N}^G e_H^G = |v^{HN/N}| \mathrm{Def}_{G/N}^G e_H^G \ ,$$

and it follows from Remark 2.5.3 that $\mathrm{Def}^G_{G/N} e^G_H$ is equal to a scalar multiple of $e^{G/N}_{HN/N}$. In particular, there is a rational number m such that

$$\mathrm{Def}^G_{G/N} e^G_G = m e^{G/N}_{G/N} \ .$$

Since $\mathrm{Def}^G_{G/N}[G/K] = [G/KN]$ for any subgroup K of G, this yields

$$\frac{1}{|G|}\sum_{X\leq G}|X|\mu(X,G)[G/XN] = \frac{m}{|G/N|}\sum_{N\leq Y\leq G}|Y/N|\mu(Y,G)[(G/N)/(Y/N)] \ ,$$

since $\mu(Y/N, G/N) = \mu(Y,G)$ for $N \leq Y \leq G$, as the poset of subgroups of G/N containing Y/N is isomorphic to the poset of subgroups of G containing Y. The coefficient of $[(G/N)/(G/N)]$ is equal to m in the right hand side, and equal to

$$\frac{1}{|G|}\sum_{XN=G}|X|\mu(X,G) = m_{G,N}$$

in the left hand side. It follows that $m = m_{G,N}$, as claimed.

The second part of Assertion 4 follows easily from the equality

$$e^G_H = \frac{1}{|N_G(H)/H|}\mathrm{Ind}^G_H e^H_H \ ,$$

and from the identity (Condition 2.(e) of 1.1.3)

$$\mathrm{Def}^G_{G/N}\mathrm{Ind}^G_H = \mathrm{Ind}^{G/N}_{HN/N}\mathrm{Iso}^{HN/N}_{H/H\cap N}\mathrm{Def}^H_{H/H\cap N} \ ,$$

which is a consequence of a straightforward isomorphism of $(G/N, H)$-bisets

$$G/N \cong (G/N)\times_{HN/N}(HN/N)\times_{H/H\cap N}(H/H\cap N) \ .$$

Assertion 5 is obvious, and this completes the proof of Theorem 5.2.4. □

5.3. Properties of the $m_{G,N}$'s

5.3.1. Proposition : *Let G be a finite group. If M and N are normal subgroups of G with $N \leq M$, then*

$$m_{G,M} = m_{G,N}m_{G/N,M/N} \ .$$

Proof: This follows from the transitivity property of deflation maps (see Sect. 1.1.3):

$$\begin{aligned}\mathrm{Def}^G_{G/M} e^G_G &= m_{G,M} e^{G/M}_{G/M} \\ &= \mathrm{Def}^{G/N}_{G/M} \mathrm{Def}^G_{G/N} e^G_G = m_{G,N} \mathrm{Def}^{G/N}_{G/M} e^{G/N}_{G/N} \\ &= m_{G,N} m_{G/N,M/N} e^{G/M}_{G/M} \ .\end{aligned}$$

Hence $m_{G,M} = m_{G,N} m_{G/N,M/N}$. □

5.3.2. Lemma : *Let G and H be finite groups, and let L be a subgroup of $H \times G$.*

1. *If $\widetilde{e}^H_H \times_H [(H \times G)/L] \neq 0$ in $\mathbb{Q}B(H,G)$, then $p_1(L) = H$.*
2. *If $[(H \times G)/L] \times_G \widetilde{e}^G_G \neq 0$ in $\mathbb{Q}B(H,G)$, then $p_2(L) = G$.*

Proof: Let $u \in \mathbb{Q}B(H,G) = \mathbb{Q}B(H \times G^{op})$. Then

$$u = \sum_M |u^M| e^{H \times G^{op}}_M \ ,$$

where M runs through a set of representatives of conjugacy classes of subgroups of $H \times G^{op}$. Now if X is an H-set and U is an (H,G)-biset, it follows from Assertion 3 of Lemma 2.5.8 that

$$(\tilde{X} \times_H U)^M \cong X^{p_1(M)} \times U^M \ .$$

By linearity, it follows that

$$|(\widetilde{e}^H_H \times_H [(H \times G)/L])^M| = |(e^H_H)^{p_1(M)}| |((H \times G)/L)^M| \ .$$

This is equal to 0, unless $p_1(M) = H$, and some conjugate of M in the group $H \times G^{op}$ is contained in L. This implies $p_1(L) = H$, and Assertion 1 follows. Assertion 2 follows by considering opposite bisets, and using Assertion 2 of Proposition 2.4.11. □

5.3.3. Proposition : *Let G be a finite group. If M and N are normal subgroups of G, then*

$$m_{G,M} = \frac{1}{|G|} \sum_{YN=YM=G} |Y| \mu(Y,G) m_{G/N,(Y \cap M)N/N} \ .$$

Proof: Let $f \in \mathbb{Q}B(G/M, G/N)$ denote the element defined by

$$f = \tilde{e}_{G/M}^{G/M} \times_{G/M} \mathrm{Def}_{G/M}^{G} \times_G \tilde{e}_G^G \times_G \mathrm{Inf}_{G/N}^{G} \times_{G/N} \tilde{e}_{G/N}^{G/N} .$$

Then by Lemma 2.5.8, the element $v = f \times_{G/N} e_{G/N}^{G/N}$ of $\mathbb{Q}B(G/M)$ is equal to

$$\begin{aligned}
v &= \tilde{e}_{G/M}^{G/M} \times_{G/M} \mathrm{Def}_{G/M}^{G} \times_G \tilde{e}_G^G \times_G \mathrm{Inf}_{G/N}^{G} \times_{G/N} \tilde{e}_{G/N}^{G/N} \times_{G/N} e_{G/N}^{G/N} \\
&= \tilde{e}_{G/M}^{G/M} \times_{G/M} \mathrm{Def}_{G/M}^{G} \times_G \tilde{e}_G^G \times_G \mathrm{Inf}_{G/N}^{G} \times_{G/N} e_{G/N}^{G/N} \\
&= \sum_{\substack{X \bmod. G \\ XN=G}} \tilde{e}_{G/M}^{G/M} \times_{G/M} \mathrm{Def}_{G/M}^{G} \times_G \tilde{e}_G^G \times_G e_X^G \\
&= \tilde{e}_{G/M}^{G/M} \times_{G/M} \mathrm{Def}_{G/M}^{G} \times_G e_G^G \\
&= m_{G,M} \tilde{e}_{G/M}^{G/M} \times_{G/M} e_{G/M}^{G/M} \\
&= m_{G,M} e_{G/M}^{G/M} .
\end{aligned}$$

On the other hand, the element f of $\mathbb{Q}B(G/M, G/N)$ is equal to

$$f = \frac{1}{|G|} \sum_{Y \leq G} |Y| \mu(Y, G) \tilde{e}_{G/M}^{G/M} \times_{G/M} \mathrm{Def}_{G/M}^{G} \times_G \widetilde{G/Y} \times_G \mathrm{Inf}_{G/N}^{G} \times_{G/N} \tilde{e}_{G/N}^{G/N} .$$

But $\mathrm{Def}_{G/M}^{G} \cong \big((G/M) \times G\big)/\Delta_M(G)$, where $\Delta_M(G) = \{(gM, g) \mid g \in G\}$, and $\mathrm{Inf}_{G/N}^{G} \cong \big(G \times (G/N)\big)/\Delta_N^o(G)$, where $\Delta_N^o(G) = \{(g, gN) \mid g \in G\}$. Moreover $\widetilde{G/Y} \cong (G \times G)/\Delta(Y)$, where $\Delta(Y) = \{(y, y) \mid y \in Y\}$. By Lemma 2.3.24, it follows that

$$\mathrm{Def}_{G/M}^{G} \times_G \widetilde{G/Y} \times_G \mathrm{Inf}_{G/N}^{G} = \big((G/M) \times (G/N)\big)/\big(\Delta_M(G) * \Delta(Y) * \Delta_N^o(G)\big) .$$

Now $\Delta_M(G) * \Delta(Y) * \Delta_N^o(G)$ is equal to the subgroup

$$\Delta_{Y,M,N} = \{(yM, yN) \mid y \in Y\}$$

of $(G/M) \times (G/N)$. The first projection of this subgroup is equal to G/M if and only if $YM = G$, and the second one is equal to G/N if and only if $YM = G$.

Moreover $k_1(\Delta_{Y,M,N}) = (Y \cap N)M/M$, and $k_2(\Delta_{Y,M,N}) = (Y \cap M)N/N$. Hence if $YN = YM = G$, there is an isomorphism of $(G/M, G/N)$-bisets

$$\Big((G/M) \times (G/N)\Big)/\Delta_{Y,M,N} \cong \mathrm{Inf}_{G/A_Y}^{G/M} \circ \mathrm{Iso}(\varphi_Y) \circ \mathrm{Def}_{G/B_Y}^{G/N} ,$$

where $A_Y = (Y \cap N)M$ and $B_Y = (Y \cap M)N$, and $\varphi_Y : G/B_Y \to G/A_Y$ is the group isomorphism sending yB_Y to yA_Y, for $y \in Y$. It follows from Lemma 5.3.2 that

$$f = \frac{1}{|G|} \sum_{YN=YM=G} |Y|\mu(Y,G)\tilde{e}_{G/M}^{G/M} \circ \mathrm{Inf}_{G/A_Y}^{G/M} \circ \mathrm{Iso}(\varphi_Y) \circ \mathrm{Def}_{G/B_Y}^{G/N} \circ \tilde{e}_{G/N}^{G/N} .$$

Hence, the image v of $e_{G/N}^{G/N}$ by this morphism is equal to

$$\begin{aligned}
v &= \frac{1}{|G|} \sum_{YN=YM=G} |Y|\mu(Y,G)\tilde{e}_{G/M}^{G/M} \circ \mathrm{Inf}_{G/A_Y}^{G/M} \circ \mathrm{Iso}(\varphi_Y) \circ \mathrm{Def}_{G/B_Y}^{G/N}(e_{G/N}^{G/N})\\
&= \frac{1}{|G|} \sum_{YN=YM=G} |Y|\mu(Y,G)m_{G/N,B_Y/N}\tilde{e}_{G/M}^{G/M} \circ \mathrm{Inf}_{G/A_Y}^{G/M} \circ \mathrm{Iso}(\varphi_Y)(e_{G/B_Y}^{G/B_Y})\\
&= \frac{1}{|G|} \sum_{YN=YM=G} |Y|\mu(Y,G)m_{G/N,B_Y/N}\tilde{e}_{G/M}^{G/M} \circ \mathrm{Inf}_{G/A_Y}^{G/M}(e_{G/A_Y}^{G/A_Y})\\
&= \frac{1}{|G|} \sum_{YN=YM=G} |Y|\mu(Y,G)m_{G/N,B_Y/N}e_{G/M}^{G/M}
\end{aligned}$$

Since $v = m_{G,M}e_{G/M}^{G/M}$, it follows that

$$\begin{aligned}
m_{G,M} &= \frac{1}{|G|} \sum_{YN=YM=G} |Y|\mu(Y,G)m_{G/N,B_Y/N}\\
&= \frac{1}{|G|} \sum_{YN=YM=G} |Y|\mu(Y,G)m_{G/N,(Y\cap M)N/N} ,
\end{aligned}$$

as was to be shown. □

5.3.4. Proposition : *Let G be a finite group. If N and M are normal subgroups of G such that $G/N \cong G/M$, then $m_{G,N} = m_{G,M}$.*

Proof: There is nothing to prove if $G = \mathbf{1}$. By induction on $|G|$, one can assume that the result holds for all finite groups H with $|H| < |G|$. If $N = \mathbf{1}$, then $M = \mathbf{1}$, and $m_{G,N} = m_{G,M} = 1$ in this case, by Example 5.2.3. And if $N \neq \mathbf{1}$ then $M \neq \mathbf{1}$ also, and by Proposition 5.3.3

$$m_{G,M} = \frac{1}{|G|} \sum_{YN=YM=G} |Y|\mu(Y,G)m_{G/N,(Y\cap M)N/N} \tag{5.3.5}$$

$$m_{G,N} = \frac{1}{|G|} \sum_{YN=YM=G} |Y|\mu(Y,G)m_{G/M,(Y\cap N)M/M} . \tag{5.3.6}$$

But if $YN = YM = G$, there are group isomorphisms

$$\begin{aligned}(G/N)/\big((Y\cap M)N/N\big) &\cong G/(Y\cap M)N = YN/(Y\cap M)N\\ &\cong Y/\big(Y\cap (Y\cap M)N\big) = Y/(Y\cap M)(Y\cap N)\\ &\cong (G/M)/\big((Y\cap N)M/M\big)\ .\end{aligned}$$

... $G/N \to G/M$ be a group isomorphism. Then it is clear from the ... that $m_{G/N,L} = m_{G/M,\varphi(L)}$, for any normal subgroup L of G/N.

$$m_{G/N,(Y\cap M)N/N} = m_{G/M,\varphi((Y\cap M)N/N)}\ .$$

$$\ldots N/N) \cong (G/N)/\big((Y\cap M)N/N\big) \cong (G/M)/\big((Y\cap N)M/M\big)\ ,$$

... the group G/M implies that

$$\ldots \cap M)N/N)\ .$$

... bgroup Y in the right □

... ic

... s

... hic to

$\mathrm{Def}^G_{G/N} e^G_G \in$

$N \trianglelefteq G$

... 2.9 that for any object

... tions of transitive (H,G)-

... uch a transitive biset factors

$\mathrm{Def}^Y_{Y/X} \circ \mathrm{Res}^G_Y$,

... $X)$ of G, such that there exists a

... ere exists a morphism φ of the above

... orem 5.2.4, the restriction of e^G_G to any

Hence $Y = G$. Moreover $\mathrm{Def}^G_{G/X} e^G_G =$

and $X = \mathbf{1}$, since Condition 3 holds. It

$\not\cong G$, so G is isomorphic to a subquotient

□

... ly implies Condition 1.

... he equivalent conditions of Proposition 5.4.5 if

... over k, according to the following definition:

... $,N\cdot$

5.4.3. Notation : *If G is an object of $\mathcal{D}$, let e_G denote the subfunctor of kB generated by $e_G^G \in kB(G)$.*

5.4.4. Remark : With the notation of Remark 3.2.9, the set $\mathcal{G}$ is equal to $\{G\}$, and $\Gamma_G = \{e_G^G\}$.

5.4.5. Proposition : *Let G be an object of $\mathcal{D}$. The following condition are equivalent:*

1. *If H is any object of $\mathcal{D}$ such that $|H| < |G|$, then $\mathsf{e}_G(H) = \{0\}$.*
2. *If H is an object of $\mathcal{D}$ such that $\mathsf{e}_G(H) \neq \{0\}$, then G is isomor a subquotient of H.*
3. *If $\mathbf{1} \neq N \trianglelefteq G$, then $m_{G,N} = 0$ in k.*
4. *If $\mathbf{1} \neq N \trianglelefteq G$, then $\mathrm{Def}_{G/N}^G e_G^G = 0$ in $kB(G/N)$.*

Proof: Suppose that Condition 1 holds. If $\mathbf{1} \neq N \trianglelefteq G$, then $\mathsf{e}_G(G/N) = \{0\}$, so Condition 4 holds.

Condition 4 holds is equivalent to Condition 3, since for

$$\mathrm{Def}_{G/N}^G e_G^G = m_{G,N} e_{G/N}^{G/N} .$$

Suppose that Condition 3 holds. Recall from Remark 3. H of $\mathcal{D}$

$$\mathsf{e}_G(H) = \mathrm{Hom}_{k\mathcal{D}}(G,H)(e_G^G)$$

Now $\mathrm{Hom}_{k\mathcal{D}}(G,H)$ is the set of k-linear combina bisets. By 4.1.5, a morphism φ corresponding to s as the composition

$$\varphi = \mathrm{Ind}_T^H \circ \mathrm{Inf}_{T/S}^T \circ \mathrm{Iso}(f) \circ$$

for suitable sections (T,S) of H and $(Y,$ group isomorphism $f : Y/X \to T/S$.

Suppose that $\mathsf{e}_G(H) \neq \{0\}$. Then th form such that $\varphi(e_G^G) \neq 0$. Now by Th proper subgroup of G is equal to 0 $m_{G,X} e_{G/X}^{G/X} \neq 0$, hence $m_{G,X} \neq 0$ follows that $T/S \cong Y/X = G/\mathbf{1}$ of H, and Condition 2 holds.

Finally, Condition 2 obviou

An object of $\mathcal{D}$ satisfies t and only if it is a B-group

5.4.6. Definition : *A finite group G is called* a B-group over k *if* $|G| \neq 0$ *in k, and if for any non trivial normal subgroup N of G, the constant $m_{G,N}$ is equal to zero in k, or equivalently* $\mathrm{Def}^G_{G/N} e^G_G = 0$ *in* $kB(G/N)$.

The class of all B-groups over k is denoted by $B\text{-}\mathrm{gr}_k$.

5.4.7. Remark : These conditions depend only on the characteristic q of k, so a B-group over k will also be called *a B-group in characteristic* q, and the class $B\text{-}\mathrm{gr}_k$ will also be denoted by $B\text{-}\mathrm{gr}_q$. Moreover, if N and M are normal subgroups of G and if $N \leq M$, then $\mathrm{Def}^G_{G/M} = \mathrm{Def}^{G/N}_{G/M} \circ \mathrm{Def}^G_{G/N}$. This shows that a group G is a B-group if and only if $\mathrm{Def}^G_{G/N} e^G_G = 0$ in $kB(G/N)$ (or equivalently $m_{G,N} = 0_k$) for any *minimal* normal subgroup N of G.

5.4.8. Proposition : *Let G and H be objects of $\mathcal{D}$.*

1. *If H is isomorphic to a quotient of G, then $\mathsf{e}_G \subseteq \mathsf{e}_H$.*
2. *If H is a B-group over k and if $\mathsf{e}_G \subseteq \mathsf{e}_H$, then H is isomorphic to a quotient of G.*

Proof: Since $\mathcal{D}$ contains group isomorphisms, and since $\mathrm{Iso}(f)(e^H_H) = e^{H'}_{H'}$ for any group isomorphism $f : H \to H'$, it follows that $\mathsf{e}_H = \mathsf{e}_{H'}$ if $H \cong H'$. So for Assertion 1, one can assume that $H = G/N$, for some $N \trianglelefteq G$. Since

$$e^G_G \, \mathrm{Inf}^G_{G/N} e^H_H = e^G_G$$

by Theorem 5.2.4, it follows that $e^G_G \in \mathsf{e}_H(G)$, hence $\mathsf{e}_G \subseteq \mathsf{e}_H$.

For Assertion 2, assume that H is a B-group over k, and that $\mathsf{e}_G \subseteq \mathsf{e}_H$. Equivalently

$$e^G_G \in \mathsf{e}_H(G) = \mathrm{Hom}_{k\mathcal{D}}(H,G)(e^H_H) \; .$$

So there exist a section (T,S) of G, a section (Y,X) of H, and a group isomorphism $f : Y/X \to T/S$ such that

$$e^G_G \, \mathrm{Ind}^G_T \mathrm{Inf}^T_{T/S} \mathrm{Iso}(f) \mathrm{Def}^Y_{Y/X} \mathrm{Res}^H_Y e^H_H \neq 0$$

in $kB(G)$. Now the restriction of e^H_H to any proper subgroup of H is equal to 0, by Theorem 5.2.4. Thus $Y = H$. Since $\mathrm{Def}^H_{H/X} e^H_H = m_{H,X} e^{H/X}_{H/X} \neq 0$, and since H is a B-group over k, it follows that $X = \mathbf{1}$. Finally $e^G_G[G/L] = |(G/L)^G| e^G_G = 0$ for any proper subgroup L of G, so $T = G$. It follows that $G/S = T/S \cong Y/X = H$, as was to be shown. □

5.4.9. Proposition : *Let F be subfunctor of kB. If H is a minimal group for F, then H is a B-group over k. Moreover $F(H) = ke^H_H$, and $\mathsf{e}_H \subseteq F$.*

In particular, if H is a B-group over k, then $\mathsf{e}_H(H) = ke^H_H$.

Proof: If $F = \{0\}$, there is nothing to prove. Otherwise let $H \in \mathrm{Min}(F)$. Then $F(H) \neq \{0\}$, and $F(K) = \{0\}$ for any group K with $|K| < |H|$. Let $u \in F(H)$. Then $u \in kB(H)$, so

$$u = \sum_{\substack{K \leq H \\ K \mathrm{mod}.H}} e_K^H u = \sum_{\substack{K \leq H \\ K \mathrm{mod}.H}} |u^H| e_K^H \ .$$

But if $|K| < |H|$, then $\mathrm{Res}_K^H u \in F(K) = \{0\}$, so $|u^K| = 0$. It follows that u is a scalar multiple of e_H^H. Thus $F(H) = ke_H^H$ since $F(H) \neq \{0\}$, and $e_H^H \in F(H)$, so $\mathsf{e}_H \subseteq F$. If K is an object of $\mathcal{D}$ such that $|K| < |H|$, then $\mathsf{e}_H(K) \subseteq F(K) = \{0\}$, thus $\mathsf{e}_H(K) = \{0\}$, so H is a B-group over k, by Proposition 5.4.5. Finally $\mathsf{e}_H(H) \supseteq ke_H^H$, for any object H of $\mathcal{D}$. If H is moreover a B-group over k, then H is minimal for e_H, by Proposition 5.4.5, so $\mathsf{e}_H(H) = ke_H^H$. □

5.4.10. Proposition : *Let G be an object of $\mathcal{D}$. Then $\mathrm{Min}(\mathsf{e}_G)$ consists of the isomorphism class of a single finite group $\beta_k(G)$. Moreover $\mathsf{e}_G = \mathsf{e}_{\beta_k(G)}$, the group $\beta_k(G)$ is isomorphic to a quotient of G, and $m_{G,N} \neq 0$ in k for any normal subgroup N of G such that $\beta_k(G) \cong G/N$.*

Proof: Let H be a minimal group for e_G. Then H is a B-group, and $\mathsf{e}_H \subseteq \mathsf{e}_G$, by Proposition 5.4.9. Since $\mathsf{e}_G(H) \neq \{0\}$, there exists a section (T, S) of H, a section (Y, X) of G, and a group isomorphism $f : Y/X \to T/S$ such that

$$u_{(T,S),f,(Y,X)} = e_H^H \,\mathrm{Ind}_T^H \mathrm{Inf}_{T/S}^T \mathrm{Iso}(f) \mathrm{Def}_{Y/X}^Y \mathrm{Res}_Y^G(e_G^G) \neq 0$$

in $kB(H)$. It follows that $\mathrm{Iso}(f)\mathrm{Def}_{Y/X}^Y \mathrm{Res}_Y^G(e_G^G)$ is a non zero element of $\mathsf{e}_G(T/S)$. Moreover T/S is an object of $\mathcal{D}$ since it is a subquotient of H. The minimality of H implies that $T = H$ and $S = \mathbf{1}$. Finally, since the restriction of e_G^G to any proper subgroup of G is equal to 0, it follows that $Y = G$. Thus

$$u_{(T,S),f,(Y,X)} = \mathrm{Iso}(f)\mathrm{Def}_{G/X}^G e_G^G = m_{G,X} \mathrm{Iso}(f)(e_{G/X}^{G/X}) = m_{G,X} e_H^H \ .$$

Hence H is isomorphic to G/X, for some normal subgroup X of G such that $m_{G,X} \neq 0$ in k. Thus $m_{G,N} \neq 0$ for any normal subgroup N of G such that $G/N \cong G/X$, since $m_{G,N} = m_{G,X}$ by Proposition 5.3.4. Now $\mathsf{e}_G \subseteq \mathsf{e}_H$ by Proposition 5.4.8, so $\mathsf{e}_G = \mathsf{e}_H$.

Finally if H and H' are minimal groups for e_G, then H and H' are B-groups over k, and $\mathsf{e}_G = \mathsf{e}_H = \mathsf{e}_{H'}$. Then by Proposition 5.4.8, the group H is a quotient of H', and H' is a quotient of H, so $H \cong H'$. This proves the unicity assertion, and completes the proof of the proposition. □

5.4.11. Theorem : *Let G be an object of $\mathcal{D}$.*

1. *If H is a isomorphic to a quotient of G, and if H is a B-group over k, then H is isomorphic to a quotient of $\beta_k(G)$.*
2. *Let N be a normal subgroup of G. Then the following conditions are equivalent:*

   a. *$m_{G,N} \neq 0$ in k.*
   b. *The group $\beta_k(G)$ is isomorphic to a quotient of G/N.*
   c. *$\beta_k(G) \cong \beta_k(G/N)$.*

3. *In particular, if $N \trianglelefteq G$, then $G/N \cong \beta_k(G)$ if and only if G/N is a B-group over k and $m_{G,N} \neq 0_k$.*

Proof: Set $K = \beta_k(G)$, so that $\mathsf{e}_G = \mathsf{e}_K$, by Proposition 5.4.10.

If H is isomorphic to a quotient of G, then $\mathsf{e}_G \subseteq \mathsf{e}_H$ by Proposition 5.4.8, hence $\mathsf{e}_K \subseteq \mathsf{e}_H$, so H is a quotient of K, if H is moreover a B-group, by Proposition 5.4.8 again. This proves Assertion 1.

For Assertion 2, let $H = G/N$ be a quotient of G with $m_{G,N} \neq 0_k$. Then since $\mathrm{Def}^G_{G/N} e^G_G = m_{G,N} e^H_H$, it follows that $e^H_H \in \mathsf{e}_G(H) = \mathsf{e}_K(H)$, so $\mathsf{e}_H \subseteq \mathsf{e}_K$. Thus K is isomorphic to a quotient of H, by Proposition 5.4.8, and (a) implies (b).

Now if (b) holds, since K is a B-group, it follows that K is isomorphic to a quotient of $\beta_k(G/N)$. But $\beta_k(G/N)$ is a B-group, isomorphic to a quotient of G/N, hence to a quotient of G. So it is a quotient of $\beta_k(G)$, and $\beta_k(G) \cong \beta_k(G/N)$. Thus (b) implies (c).

Finally if (c) holds, then choose a normal subgroup M/N of G/N such that $\beta_k(G/N) \cong (G/N)/(M/N)$ and $m_{G/N,M/N} \neq 0_k$. Such a subgroup exists by Proposition 5.4.10, and $G/M \cong (G/N)/(M/N) \cong \beta_k(G/N) \cong \beta_k(G)$. Similarly, there exists a normal subgroup M' of G such that $G/M' \cong \beta_k(G)$ and $m_{G,M'} \neq 0_k$. Since $G/M \cong G/M'$, it follows from Proposition 5.3.4 that $m_{G,M} = m_{G,M'}$, thus $m_{G,M} \neq 0_k$. Now $m_{G,M} = m_{G,N} m_{G/N,M/N}$, so $m_{G,N} \neq 0_k$, and (a) holds.

If $N \trianglelefteq G$ and $G/N \cong \beta_k(G)$, then G/N is a B-group over k and $m_{G,N} \neq 0$, by Proposition 5.4.10, since $m_{G,N}$ does not depend on the normal subgroup N such that $G/N \cong \beta_k(G)$, by Proposition 5.3.4. Conversely, if G/N is a B-group and $m_{G,N} \neq 0$, then $\beta_k(G) \cong \beta_k(G/N) = G/N$, by Assertion 2.

□

5.4.12. Remark : Theorem 5.4.11 shows that the group $\beta_k(G)$ depends only on G and q, and not on the category $\mathcal{D}$ containing G and satisfying Hypothesis 5.4.1. So it will also be denoted by $\beta_q(G)$. It is uniquely determined, up to isomorphism, by the fact that it is a B-group over k, which is isomorphic to a quotient G/N of G by a normal subgroup N with $m_{G,N} \neq 0_k$. But in general, such a normal subgroup N is far from being unique (see Example 5.6.9).

5.4.13. Definition and Notation : *Let $[B\text{-gr}_k]$ denote a set of representatives of isomorphism classes of B-groups over k.*

If $\mathcal{D}$ is a replete subcategory of $\mathcal{C}$, let $B\text{-gr}_k(\mathcal{D})$ denote the subclass of the class of objects of $\mathcal{D}$ consisting of B-groups over k, and by $[B\text{-gr}_k(\mathcal{D})]$ the intersection $[B\text{-gr}_k] \cap B\text{-gr}_k(\mathcal{D})$.

Define a relation $\gg$ on $B\text{-gr}_k(\mathcal{D})$, by $G \gg H$ if and only if H is isomorphic to a quotient of G. Then $([B\text{-gr}_k(\mathcal{D})], \gg)$ is a poset. A subclass $\mathcal{M}$ of $B\text{-gr}_k(\mathcal{D})$ is said to be closed *if*

$$\forall G, H \in B\text{-gr}_k(\mathcal{D}), \quad G \gg H \in \mathcal{M} \Rightarrow G \in \mathcal{M} .$$

A subset A of $[B\text{-gr}_k(\mathcal{D})]$ is a closed subset *if there exists a closed subclass $\mathcal{M}$ of $B\text{-gr}_k(\mathcal{D})$ such that $A = [B\text{-gr}_k(\mathcal{D})] \cap \mathcal{M}$.*

5.4.14. Theorem : *Let $\mathcal{S}_{k\mathcal{D}}$ denote the set of subfunctors of the Burnside functor kB on $k\mathcal{D}$, ordered by inclusion of subfunctors, and let $\mathcal{T}_{k\mathcal{D}}$ denote the set of closed subsets of $[B\text{-gr}_k(\mathcal{D})]$, ordered by inclusion of subsets.*

Then the map

$$\Theta : F \mapsto \{H \in [B\text{-gr}_k(\mathcal{D})] \mid \mathsf{e}_H \subseteq F\}$$

is an isomorphism of posets from $\mathcal{S}_{k\mathcal{D}}$ to $\mathcal{T}_{k\mathcal{D}}$. The inverse isomorphism is the map

$$\Psi : A \mapsto \sum_{H \in A} \mathsf{e}_H .$$

Proof: Let F be a subfunctor of kB. The set $\Theta(F)$ is closed, by Proposition 5.4.8, so Θ is indeed a map from $\mathcal{S}_{k\mathcal{D}}$ to $\mathcal{T}_{k\mathcal{D}}$. It is obviously a map of posets: if $F \subseteq F'$ are subfunctors of kB, then $\mathsf{e}_H \subseteq F$ implies $\mathsf{e}_H \subseteq F'$. The map Ψ is also obviously a map of posets.

Let $F \in \mathcal{S}_{k\mathcal{D}}$, and let G be an object of $\mathcal{D}$. Then $F(G)$ is the sum in $kB(G)$ of those one dimensional subspaces ke_H^G for which $e_H^G \in F(G)$. But if $e_H^G \in F(G)$, then $\mathrm{Res}_H^G e_H^G = e_H^H \in F(H)$, i.e. $\mathsf{e}_H \subseteq F$. Conversely, if $\mathsf{e}_H \subseteq F$, i.e. if $e_H^H \in F(H)$, then $e_H^G \in F(G)$, because e_H^G is a non zero scalar multiple of $\mathrm{Ind}_H^G e_H^H$, by Theorem 5.2.4. It follows that

$$F(G) = \sum_{\substack{H \leq G \\ \mathsf{e}_H \subseteq F}} \mathsf{e}_H(G) ,$$

for any object G of $\mathcal{D}$. Thus

$$F = \sum_{\substack{H \in \mathcal{O} \\ \mathsf{e}_H \subseteq F}} \mathsf{e}_H ,$$

where $\mathcal{O}$ is any set of representatives of objects of $\mathcal{D}$ up to group isomorphism. But $\mathsf{e}_H = \mathsf{e}_{\beta_k(H)}$, for any object H of $\mathcal{D}$, and $\beta_k(H)$ is a B-group over k. It follows that

$$F = \sum_{\substack{H\in[B\text{-gr}_k(\mathcal{D})]\\ \mathsf{e}_H\subseteq F}} \mathsf{e}_H \ ,$$

i.e. that $\Psi\Theta(F) = F$.

Conversely, let $A \in \mathcal{T}_{k\mathcal{D}}$. Then

$$\Theta\Psi(A) = \{H \in [B\text{-gr}_k(\mathcal{D})] \mid \mathsf{e}_H \subseteq \sum_{K\in A} \mathsf{e}_K\} \ ,$$

so obviously $\Theta\Psi(A) \supseteq A$. Conversely, if $H \in \Theta\Psi(A)$, then $\mathsf{e}_H \subseteq \sum_{K\in A} \mathsf{e}_K$, i.e.

$$e_H^H \in \sum_{K\in A} \mathsf{e}_K(H) \ .$$

So there is $K \in A$ such that $e_H^H \mathsf{e}_K(H) \neq \{0\}$, i.e. $e_H^H \mathsf{e}_K(H) = ke_H^H$. But $\mathsf{e}_K(H)$ is an ideal of $kB(H)$, so $e_H^H \in \mathsf{e}_K(H)$, i.e. $\mathsf{e}_H \subseteq \mathsf{e}_K$. Since K is a B-group over k, it follows from Proposition 5.4.8 that K is isomorphic to a quotient of H, and then $H \in A$, since $K \in A$ and A is a closed subset of $[B\text{-gr}_k(\mathcal{D})]$. This shows that $\Theta\Psi(A) = A$, and completes the proof of Theorem 5.4.14. □

5.5. Application: Some Simple Biset Functors

5.5.1. Proposition :

1. *Let G be a B-group over k. Then the subfunctor e_G of kB has a unique maximal subfunctor, equal to*

$$\mathsf{j}_G = \sum_{\substack{H\in[B\text{-gr}_k(\mathcal{D})]\\ H\gg G,\ H\not\cong G}} \mathsf{e}_H \ ,$$

and the quotient functor $\mathsf{e}_G/\mathsf{j}_G$ is isomorphic to the simple functor $S_{G,k}$.

2. *If $F \subset F'$ are subfunctors of kB such that F'/F is simple, then there exists a unique $G \in [B\text{-gr}_k(\mathcal{D})]$ such that $\mathsf{e}_G \subseteq F'$ and $\mathsf{e}_G \not\subseteq F$. In particular $\mathsf{e}_G + F = F'$, $\mathsf{e}_G \cap F = \mathsf{j}_G$, and $F'/F \cong S_{G,k}$.*

Proof: By Theorem 5.4.14, the poset of subfunctors of e_G is isomorphic to the poset of closed subsets of

$$\Theta(\mathsf{e}_G) = \{H \in [B\text{-gr}_k(\mathcal{D})] \mid \mathsf{e}_H \subseteq \mathsf{e}_G\} .$$

But $\mathsf{e}_H \subseteq \mathsf{e}_G$ if and only if $H \gg G$, since G is a B-group over k, by Proposition 5.4.8. Thus

$$\Theta(\mathsf{e}_G) = \{H \in [B\text{-gr}_k(\mathcal{D})] \mid H \gg G\} .$$

If A is a closed subset of $\Theta(\mathsf{e}_G)$, then either $A \ni G$, and then $A = \Theta(\mathsf{e}_G)$, or A is contained in the set

$$\{H \in [B\text{-gr}_k(\mathcal{D}] \mid H \gg G,\ H \ncong G\} ,$$

which is a closed subset of $\Theta(\mathsf{e}_G)$. It follows that e_G has a unique maximal (proper) subfunctor, equal to j_G.

The quotient functor $S = \mathsf{e}_G/\mathsf{j}_G$ is a simple functor. Moreover $S(G) \cong \mathsf{e}_G(G) \cong k$, since $\mathsf{e}_G(G) = ke_G^G$ by Proposition 5.4.9, and $\mathsf{j}_G(G) = \{0\}$ by Proposition 5.4.5. And if K is an object of $\mathcal{D}$ with $|K| < |G|$, then $\mathsf{e}_G(K) = \{0\}$ by Proposition 5.4.5, so $S(K) = \{0\}$. Thus G is a minimal object for S, and $S(G)$ is one dimensional, generated by the image of e_G^G, which is invariant by any automorphism of G. Hence $S(G)$ is the trivial $k\mathrm{Out}(G)$-module, so $S \cong S_{G,k}$, and Assertion 1 holds.

Now if $F \subset F' \subseteq kB$ and F'/F is simple, let G and G' be elements of $\Theta(F') - \Theta(F)$. Since F'/F is simple, any closed subset of $\Theta(F')$ containing $\Theta(F)$ and G is equal to $\Theta(F')$, hence contains G'. Thus

$$G' \in \Theta(F) \cup \{H \in [B\text{-gr}_k(\mathcal{D}] \mid H \gg G\} ,$$

hence $G' \gg G$, and $G \gg G'$ by symmetry, so $G = G'$. Now $\mathsf{e}_G \subseteq F'$, but $\mathsf{e}_G \nsubseteq F$, so $\mathsf{e}_G + F = F'$, since F'/F is simple. It follows that $F'/F \cong \mathsf{e}_G/(\mathsf{e}_G \cap F)$ is simple, hence $\mathsf{e}_G \cap F = \mathsf{j}_G$ and $F'/F \cong S_{G,k}$ by Assertion 1, as was to be shown. □

5.5.2. Remark : Proposition 5.5.1 means that the "composition factors" (i.e. the simple subquotients) of the Burnside functor kB on $k\mathcal{D}$ are the exactly the functors $S_{G,k}$, where G is an object of $\mathcal{D}$ which is a B-group over k.

5.5.3. Proposition : *Let G and H be objects of $\mathcal{D}$. Then $\mathsf{e}_G(H)$ is the subspace of $kB(H)$ with basis the set of idempotents e_K^H, where K runs through a set of representatives of conjugacy classes of subgroups K of H such that $K \gg \beta_k(G)$.*

Proof: Since $\mathsf{e}_G = \mathsf{e}_{\beta_k(G)}$, one can suppose that $G = \beta_k(G)$, in other words that G is a B-group over k.

Since $\mathsf{e}_G(H)$ is an ideal of $kB(H)$ by Lemma 5.2.1, it has a k-basis consisting of the different idempotents e_K^H it contains. Now if $e_K^H \in \mathsf{e}_G(H)$, it follows that $e_K^K = \mathrm{Res}_K^H e_K^H \in \mathsf{e}_G(K)$, thus $\mathsf{e}_K \subseteq \mathsf{e}_G$, and $K \gg G$ by Proposition 5.4.8, since G is a B-group over k. Conversely, if $K \gg G$, then $\mathsf{e}_K \subseteq \mathsf{e}_G$, so $e_K^K \in \mathsf{e}_G(K)$. Since e_K^H is a non zero multiple of $\mathrm{Ind}_K^H e_K^K$, it follows that $e_K^H \in \mathsf{e}_G(H)$. □

5.5.4. Theorem : *Let k be a field of characteristic q, and let G be a B-group in characteristic q. If H is a finite group of order coprime to q, then $\dim_k S_{G,k}(H)$ is equal to the number of conjugacy classes of subgroups K of H such that $\beta_q(K) \cong G$.*

Proof: Let $\mathcal{D}$ denote the full subcategory of $\mathcal{C}$ consisting of groups of order coprime to q. Then Proposition 5.5.3 shows that $\mathsf{e}_G(H)$ is the subspace of $kB(H)$ generated by the idempotents e_K^H, where K is a subgroup of H such that $K \gg \beta_k(G) = G$. By Theorem 5.4.11, this condition is equivalent to the condition $\beta_k(K) \gg G$. Now if $\beta_k(K) \gg G$ and $\beta_k(K) \not\cong G$, then

$$e_K^H \in \mathsf{e}_{\beta_k(K)}(H) \subseteq \mathsf{j}_G(H) .$$

This shows that the quotient $\mathsf{e}_G(H)/\mathsf{j}_G(H) \cong S_{G,k}(H)$ is generated by the images of the idempotents e_K^H, for $\beta_k(K) \cong G$, up to H-conjugation. Conversely, these images are linearly independent over k: indeed $\mathsf{j}_G(H)$ has a k-basis consisting of the different idempotents e_K^H for which $\beta_k(K) \gg G$, but $\beta_k(K) \not\cong G$. □

5.6. Examples

5.6.1. Proposition : *Let G be a finite group. Then $m_{G,G} = 0$ if G is not cyclic, and $m_{G,G} = \varphi(n)/n$ if G is cyclic of order n, where φ is the Euler totient function.*

Proof: Indeed

$$\begin{aligned} m_{G,G} &= \frac{1}{|G|} \sum_{XG=G} |X|\mu(X,G) = \frac{1}{|G|} \sum_{X \leq G} |X|\mu(X,G) \\ &= \frac{1}{|G|} \sum_{x \in G} \Big(\sum_{<x> \leq X \leq G} \mu(X,G) \Big) \\ &= \frac{1}{|G|} |\{x \in G \mid <x> = G\}| . \end{aligned}$$

The proposition follows. □

5.6.2. Remark : This shows in particular that $\beta_0(G) = \mathbf{1}$ if and only if G is cyclic. So Theorem 4.4.8 is a special case of Theorem 5.5.4.

5.6.3. Proposition : *Let G be a non cyclic finite simple group. Then G is a B-group in characteristic q, for any q not dividing $|G|$.*

Proof: Indeed, the only non-trivial normal subgroup of G is G itself. But $m_{G,G} = 0$ if G is not cyclic. □

5.6.4. Proposition : *Let G be a finite group. If N is a minimal abelian normal subgroup of G, then*

$$m_{G,N} = 1 - \frac{|K_G(N)|}{|N|} ,$$

where $K_G(N)$ is the set of complements of N in G.

Hence if G is solvable, then G is a B-group in characteristic q if and only if q does not divide $|G|$, and if $|K_G(N)| \equiv |N| \pmod{q}$, for any minimal normal subgroup N of G.

Proof: Let X be a subgroup of G such that $XN = G$. Then $X \cap N$ is normalized by X, and by N since N is abelian. It follows that $X \cap N \trianglelefteq G$, thus $X \cap N = N$ or $X \cap N = \mathbf{1}$ by minimality of N. In the first case $X = G$, and in the second case $X \in K_G(N)$, so $|X| = |G : N|$, and moreover X is a maximal subgroup of G, so $\mu_G(X, G) = -1$. This shows that

$$m_{G,N} = 1 - \frac{|K_G(N)|}{|N|} .$$

The second assertion of the proposition is straightforward. □

5.6.5. Definition : *Two finite groups G and H are said to have* no non trivial common quotient *if $G \gg K$ and $H \gg K$ implies $K \cong \mathbf{1}$.*

This occurs in particular if G and H have coprime orders.

5.6.6. Proposition : *Let G and H be finite groups having no non trivial common quotient, and let q be an integer equal to 0 or a prime number coprime to both $|G|$ and $|H|$.*

1. *If $M \trianglelefteq G$ and $N \trianglelefteq H$, then $m_{G\times H, M\times N} = m_{G,M}m_{H,N}$.*
2. *The group $\beta_q(G \times H)$ is isomorphic to $\beta_q(G) \times \beta_q(H)$.*
3. *In particular the group $G \times H$ is a B-group in characteristic q if and only if G and H are B-groups in characteristic q.*

Proof: The subgroups $\mathbf{1} \times N$ and $M \times N$ are normal subgroups of $G \times H$. By Lemma 5.3.1

$$\begin{aligned} m_{G\times H, M\times N} &= m_{G\times H,\mathbf{1}\times N} m_{(G\times H)/(\mathbf{1}\times N),(M\times N)/(\mathbf{1}\times N)} \\ &= m_{G\times H,\mathbf{1}\times N} m_{G\times (H/N), M\times \mathbf{1}} \ . \end{aligned} \tag{5.6.7}$$

Now to compute $m_{G\times H,\mathbf{1}\times N}$, observe that $G \times \mathbf{1}$ is a normal subgroup of $G \times H$. Setting $K = G \times H$, $P = \mathbf{1} \times N$, and $Q = G \times \mathbf{1}$, Proposition 5.3.3 gives

$$m_{K,P} = \frac{1}{|G||H|} \sum_{Y} |Y|\mu(Y, K) m_{K/Q,(Y\cap P)Q/Q} \ ,$$

where Y runs through the set of subgroups of $G \times H$ such that

$$Y(\mathbf{1} \times N) = G \times H = Y(G \times \mathbf{1}) \ .$$

The first of these equalities implies $p_1(Y) = G$, and the second one implies $p_2(Y) = H$. Then $q(Y)$ is isomorphic to a quotient of both G and H, so $q(Y) \cong \mathbf{1}$. It follows that $Y = G \times H$, and

$$m_{G\times H,\mathbf{1}\times N} = m_{K/Q,PQ/Q} = m_{H,N} \ .$$

Since the groups G and H/N also have no non trivial quotient, a similar argument shows that $m_{G\times(H/N),M\times\mathbf{1}} = m_{G,M}$, thus by 5.6.7

$$m_{G\times H,M\times N} = m_{G,M} m_{H,N} \ ,$$

which proves Assertion 1.

Let $M \trianglelefteq G$ and $N \trianglelefteq H$ such that $\beta_q(G) \cong G/M$ and $\beta_q(H) \cong H/N$. Since $m_{G\times H,M\times N} = m_{G,M} m_{H,N} \neq 0_k$, it follows from Proposition 5.4.11 that $\beta_q(G \times H)$ is a quotient of $(G \times H)/(M \times N) \cong \beta_q(G) \times \beta_q(H)$.

Conversely, the groups $\beta_q(G)$ and $\beta_q(H)$ are B-groups in characteristic q, and they are both isomorphic to a quotient of $G \times H$. Hence by Proposition 5.4.11, there is a surjective morphism $s : \beta_q(G\times H) \to \beta_q(G)$ and a surjective morphism $t : \beta_q(G\times H) \to \beta_q(H)$. The image Y of $\beta_q(G\times H)$ in $\beta_q(G)\times\beta_q(H)$ by the map $s \times t$ is such that $p_1(Y) = \beta_q(G)$ and $p_2(Y) = \beta_q(H)$. Since the groups $\beta_q(G)$ and $\beta_q(H)$ also have no non trivial common quotient, it follows that $Y = \beta_q(G) \times \beta_q(H)$, so $\beta_q(G) \times \beta_q(H)$ is isomorphic to a quotient of $\beta_q(G \times H)$. Assertion 2 follows.

Finally, if G and H are B-groups in characteristic q, then $\beta_q(G \times H) \cong \beta_q(G)\times\beta_q(H) \cong G\times H$, so $G\times H$ is a B-group in characteristic q. Conversely, if $G \times H$ is a B-group in characteristic q, then

$$\beta_q(G \times H) \cong G \times H \cong \beta_q(G) \times \beta_q(H) \ .$$

Since $G \gg \beta_q(G)$ and $H \gg \beta_q(H)$, it follows that $G \cong \beta_q(G)$ and $H \cong \beta_q(H)$, so G and H are B-groups in characteristic q. □

5.6.8. The Case $q = 0$. Some examples of B-groups in characteristic 0 can be found in [6], where it is shown in particular that the symmetric groups S_n are B-groups in characteristic 0, for any $n \in \mathbb{N}$, and that a nilpotent finite group is a B-group in characteristic 0 if and only if it is isomorphic to $(\mathbb{Z}/n\mathbb{Z})^2$, for some *square free* integer n. This result will follow from Proposition 5.6.6, and from the discussion of the case of p-groups in the next section.

5.6.9. The Case of p-Groups. [Bouc–Thévenaz [19] Sect. 8] If G is a p-group, and if $N \trianglelefteq G$, then $m_{G,N} = m_{G/\Phi(G),N\Phi(G)/\Phi(G)}$ by Remark 5.2.3, so the computation can be made for the elementary abelian p-group $G/\Phi(G)$. In particular, if G is also a B-group in characteristic q (so $q \neq p$ by Definition 5.4.6), it follows that G is elementary abelian.

If G is elementary abelian of order $p^n > 1$, and if N is a minimal normal subgroup of G, then N has order p, and N has p^{n-1} complements in G, thus

$$m_{G,N} = 1 - p^{n-1}/p = 1 - p^{n-2}$$

by Proposition 5.6.4. It follows that G is a B-group modulo q if and only if $p^{n-2} = 1$ modulo q.

- If $q = 0$, this implies $n = 2$. So a p-group G is a B-group in characteristic 0 if and only if G is trivial or isomorphic to $E = (\mathbb{Z}/p\mathbb{Z})^2$. Together with Proposition 5.6.6, this gives the structure of nilpotent B-groups in characteristic 0 mentioned in Remark 5.6.8.
  If $\mathcal{C}_p$ is the full subcategory of $\mathcal{C}$ consisting of p-groups, then the poset $([B\text{-gr}_k(\mathcal{C}_p)], \gg)$ is the poset $\{\mathbf{1}, E\}$, with $E \gg \mathbf{1}$. The set of closed subsets of this poset is $\{\emptyset, \{E\}, \{\mathbf{1}, E\}\}$.
  It follows that if k is a field of characteristic 0, then the Burnside functor on $k\mathcal{C}_p$ has a unique proper non-zero subfunctor J, generated by e_E^E. This functor J is equal to the kernel of the linearization morphism $kB \to kR_{\mathbb{Q}}$, and the functor $kR_{\mathbb{Q}}$ is isomorphic to the simple functor $S_{\mathbf{1},k}$. The functor J is simple, isomorphic to $S_{E,k}$, since E is a minimal group for J, and $J(E) \cong k$. This simple functor is closely related to *the Dade functor*, to be discussed in Chap. 12 (see Corollary 12.9.11, and also Theorems 10.1 and 10.4 of [19] for details).
  Finally, if P is a finite p-group, then $\beta_0(P)$ is trivial if P is cyclic, and isomorphic to $(\mathbb{Z}/p\mathbb{Z})^2$ otherwise. This illustrates Remark 5.4.12, for if $|P/\Phi(P)| \geq p^3$, there are many different normal subgroups N in P such that $P/N \cong (\mathbb{Z}/p\mathbb{Z})^2$.
- If $q \mid p-1$, then any elementary abelian p-group is a B-group in characteristic p. In this case, the poset $([B\text{-gr}_k(\mathcal{C}_p)], \gg)$ is isomorphic to $(\mathbb{N}, \geq)$, and the poset of its closed subsets is isomorphic to $(\overline{\mathbb{N}}, \geq)$, where $\overline{\mathbb{N}} = \mathbb{N} \sqcup \{\infty\}$,

with $\infty \geq n$ for any $n \in \mathbb{N}$. So the Burnside functor on $k\mathcal{C}_p$ is uniserial, and the term F_n in its unique filtration

$$kB = F_0 \supset F_1 \supset \ldots \supset F_n \supset \ldots$$

by subfunctors is the subfunctor generated by the elements e_E^E, where E is an elementary abelian group of order at least p^n. It is easy to see that if P is a finite p-group, then $F_n(P)$ is the subspace of $kB(P)$ generated by the idempotents e_Q^P, for $|Q/\Phi(Q)| \geq p^n$.
In this case $\beta_q(P)$ is isomorphic to $P/\Phi(P)$, so there is a unique normal subgroup N of P such that $P/N \cong \beta_q(P)$.

- If $0 < q \nmid p-1$, let $m \geq 2$ denote the order of p modulo q. Then a p-group E is a B-group in characteristic q if and only if it is trivial, or elementary abelian of order p^n, with $n \equiv 2$ modulo m. The poset $([B\text{-gr}_k(\mathcal{C}_p)], \gg)$ is isomorphic to $(\{0\} \sqcup (2+m\mathbb{N}), \geq)$, and the poset of its closed subsets is isomorphic to the poset $(\{0\} \sqcup (2+m\mathbb{N}) \sqcup \{\infty\}, \geq)$ obtained by adding a maximal element ∞.
The functor kB on $k\mathcal{C}_p$ is again uniserial in this case, and for $n > 0$, the subfunctor F_n in its unique filtration

$$kB = F_0 \supset F_1 \supset \ldots \supset F_n \supset \ldots$$

by subfunctors is the subfunctor generated by the elements e_E^E, where E is an elementary abelian group of order p^{2+lm}, with $l \geq n-1$. If P is a p-group, then $F_n(P)$ is the subspace of $kB(P)$ generated by the idempotents e_Q^P, for $|Q/\Phi(Q)| \geq p^{2+(n-1)m}$.
In this case, if $|P/\Phi(P)| = p^h$, then $\beta_q(P)$ is trivial if $h \leq 1$, i.e. if P is cyclic, and $\beta_q(P)$ is isomorphic to $(\mathbb{Z}/p\mathbb{Z})^{2+lm}$ otherwise, where l is the largest integer such that $2 + lm \leq h$.

Chapter 6
Endomorphism Algebras

This chapter is devoted to some results on the endomorphism algebra $\mathrm{End}_{R\mathcal{D}}(G)$ of an object G of $\mathcal{D}$. Since $\mathcal{D}$ is a full subcategory of the biset category, it follows that for any objects G and H of $\mathcal{D}$

$$\mathrm{Hom}_{R\mathcal{D}}(G,H) = RB(H,G) .$$

In particular, for any object G of $\mathcal{D}$

$$\mathrm{End}_{R\mathcal{D}}(G) = RB(G,G) ,$$

and both symbols will be used to denote this algebra in the present chapter. Recall in particular from Proposition 4.3.2 that there is a surjective algebra homomorphism

$$\varpi_G : RB(G,G) \to R\mathrm{Out}(G) ,$$

whose kernel I_G consists of linear combinations of bisets $[(G \times G)/L]$, for $L \leq G \times G$, which "factor trough a group strictly smaller than G", i.e. such that $|q(L)| < |G|$. This algebra homomorphism has a section, induced by the map $\varphi \mapsto \mathrm{Iso}(\varphi)$ from $\mathrm{Aut}(G)$ to $B(G,G)$.

6.1. Simple Modules and Radical

Fix an object G of $\mathcal{D}$. It was shown in Corollary 4.2.4 that if V is a simple $\mathrm{End}_{R\mathcal{D}}(G)$-module, then the functor $L_{G,V}$ has a unique simple quotient $F = S_{G,V}$, and V is isomorphic to $S_{G,V}(G)$ as an $\mathrm{End}_{R\mathcal{D}}(G)$-module. Moreover, since the category $\mathcal{D}$ is admissible (see Example 4.1.8), the simple objects of $\mathcal{F}_{\mathcal{D},R}$ are parametrized by seeds of $R\mathcal{D}$: the seed corresponding to the simple functor F is any pair (H,W) consisting of a minimal object H for F, and $W = F(H)$, viewed as an $R\mathrm{Out}(H)$-module.

This shows that there is a seed (H,W) of $R\mathcal{D}$ such that $V \cong S_{H,W}(G)$. Conversely, if (H,W) is a seed of $R\mathcal{D}$ such that $S_{H,W}(G) \neq \{0\}$, then $V = S_{H,W}(G)$ is a simple $\mathrm{End}_{R\mathcal{D}}(G)$-module, by Corollary 4.2.4. In this case, Lemma 4.3.9 shows that H is isomorphic to a subquotient of G.

S. Bouc, *Biset Functors for Finite Groups*, Lecture Notes in Mathematics 1990,
DOI 10.1007/978-3-642-11297-3_6,

By Remark 4.2.6, the evaluation $S_{H,W}(G)$ is equal to $\{0\}$ if and only if $(\psi\varphi).w = 0$, for any $\varphi \in \mathrm{Hom}_{R\mathcal{D}}(H,G)$, any $\psi \in \mathrm{Hom}_{R\mathcal{D}}(G,H)$, and any $w \in W$, i.e. if

$$\varpi_H\big(\mathrm{Hom}_{R\mathcal{D}}(G,H) \circ \mathrm{Hom}_{R\mathcal{D}}(H,G)\big)W = \{0\} .$$

6.1.1. Notation : *Let $R[H,G,H]$ denote the image of the two sided ideal $RB(H,G) \times_G RB(G,H)$ of $RB(H,H)$ in the algebra $R\mathrm{Out}(H)$ by the projection map $\varpi_H : RB(H,H) \to R\mathrm{Out}(H)$.*

With this notation, if $S_{H,W}(G) \neq \{0\}$, then $R[H,G,H]W \neq \{0\}$, or equivalently $R[H,G,H]W = W$, since $R[H,G,H]W$ is an $R\mathrm{Out}(H)$-submodule of W. Hence any simple $R\mathrm{End}_{R\mathcal{D}}(G)$-module V yields such a seed (H,W), where H is a minimal group for $S_{G,V}$ and $W = S_{G,V}(H)$. Conversely, if (H,W) is a seed of $R\mathcal{D}$ such that $R[H,G,H]W \neq \{0\}$, then $S_{H,W}(G)$ is non zero, and it is a simple $R\mathrm{End}_{R\mathcal{D}}(G)$-module.

6.1.2. Proposition : *Let G be an object of $\mathcal{D}$. Then the correspondence*

$$(H,W) \mapsto S_{H,W}(G)$$

induces a bijection between the set of isomorphism classes of seeds (H,W) of $R\mathcal{D}$ such that $R[H,G,H]W \neq \{0\}$ and the set of isomorphism classes of simple $\mathrm{End}_{R\mathcal{D}}(G)$-modules.

6.1.3. Notation : *If H is a finite group, let $J_H \subseteq RB(H,H)$ be the two-sided ideal defined by*

$$J_H = \varpi_H^{-1}(\mathrm{Rad}\, R\mathrm{Out}(H)) ,$$

where $\mathrm{Rad}\, R\mathrm{Out}(H)$ is the Jacobson radical of $R\mathrm{Out}(H)$.

In particular $J_H \supseteq I_H = \varpi^{-1}(\{0\})$. With this notation:

6.1.4. Proposition : *Let G be a finite group. Then the Jacobson radical $\mathrm{Rad}\, RB(G,G)$ is the set of elements $u \in RB(G,G)$ such that*

$$RB(H,G) \times_G u \times_G RB(G,H) \subseteq J_H ,$$

for any group H isomorphic to a subquotient of G.

Proof: Let $\mathcal{D}$ be a replete subcategory of $\mathcal{C}$, containing G (e.g. $\mathcal{D} = \mathcal{C}$). Then $RB(G,G) = \mathrm{End}_{R\mathcal{D}}(G)$. Now the Jacobson radical of $\mathrm{End}_{R\mathcal{D}}(G)$ is the intersection of the annihilators of the simple $\mathrm{End}_{R\mathcal{D}}(G)$-modules (cf. [3] Corollary 15.5), i.e. the intersection of the annihilators of the modules $S_{H,W}(G)$, where (H,W) is a seed of $R\mathcal{D}$. It suffices to consider those seeds for which H is a subquotient of G, by Lemma 4.3.9. Now if $u \in \mathrm{End}_{R\mathcal{D}}(G)$ and (H,W) is a seed of G, then Remark 4.2.6 shows that $uS_{H,W}(G) = \{0\}$ if and only if

$$\varpi_H\big(\mathrm{Hom}_{R\mathcal{D}}(G,H) \circ u \circ \mathrm{Hom}_{R\mathcal{D}}(H,G)\big)W = \{0\} \; .$$

This holds for any simple $R\mathrm{Out}(H)$-module if and only if

$$\varpi_H\big(\mathrm{Hom}_{R\mathcal{D}}(G,H) \circ u \circ \mathrm{Hom}_{R\mathcal{D}}(H,G)\big) \subseteq \mathrm{Rad}\, R\mathrm{Out}(H) \; ,$$

and the Proposition follows. □

6.1.5. Proposition : *Let G be a finite group. If* $\mathrm{Rad}\, RB(G,G) = \{0\}$, *then* $\mathrm{Rad}\, R = \{0\}$ *and G is cyclic.*

Proof: If X is a G-set, let $\widehat{X}$ denote the (G,G)-biset equal to X as a G-set, with trivial right action of G. In other words, if X is considered as a $(G,\mathbf{1})$-biset, then

$$\widehat{X} = X \times_{\mathbf{1}} \bullet \; , \qquad \textbf{(6.1.6)}$$

where $\bullet$ is a $(\mathbf{1},G)$-biset of cardinality 1. The correspondence $X \mapsto \widehat{X}$ induces a $\mathbb{Z}$-linear map $B(G) \to B(G,G)$, denoted by $u \mapsto \widehat{u}$.

Let $K(G)$ denote the kernel of the linearization map

$$\chi_G = \chi_{\mathbb{Q},G} : B(G) \to R_{\mathbb{Q}}(G)$$

(see Remark 1.2.3). Since $R_{\mathbb{Q}}(G)$ is a finitely generated free abelian group, the image I of $\chi_{\mathbb{Q},G}$ is a free abelian group, and there is a $\mathbb{Z}$-linear map $\sigma : I \to B(G)$ such that $\chi_{\mathbb{Q},G} \circ \sigma = \mathrm{Id}_I$. It follows that $K(G)$ is a direct summand of $B(G)$, and in particular, it is a free abelian group. Moreover, the cokernel of $\chi_{\mathbb{Q},G}$ is finite by Artin's induction Theorem (cf. [29] Theorem 15.4), hence the rank of I is equal to the rank of $R_{\mathbb{Q}}(G)$, i.e. the number of conjugacy classes of cyclic subgroups of G (cf. [45] Sect. 13 Théorème 29 Corollaire 1). It follows that the rank of $K(G)$ is equal to the number of conjugacy classes of non cyclic subgroups of G. In particular $K(G) = \{0\}$ if and only if G is cyclic.

Let $v \in B(G)$, and $j \in R$. Consider the element $u = j\widehat{v}$ of $RB(G,G)$. If H is a finite group, then Equation 6.1.6 shows that for a finite G-set X, any element of

$$RB(H,G) \times_G [\widehat{X}] \times_G RB(G,H)$$

is a linear combination of (H,H)-bisets which factor through the trivial group $\mathbf{1}$. It follows that if $H \neq \mathbf{1}$, then

$$RB(H,G) \times_G u \times_G RB(G,H) \subseteq I_H \subseteq J_H \ .$$

And if $H = \mathbf{1}$, then

$$\begin{aligned} RB(\mathbf{1},G) \times_G u \times_G RB(G,\mathbf{1}) &= RjB(\mathbf{1},G) \times_G \widehat{v} \times_G B(G,\mathbf{1}) \\ &= RjB_v \ , \end{aligned}$$

where $B_v = B(\mathbf{1},G) \times_G \widehat{v} \times_G B(G,\mathbf{1}) = B(\mathbf{1},G) \times_G v \subseteq B(\mathbf{1}) = \mathbb{Z}$.

If $j \in \operatorname{Rad} R$, then $RjB_v \subseteq \operatorname{Rad} R = J_{\mathbf{1}}$, so $u = j\widehat{v} \in \operatorname{Rad} RB(G,G)$, by Proposition 6.1.4, for any $v \in B(G)$. Hence if $\operatorname{Rad} RB(G,G) = \{0\}$, it follows that $u = 0$, so $\operatorname{Rad} R = \{0\}$.

Moreover, if $v \in K(G)$, then

$$B_v = B(\mathbf{1},G) \times_G v \subseteq K(\mathbf{1}) = \{0\} \ ,$$

so $u = j\widehat{v} \in \operatorname{Rad} RB(G,G)$ in this case, for any $j \in R$. It follows that if $\operatorname{Rad} RB(G,G) = \{0\}$, then $K(G) = \{0\}$, and G is cyclic, as was to be shown. □

The following is a partial converse of Proposition 6.1.5 in the case where R is a field:

6.1.7. Proposition : *Let k be a field of characteristic $q \geq 0$, and G be a cyclic group of order m. Then $kB(G,G)$ is semisimple if and only if $q \nmid \varphi(m)$, where φ is Euler totient's function.*

Proof: Recall that if A is a finite dimensional k-algebra, then the A-module $A/\operatorname{Rad} A$ is semisimple. More precisely

$$A/\operatorname{Rad} A \cong \bigoplus_{S \in \mathcal{S}} \frac{\dim_k S}{\dim_k \operatorname{End}_A(S)} S \ ,$$

where $\mathcal{S}$ is a set of representatives of isomorphism classes of simple A-modules. Hence

$$\dim_k A \geq \sum_{S \in \mathcal{S}} \frac{(\dim_k S)^2}{\dim_k \operatorname{End}_A(S)} \ ,$$

and equality holds if and only if $\operatorname{Rad} A = \{0\}$, i.e. if A is semisimple.

In particular, the algebra $kB(G,G)$ is semisimple if and only if

$$\textbf{(6.1.8)} \qquad \dim_k kB(G,G) = \sum_{S} \frac{(\dim_k S)^2}{\dim_k \operatorname{End}_{kB(G,G)}(S)} \ ,$$

where S runs through a set of representatives of isomorphism classes of simple $kB(G,G)$-modules.

The left hand side of Equation 6.1.8 is equal to the number of conjugacy classes of subgroups of $G \times G$, i.e. to the number of subgroups of $G \times G$, since G is abelian. Let L be such a subgroup. Then there are sections $(B, A) = \big(p_1(L), k_1(L)\big)$ and $(T, S) = \big(p_2(L), k_2(L)\big)$ of G, and a group isomorphism $\theta : T/S \to B/A$, such that

$$L = \{(g, h) \in G \times G \mid gA = \theta(hS)\} .$$

Conversely, such an isomorphism $\theta : T/S \to B/A$ between subquotients of G determines a subgroup of $G \times G$.

Set $n = |T : S| = |B : A|$, $s = |S|$, and $a = |A|$. Then $n \mid m$, and a and s are divisors of m/n. Moreover, the integers a, s, and n determine the subgroups A, S, B, and T, since a cyclic group has a unique subgroup of a given order. Finally, there are $\varphi(n) = |\mathrm{Aut}(\mathbb{Z}/n\mathbb{Z})|$ choices for the isomorphism θ, where φ is the Euler function. It follows that

$$\dim_k kB(G, G) = \sum_{n|m} \varphi(n)\sigma(m/n)^2 ,$$

where $\sigma(m/n)$ is the number of divisors of m/n.

Now if $m = m'm''$, for coprime integers m' and m'', then the cyclic group G of order m is isomorphic to $M' \times M''$, where M' is cyclic of order m' and M'' is cyclic of order m''. Moreover

$$kB(G, G) \cong kB(M', M') \otimes_k kB(M'', M'') ,$$

by Proposition 2.5.14, since $B(G, G)$ is a free $\mathbb{Z}$-module, for any finite group G.

If $kB(G, G)$ is semisimple, then $kB(M', M')$ and $kB(M'', M'')$ are semisimple, for $\mathrm{Rad}\, kB(M', M') \otimes kB(M'', M'')$ and $kB(M', M') \otimes \mathrm{Rad}\, kB(M'', M'')$ are nilpotent two sided ideals of $kB(G, G)$. Conversely, if $kB(M', M')$ and $kB(M'', M'')$ are semisimple, then their tensor product is semisimple.

Since moreover $\varphi(m) = \varphi(m')\cdot\varphi(m'')$, then $q \nmid \varphi(m)$ if and only if $q \nmid \varphi(m')$ and $q \nmid \varphi(m'')$, so if suffices to consider the case where G is a cyclic p-group, of order $m = p^a$. The case $a = 0$ is trivial, since if $G = \mathbf{1}$, then $kB(G, G) \cong k$ is semisimple, and $\varphi(1) = 1$. So one can assume $a > 0$.

In the right hand side $\mathcal{R}$ of Equation 6.1.8, if $S = S_{H,W}(G)$, for some group H and some simple $k\mathrm{Out}(H)$-module W, then H is isomorphic to a subquotient of G, by Lemma 4.3.9, so H is cyclic, isomorphic to $\mathbb{Z}/p^b\mathbb{Z}$, for some $b \leq a$.

By Corollary 4.2.4, there are isomorphisms of skew fields

$$\mathrm{End}_{kB(G,G)}(S) \cong \mathrm{End}_{kB(H,H)}(W) \cong \mathrm{End}_{k\mathrm{Out}(H)}(W) .$$

Let $\sigma_H(G)$ denote the set of sections (T,S) of G such that $T/S \cong H$. Since $G = \mathbb{Z}/p^a\mathbb{Z}$ has a unique subgroup of order p^i, namely $p^{a-i}G$, the set $\sigma_H(G)$ consists of the $b-a+1$ sections $(p^iG, p^{i+b-a}G)$, for $0 \leq i \leq a-b$.

By Theorem 4.3.20, the k-vector space $S_{H,W}(G)$ is isomorphic to the image of the map

$$\Psi : \bigoplus_{(T,S)\in\sigma_H(G)} W \to \bigoplus_{(B,A)\in\sigma_H(G)} W ,$$

whose block from the component (T,S) to the component (B,A) is the endomorphism of W defined by

$$\Psi_{(B,A),(T,S)}(w) = \langle (B,A) \parallel (T,S) \rangle_H \cdot w ,$$

where $\langle (B,A) \parallel (T,S) \rangle_H$ is the element of $k\mathrm{Out}(H)$ defined in Notation 4.3.19 by

$$\langle (B,A) \parallel (T,S) \rangle_H = \sum_{\substack{g\in[B\backslash G/T] \\ (B,A) \,—\, ({}^gT,{}^gS)}} \pi_H\big(\bar{s}_{B,A}\theta_g(\bar{s}_{T,S})^{-1}\big) .$$

Moreover $({}^gT, {}^gS) = (T,S)$ since G is commutative, and $(B,A) — (T,S)$ if and only if $BS = AT$ and $B \cap S = A \cap T$.

If $H \neq \mathbf{1}$, this implies $B = T$ and $A = S$. So

$$\langle (B,A) \parallel (T,S) \rangle_H = \begin{cases} |G:T|1_{\mathrm{Out}(H)} & \text{if } (B,A) = (T,S) \\ 0 & \text{otherwise.} \end{cases}$$

It follows that the matrix of the map Ψ is a diagonal block matrix, whose diagonal block indexed by (T,S) is equal to $|G:T|\mathrm{Id}_W$. Thus

$$\dim_k S_{H,W}(G) = \begin{cases} (a-b+1)\dim_k W & \text{if } q \neq p \\ \dim_k W & \text{if } q = p\ . \end{cases}$$

Note that in any case $\dim_k S_{H,W}(G) \leq (a-b+1)\dim_k W$.

And if $H = \mathbf{1}$, then W is the trivial module k, and $\sigma_H(G)$ consists of sections (T,T), for $T \leq G$. Moreover, for any subgroups B and T of G

$$\langle (B,B) \parallel (T,T) \rangle_{\mathbf{1}} = |G:T|1 ,$$

so $\dim_k S_{\mathbf{1},k}(G)$ is equal to the rank of the matrix

$$\begin{pmatrix} 1 & 1 & 1 & \cdots & 1 \\ 1 & p & p & \cdots & p \\ 1 & p & p^2 & \cdots & p^2 \\ \vdots & \vdots & \vdots & \ddots & \vdots \\ 1 & p & p^2 & \cdots & p^a \end{pmatrix} ,$$

thus

$$\dim_k S_{\mathbf{1},k}(G) = \begin{cases} 1 & \text{if } q \mid p-1 \\ 2 & \text{if } q = p \\ a+1 & \text{otherwise.} \end{cases}$$

In this case again, note that $\dim_k S_{H,W}(G) \leq (a-b+1) \dim_k W$, with $H = \mathbf{1}$, i.e. $b = 0$, and $W = k$.

Now there are two cases:

- If $a \geq 2$, then $q \nmid \varphi(p^a) = p^{a-1}(p-1)$ if and only if $q \neq p$ and $q \nmid p-1$. In this case for any $H = \mathbb{Z}/p^b\mathbb{Z}$, for $b \leq a$, and any $k\mathrm{Out}(H)$-module W

$$\dim_k S_{H,W}(G) = (a-b+1) \dim_k W \ .$$

Conversely, if $q \mid \varphi(p^a)$, then there exists a subquotient H of G such that

$$\dim_k S_{H,W}(G) < (a-b+1) \dim_k W \ ,$$

for any simple $k\mathrm{Out}(H)$-module W.

- If $a = 1$, i.e. $G \cong \mathbb{Z}/p\mathbb{Z}$, then $S_{H,W}(G) \cong W$ if $|H| = p$, and $\dim_k S_{\mathbf{1},k}(G)$ is equal to 1 if $q \mid p-1$, and to 2 otherwise. Thus when $|H| = p^b$,

$$\dim_k S_{H,W}(G) \leq (a-b+1) \dim_k W \ ,$$

and this inequality is strict if and only if $b = 0$ and $q \mid p-1 = \varphi(p^a)$.

In both cases, if $q \nmid \varphi(p^a)$, the right hand side $\mathcal{R}$ of Equation 6.1.8 is equal to

$$\mathcal{S} = \sum_{H,W} \frac{(a-b+1)^2 (\dim_k W)^2}{\dim_k \mathrm{End}_{k\mathrm{Out}(H)} W} \ ,$$

where H is a cyclic group of order $p^b \leq p^a$, and W is a simple $k\mathrm{Out}(H)$-module (up to isomorphism), and $\mathcal{R} < \mathcal{S}$ if $q \mid \varphi(p^a)$.

Now for a given H, the sum

$$\sum_W \frac{(\dim_k W)^2}{\dim_k \mathrm{End}_{k\mathrm{Out}(H)} W} \ ,$$

when W runs through a set of representatives of simple $k\mathrm{Out}(H)$-modules, is at most equal to $\dim_k k\mathrm{Out}(H) = \varphi(p^b)$, with equality if and only if $k\mathrm{Out}(H)$ is semisimple, i.e. if $q \nmid \varphi(p^b)$. Since moreover $\sigma(p^{a-b}) = a-b+1$, it follows that

$$\mathcal{R} \leq \mathcal{S} \leq \sum_{b=0}^{a} (a-b+1)^2 \varphi(p^b) = \sum_{n \mid p^a} \varphi(n) \sigma(p^a/n)^2 = \dim_k kB(G,G) \ .$$

Equality 6.1.8 holds if and only if $\mathcal{R} = \dim_k kB(G,G)$, and this occurs if and only if $q \nmid \varphi(p^b)$, for $0 \leq b \leq a$, i.e. if $q \nmid \varphi(p^a)$. This completes the proof of Proposition 6.1.7. □

6.1.9. Remark : Similar more general algebras, in relation with the question of the semi-simplicity of categories of biset functors, have been considered by P. Webb [54,57], L. Barker [4], and O. Coşkun [28].

6.2. Idempotents

6.2.1. Notation : *Let G be a finite group. If N is a normal subgroup of G, let j_N^G be the element of $RB(G,G)$ defined by*

$$j_N^G = \mathrm{Inf}_{G/N}^G \times_{G/N} \mathrm{Def}_{G/N}^G \, .$$

In other words j_N^G is the endomorphism of G (in the biset category) corresponding to the biset G/N, endowed with the (G,G)-biset structure given by projection on G/N and multiplication inside G/N.

6.2.2. Remark : Note that if G is an object of the replete subcategory $\mathcal{D}$, then G/N is also an object of $\mathcal{D}$, for any $N \trianglelefteq G$, and that j_N^G factors as

$$j_N^G = \mathrm{Inf}_{G/N}^G \circ \mathrm{Def}_{G/N}^G$$

in the category $\mathcal{D}$.

6.2.3. Lemma : *Let G be a finite group.*

1. *If N and M are normal subgroups of G, then there is an isomorphism of $(G/N, G/M)$-bisets*

$$\mathrm{Def}_{G/N}^G \circ \mathrm{Inf}_{G/M}^G \cong \mathrm{Inf}_{G/MN}^{G/N} \circ \mathrm{Def}_{G/MN}^{G/M} \, .$$

2. *Thus $j_N^G \circ j_M^G = j_{MN}^G$, and in particular j_N^G is an idempotent of $RB(G,G)$.*

Proof: Assertion 1 follows from the obvious isomorphism of $(G/N, G/M)$-bisets

$$G/N \times_G G/M \cong G/NM \, ,$$

given by

$$\forall g, h \in G, \ \ (gN,_G hM) \mapsto ghNM \, ,$$

and Assertion 2 is a straightforward consequence. □

6.2.4. Definition : *Let G be a finite group. If N is a normal subgroup of G, let f_N^G denote the element of $RB(G,G)$ defined by*

$$f_N^G = \sum_{N \leq M \trianglelefteq G} \mu_{\trianglelefteq G}(N,M)\, j_M^G \ ,$$

where $\mu_{\trianglelefteq G}$ is the Möbius function of the poset of normal subgroups of G.

6.2.5. Lemma : *Let G be a finite group. If $N \trianglelefteq G$, then*

$$j_N^G = \sum_{N \leq M \trianglelefteq G} f_M^G \ .$$

Proof: This is a straightforward consequence of the Definition 6.2.4. □

6.2.6. Proposition : *Let G be a finite group. If M and N are normal subgroups of G, then*

1. $$\mathrm{Def}^G_{G/M} \circ f_N^G = \begin{cases} 0 & \text{if } M \not\leq N \\ f^{G/M}_{N/M} \circ \mathrm{Def}^G_{G/M} & \text{if } M \leq N \ . \end{cases}$$

2. $$f_N^G \circ \mathrm{Inf}^G_{G/M} = \begin{cases} 0 & \text{if } M \not\leq N \\ \mathrm{Inf}^G_{G/M} \circ f^{G/M}_{N/M} & \text{if } M \leq N \ . \end{cases}$$

Proof: Indeed

$$\begin{aligned} \mathrm{Def}^G_{G/M} \circ f_N^G &= \sum_{N \leq L \trianglelefteq G} \mu_{\trianglelefteq G}(N,L)\, \mathrm{Def}^G_{G/M} \circ \mathrm{Inf}^G_{G/L} \circ \mathrm{Def}^G_{G/L} \\ &= \sum_{N \leq L \trianglelefteq G} \mu_{\trianglelefteq G}(N,L)\, \mathrm{Inf}^{G/M}_{G/ML} \circ \mathrm{Def}^G_{G/ML} \\ &= \sum_{NM \leq P \trianglelefteq G} \Big(\sum_{\substack{N \leq L \trianglelefteq G \\ ML = P}} \mu_{\trianglelefteq G}(N,L) \Big)\, \mathrm{Inf}^{G/M}_{G/P} \circ \mathrm{Def}^G_{G/P} \ . \end{aligned}$$

Set $s_P = \sum_{\substack{N \leq L \trianglelefteq G \\ ML = P}} \mu_{\trianglelefteq G}(N,L)$. A classical combinatorial result states that $s_P = 0$ unless $NM = N$. It can be proved as follows: first if $NM \neq N$, then

$$s_{NM} = \sum_{\substack{L \trianglelefteq G \\ N \leq L \leq NM}} \mu_{\trianglelefteq G}(N,L) = 0 \ ,$$

by the defining property of the Möbius function. Next, for $NM \leq Q \trianglelefteq G$

$$\sum_{\substack{P \trianglelefteq G \\ NM \leq P \leq Q}} s_P = \sum_{\substack{L \trianglelefteq G \\ N \leq L \leq Q}} \mu_{\trianglelefteq G}(N, L) = 0 ,$$

so $s_Q = 0$ by an easy induction argument.

It follows that $\mathrm{Def}^G_{G/M} \circ f^G_N = 0$ unless $NM = N$, i.e. $M \leq N$. And in this case $s_P = \mu_{\trianglelefteq G}(N, P)$ for any $P \trianglelefteq G$, with $P \geq N$, thus

$$\begin{aligned}\mathrm{Def}^G_{G/M} \circ f^G_N &= \sum_{N \leq P \trianglelefteq G} \mu_{\trianglelefteq G}(N, P)\,\mathrm{Inf}^{G/M}_{G/P} \circ \mathrm{Def}^G_{G/P} \\ &= \Big(\sum_{N/M \leq P/M \trianglelefteq G/M} \mu_{\trianglelefteq G/M}(N/M, P/M)\,\mathrm{Inf}^{G/M}_{G/P} \circ \mathrm{Def}^{G/M}_{G/P}\Big) \circ \mathrm{Def}^G_{G/M} \\ &= f^{G/M}_{N/M} \circ \mathrm{Def}^G_{G/M}\end{aligned}$$

as was to be shown for Assertion 1. Assertion 2 follows by considering opposite bisets, and using Assertion 2 of Lemma 2.3.14, since $(f^G_N)^{op} = f^G_N$. □

6.2.7. Proposition : *Let G be a finite group. Then the elements f^G_N, for $N \trianglelefteq G$, are orthogonal idempotents of $RB(G,G)$, and*

$$\sum_{N \trianglelefteq G} f^G_N = \mathrm{Id}_G .$$

Proof: Indeed if N and M are normal subgroups of G, then

$$f^G_M \circ f^G_N = \sum_{M \leq L \trianglelefteq G} \mu_{\trianglelefteq G}(M, L)\mathrm{Inf}^G_{G/L} \circ \mathrm{Def}^G_{G/L} \circ f^G_N .$$

If for some $L \geq M$, the product $\mathrm{Def}^G_{G/L} \circ f^G_N$ is non zero, then $L \leq N$ by Proposition 6.2.6, hence $M \leq N$. Thus if $M \not\leq N$, then $f^G_M \circ f^G_N = 0$. But $f^G_M \circ f^G_N = (f^G_N \circ f^G_M)^{op}$, and this is equal to 0 if $N \not\leq M$. Thus $f^G_M \circ f^G_N = 0$ if $M \neq N$. Now

$$\begin{aligned}\sum_{N \trianglelefteq G} f^G_N &= \sum_{N \trianglelefteq G} \sum_{N \leq M \trianglelefteq G} \mu_{\trianglelefteq G}(N, M) j^G_M \\ &= \sum_{M \trianglelefteq G} \Big(\sum_{\substack{N \trianglelefteq G \\ 1 \leq N \leq M}} \mu_{\trianglelefteq G}(N, M)\Big) j^G_M .\end{aligned}$$

The sum $\sum_{\substack{N \trianglelefteq G \\ \mathbf{1} \le N \le M}} \mu_{\trianglelefteq G}(N, M)$ is equal to 0, unless $M = \mathbf{1}$. Thus

$$\sum_{N \trianglelefteq G} f_N^G = j_{\mathbf{1}}^G = \mathrm{Inf}_{G/\mathbf{1}}^G \circ \mathrm{Def}_{G/\mathbf{1}}^G = \mathrm{Id}_G \ .$$

It follows that for $M \trianglelefteq G$

$$f_M^G = f_M^G \circ \big(\sum_{N \trianglelefteq G} f_N^G\big) = (f_M^G)^2 \ ,$$

so the elements f_M^G, for $M \trianglelefteq G$, are orthogonal idempotents, whose sum is equal to Id_G. □

6.2.8. Proposition : *Let G be a finite group, and let N be a normal subgroup of G. Then the maps*

$$a : u \in RB(G, G) \mapsto \mathrm{Def}_{G/N}^G \circ u \circ \mathrm{Inf}_{G/N}^G \in RB(G/N, G/N)$$

and

$$b : v \in RB(G/N, G/N) \mapsto \mathrm{Inf}_{G/N}^G \circ v \circ \mathrm{Def}_{G/N}^G \in RB(G, G)$$

restrict to mutual inverse isomorphisms of unital algebras between $f_N^G RB(G, G) f_N^G$ and $f_{\mathbf{1}}^{G/N} RB(G/N, G/N) f_{\mathbf{1}}^{G/N}$.

Proof: First if $u \in f_N^G RB(G, G) f_N^G$, then by Lemma 6.2.5

$$\begin{aligned} ba(u) &= \mathrm{Inf}_{G/N}^G \circ \mathrm{Def}_{G/N}^G \circ u \circ \mathrm{Inf}_{G/N}^G \circ \mathrm{Def}_{G/N}^G \\ &= j_N^G \circ u \circ j_N^G \\ &= \big(\sum_{N \le M \trianglelefteq G} f_M^G\big) \circ f_N^G \circ u \circ f_N^G \circ \big(\sum_{N \le M \trianglelefteq G} f_M^G\big) \\ &= f_N^G \circ u \circ f_N^G = u \ . \end{aligned}$$

Moreover by Proposition 6.2.6

$$\begin{aligned} f_{\mathbf{1}}^{G/N} \circ a(u) \circ f_{\mathbf{1}}^{G/N} &= f_{\mathbf{1}}^{G/N} \circ \mathrm{Def}_{G/N}^G \circ u \circ \mathrm{Inf}_{G/N}^G \circ f_{\mathbf{1}}^{G/N} \\ &= \mathrm{Def}_{G/N}^G \circ f_N^G \circ u \circ f_N^G \circ \mathrm{Inf}_{G/N}^G \\ &= \mathrm{Def}_{G/N}^G \circ u \circ \mathrm{Inf}_{G/N}^G \\ &= a(u) \ . \end{aligned}$$

Similarly, if $v \in RB(G/N, G/N)$, then

$$\begin{aligned} ab(v) &= \mathrm{Def}^G_{G/N} \circ \mathrm{Inf}^G_{G/N} \circ v \circ \mathrm{Def}^G_{G/N} \circ \mathrm{Inf}^G_{G/N} \\ &= v \ , \end{aligned}$$

since $\mathrm{Def}^G_{G/N} \circ \mathrm{Inf}^G_{G/N} = \mathrm{Id}_{G/N}$. Moreover, if $v \in f_{\mathbf{1}}^{G/N} RB(G/N, G/N) f_{\mathbf{1}}^{G/N}$, then by Proposition 6.2.6

$$\begin{aligned} f_N^G \circ b(v) \circ f_N^G &= f_N^G \circ \mathrm{Inf}^G_{G/N} \circ v \circ \mathrm{Def}^G_{G/N} \circ f_N^G \\ &= \mathrm{Inf}^G_{G/N} \circ f_{\mathbf{1}}^{G/N} \circ v \circ f_{\mathbf{1}}^{G/N} \circ \mathrm{Def}_{\mathbf{1}}^{G/N} \\ &= \mathrm{Inf}^G_{G/N} \circ v \circ \mathrm{Def}_{\mathbf{1}}^{G/N} \\ &= b(v) \ . \end{aligned}$$

Finally, if $v, v' \in RB(G/N, G/N)$, then

$$\begin{aligned} b(v) \circ b(v') &= \mathrm{Inf}^G_{G/N} \circ v \circ \mathrm{Def}^G_{G/N} \circ \mathrm{Inf}^G_{G/N} \circ v' \circ \mathrm{Def}^G_{G/N} \\ &= \mathrm{Inf}^G_{G/N} \circ v \circ v' \circ \mathrm{Def}^G_{G/N} \\ &= b(v \circ v') \ . \end{aligned}$$

Finally

$$\begin{aligned} b(f_{\mathbf{1}}^{G/N}) &= \mathrm{Inf}^G_{G/N} \circ f_{\mathbf{1}}^{G/N} \circ \mathrm{Def}^G_{G/N} \\ &= f_N^G \circ \mathrm{Inf}^G_{G/N} \circ \mathrm{Def}^G_{G/N} = f_N^G \circ j_N^G \\ &= f_N^G \ . \end{aligned}$$

Hence a and b are mutual inverse isomorphisms of unital algebras between $f_N^G RB(G, G) f_N^G$ and $f_{\mathbf{1}}^{G/N} RB(G/N, G/N) f_{\mathbf{1}}^{G/N}$. □

6.2.9. Remark : It follows in particular that for any $N \trianglelefteq G$

$$f_N^G = \mathrm{Inf}^G_{G/N} \circ f_{\mathbf{1}}^{G/N} \circ \mathrm{Def}^G_{G/N} \ ,$$

so the idempotents f_N^G can be recovered from the idempotents $f_{\mathbf{1}}^H$, for factor groups H of G. These idempotents indexed by the trivial normal subgroup play a special rôle, leading to the notion of *faithful elements* defined in the next section.

When G is a p-group, for some prime number p, the idempotent $f_{\mathbf{1}}^G$ can be computed easily:

6.2.10. Lemma : *Let p be a prime number, and G be a finite p-group. Then*

$$f_{\mathbf{1}}^G = \sum_{\mathbf{1} \leq Q \leq \Omega_1 Z(G)} \mu(\mathbf{1}, Q) \operatorname{Inf}_{G/Q}^G \circ \operatorname{Def}_{G/Q}^G ,$$

where $\Omega_1 Z(G)$ is the largest elementary abelian central subgroup of G, and μ is the Möbius function of the poset of subgroups of Q, given by

$$\mu(\mathbf{1}, Q) = (-1)^k p^{\binom{k}{2}}$$

if Q is elementary abelian of rank k.

Proof: Let $Q \trianglelefteq G$. Then the value $\mu_{\trianglelefteq G}(\mathbf{1}, Q)$ is equal to the reduced Euler–Poincaré characteristic (see Sect. 11.2.6) of the poset $]\mathbf{1}, Q[^G$ of non trivial normal subgroups of G, properly contained in Q.

If Q is not elementary abelian, this poset is contractible via the maps $S \mapsto S\Phi(Q) \mapsto \Phi(Q)$, where $\Phi(Q)$ is the Frattini subgroup of Q. And if Q is elementary abelian, but not central in G, then $]\mathbf{1}, Q[^G$ is contractible via the maps $S \mapsto S \cap Z(G) \mapsto Q \cap Z(G)$.

Now if Q is central in G, then $\mu_{\trianglelefteq G}(\mathbf{1}, Q) = \mu(\mathbf{1}, Q)$. The value $\mu(\mathbf{1}, Q) = (-1)^k p^{\binom{k}{2}}$ when $Q \cong (\mathbb{Z}/p\mathbb{Z})^k$ is well known (see [38] 2.4, [46] 3.9–3.10, or [5] Proposition 5). □

6.3. Faithful Elements

If F is a biset functor, then for any object G of $\mathcal{D}$, the R-module $F(G)$ has a natural structure of $\operatorname{End}_{R\mathcal{D}}(G)$-module. In particular, by Proposition 6.2.7, there is a decomposition

$$F(G) \cong \bigoplus_{N \trianglelefteq G} f_N^G F(G) ,$$

as a direct sum of R-modules. The summand corresponding to $N = \mathbf{1}$ in this decomposition is called the set of *faithful elements* of $F(G)$:

6.3.1. Definition and Notation : *Let F be a biset functor, and G be an object of $\mathcal{D}$. The set of* faithful elements *of $F(G)$ is the submodule*

$$\partial F(G) = f_{\mathbf{1}}^G F(G)$$

of $F(G)$.

The rationale behind this terminology is the following lemma (see also Proposition 9.1.3):

6.3.2. Lemma : *Let F be a biset functor, and G be an object of $\mathcal{D}$.*

1. *If $\mathbf{1} < N \trianglelefteq G$, and $u \in F(G/N)$, then $f_{\mathbf{1}}^G \mathrm{Inf}_{G/N}^G u = 0$.*
2. *$F(G) = \partial F(G) \oplus \sum\limits_{\mathbf{1}<N\leq G} \mathrm{Im}\,\mathrm{Inf}_{G/N}^G$.*
3. *$\partial F(G) = \bigcap\limits_{\mathbf{1}<N\trianglelefteq G} \mathrm{Ker}\,\mathrm{Def}_{G/N}^G$.*

Proof: Assertion 1 is a special case of Assertion 1 of Proposition 6.2.6. It follows that $\sum\limits_{\mathbf{1}<N\leq G} \mathrm{Im}\,\mathrm{Inf}_{G/N}^G \subseteq \mathrm{Ker}\, f_{\mathbf{1}}^G$. Conversely, Proposition 6.2.7 implies that $F(G) = \partial F(G) \oplus \mathrm{Ker}\, f_{\mathbf{1}}^G$, and that $\mathrm{Ker}\, f_{\mathbf{1}}^G$ is equal to the sum of the images of the idempotents f_N^G, for $\mathbf{1} < N \trianglelefteq G$.

Moreover $f_N^G = \mathrm{Inf}_{G/N}^G \circ f_{\mathbf{1}}^{G/N} \circ \mathrm{Def}_{G/N}^G$ by Proposition 6.2.8, so $\mathrm{Im}\, f_N^G \subseteq \mathrm{Im}\,\mathrm{Inf}_{G/N}^G$, thus $\mathrm{Ker}\, f_{\mathbf{1}}^G \subseteq \sum\limits_{\mathbf{1}<N\trianglelefteq G} \mathrm{Im}\,\mathrm{Inf}_{G/N}^G$, so equality holds, and Assertion 2 follows.

Finally, if $\mathbf{1} < N \trianglelefteq G$, then $\mathrm{Def}_{G/N}^G \circ f_{\mathbf{1}}^G = 0$ by Proposition 6.2.6. Conversely, if $u \in \bigcap\limits_{\mathbf{1}<N\trianglelefteq G} \mathrm{Ker}\,\mathrm{Def}_{G/N}^G$, then $f_N^G u = 0$ for $\mathbf{1} < N \trianglelefteq G$, since $f_N^G = \mathrm{Inf}_{G/N}^G \circ f_{\mathbf{1}}^{G/N} \circ \mathrm{Def}_{G/N}^G$, thus $u = f_{\mathbf{1}}^G u \in \partial F(G)$ by Proposition 6.2.7. Assertion 3 follows, and this completes the proof of the proposition. □

6.3.3. Proposition : *Let F be a biset functor, and G be an object of $\mathcal{D}$. Then the map $\delta : F(G) \to \bigoplus\limits_{N\trianglelefteq G} \partial F(G/N)$ defined by*

$$\delta(u) = \bigoplus_{N\trianglelefteq G} f_{\mathbf{1}}^{G/N} \mathrm{Def}_{G/N}^G u$$

is an isomorphism of R-modules. The inverse isomorphism is the map ι defined by

$$\iota(\bigoplus_{N\trianglelefteq G} v_N) = \sum_{N\trianglelefteq G} \mathrm{Inf}_{G/N}^G v_N \ .$$

Proof: Let $u \in F(G)$. Then by Propositions 6.2.6 and 6.2.7

$$\begin{aligned}\iota\delta(u) &= \sum_{N\trianglelefteq G} \mathrm{Inf}_{G/N}^G f_{\mathbf{1}}^{G/N} \mathrm{Def}_{G/N}^G u \\ &= \sum_{N\trianglelefteq G} \mathrm{Inf}_{G/N}^G \mathrm{Def}_{G/N}^G f_N^G u\end{aligned}$$

$$= \sum_{N \trianglelefteq G} j_N^G f_N^G u$$
$$= \sum_{N \trianglelefteq G} f_N^G u$$
$$= u \ .$$

Similarly, if $v = \bigoplus_N v_N \in \bigoplus_{N \trianglelefteq G} \partial F(G/N)$, then by Proposition 6.2.6

$$\begin{aligned}
\delta\iota(v) &= \bigoplus_{N \trianglelefteq G} f_{\mathbf{1}}^{G/N} \mathrm{Def}_{G/N}^G \big(\sum_{M \trianglelefteq G} \mathrm{Inf}_{G/M}^G v_M \big) \\
&= \bigoplus_{N \trianglelefteq G} f_{\mathbf{1}}^{G/N} \big(\sum_{M \trianglelefteq G} \mathrm{Def}_{G/N}^G \mathrm{Inf}_{G/M}^G v_M \big) \\
&= \bigoplus_{N \trianglelefteq G} \big(\sum_{M \trianglelefteq G} f_{\mathbf{1}}^{G/N} \mathrm{Inf}_{G/MN}^{G/N} \mathrm{Def}_{G/NM}^{G/M} v_M \big) \\
&= \bigoplus_{N \trianglelefteq G} \big(\sum_{\substack{M \trianglelefteq G \\ M \leq N}} f_{\mathbf{1}}^{G/N} \mathrm{Def}_{G/N}^{G/M} v_M \big) \\
&= \bigoplus_{N \trianglelefteq G} \big(\sum_{\substack{M \trianglelefteq G \\ M \leq N}} f_{\mathbf{1}}^{G/N} \mathrm{Def}_{G/N}^{G/M} f_{\mathbf{1}}^{G/M} v_M \big) \\
&= \bigoplus_{N \trianglelefteq G} f_{\mathbf{1}}^{G/N} v_N = v \ ,
\end{aligned}$$

as was to be shown. □

6.4. More Idempotents

This section is devoted to a technical result, that will prove very useful later in Part III for studying p-biset functors. The idea is to generalize the result of the previous section, by finding a sufficient condition on a not necessarily normal subgroup H of G, ensuring that the map

$$\mathrm{Indinf}_{N_G(H)/H}^G : \partial F\big(N_G(H)/H\big) \to F(G)$$

is split injective.

6.4.1. Notation : *Let G be a group. If H and K are subgroups of G, denote by $H \overset{G}{\uparrow} K$ the subgroup of $N_G(K)$ defined by*

$$H \overset{G}{\uparrow} K = \bigcap_{g \in N_G(K)} \big(H^g \cap N_G(K)\big)K \ .$$

6.4.2. Remark : The group $H \overset{G}{\uparrow} K$ is a normal subgroup of $N_G(K)$, containing K. Note that for $g \in N_G(K)$

$$\big(H^g \cap N_G(K)\big)K = H^g{\cdot}K \cap N_G(K) = \{t \in N_G(K) \mid HgKt = HgK\} ,$$

so $H \overset{G}{\uparrow} K$ is the subgroup of $N_G(K)$ consisting of elements which stabilize each double coset HgK, for $g \in N_G(K)$. More generally, if $x \in G$, then $H^x \overset{G}{\uparrow} K$ is the set of elements of $N_G(K)$ which permute trivially the double cosets $HxgK$, for $g \in N_G(K)$, by multiplication on the right.

6.4.3. Definition : *A subgroup H of G is called* weakly expansive *if for any $x \in G$, the equalities $H^x \overset{G}{\uparrow} H = H = {}^xH \overset{G}{\uparrow} H$ imply $x \in N_G(H)$. It is called* expansive *if for any $x \in G$, the equality $H^x \overset{G}{\uparrow} H = H$ implies $x \in N_G(H)$.*

If H and K are subgroups of G, write $H \mathrel{\widehat{=}_G} K$ if there exists $x \in G$ such that ${}^xH \overset{G}{\uparrow} K = K$ and $K^x \overset{G}{\uparrow} H = H$, and $H \mathrel{\not\widehat{=}_G} K$ otherwise.

In particular, a normal subgroup of G is an expansive subgroup. If H and K are normal subgroups of G, then $H \mathrel{\widehat{=}_G} K$ if and only if $H = K$.

6.4.4. Proposition : *Let G be a finite group. If H is a subgroup of G, let $\mathcal{I}_H$ denote the element of $RB\big(G, N_G(H)/H\big)$ defined by*

$$\mathcal{I}_H = \mathrm{Indinf}_{N_G(H)/H}^{G} \circ f_{\mathbf{1}}^{N_G(H)/H} = (G/H) \times_{N_G(H)/H} f_{\mathbf{1}}^{N_G(H)/H} .$$

Set moreover $\mathcal{D}_H = \mathcal{I}_H^{op} = f_{\mathbf{1}}^{N_G(H)/H} \circ \mathrm{Defres}_{N_G(H)/H}^{G} \in B\big(N_G(H)/H, G\big)$.

1. *Let H and K be subgroups of G such that $H \mathrel{\not\widehat{=}_G} K$. Then*

$$\mathcal{D}_K \circ \mathcal{I}_H = 0 .$$

2. *Suppose that H is a weakly expansive subgroup G. Then*

$$\mathcal{D}_H \circ \mathcal{I}_H = f_{\mathbf{1}}^{N_G(H)/H} .$$

3. *Suppose that H is an expansive subgroup of G. Then*

$$\mathrm{Defres}_{N_G(H)/H}^{G} \circ \mathcal{I}_H = f_{\mathbf{1}}^{N_G(H)/H} .$$

Proof: Let H and K be any subgroups of G, and set $S = N_G(H)$ and $T = N_G(K)$. Then

$$\mathcal{D}_K \circ \mathcal{I}_H = f_{\mathbf{1}}^{T/K} \times_{T/K} (K\backslash G/H) \times_{S/H} f_{\mathbf{1}}^{S/H} \in B(T/K, S/H) .$$

Now the $(T/K, S/H)$-biset $K\backslash G/H$ is isomorphic to

$$K\backslash G/H = \bigsqcup_{x\in[T\backslash G/S]} K\backslash TxS/H \; .$$

For $x \in G$, set $N_x = K^x \stackrel{G}{\uparrow} H$. Then N_x/H acts trivially on the right on $K\backslash TxS/H$, by Remark 6.4.2. It follows that

$$K\backslash TxS/H = (K\backslash TxS/H) \times_{S/H} \mathrm{Inf}_{S/N_x}^{S/H} \times_{S/N_x} \mathrm{Def}_{S/N_x}^{S/H} \; .$$

But $\mathrm{Def}_{S/N_x}^{S/H} \times_{S/H} f_{\mathbf{1}}^{S/H} = 0$ if $N_x/H \neq \mathbf{1}$, by Proposition 6.2.6. Thus $(K\backslash TxS/H) \times_{S/H} f_{\mathbf{1}}^{S/H} = 0$ if $K^x \stackrel{G}{\uparrow} H \neq H$. A similar argument shows that $f_{\mathbf{1}}^{T/K} \times_{T/K} (K\backslash TxS/H) = 0$ if ${}^xH \stackrel{G}{\uparrow} K \neq K$. Thus

$$f_{\mathbf{1}}^{T/K} \times_{T/K} (K\backslash TxS/H) \times_{S/H} f_{\mathbf{1}}^{S/H} = 0 \; ,$$

unless ${}^xH \stackrel{G}{\uparrow} K = K$ and $K^x \stackrel{G}{\uparrow} H = H$. Assertion 1 follows.

For Assertion 2, suppose that $K = H$, and that H is a weakly expansive subgroup of G. Then as above

$$f_{\mathbf{1}}^{S/H} \times_{S/H} (H\backslash SxS/H) \times_{S/H} f_{\mathbf{1}}^{S/H} = 0 \; ,$$

unless ${}^xH \stackrel{G}{\uparrow} H = H = H^x \stackrel{G}{\uparrow} H$. This implies $x \in N_G(H) = S$, and it follows that

$$\mathcal{D}_H \circ \mathcal{I}_H = f_{\mathbf{1}}^{S/H} \circ (H\backslash S/H) \times_{S/H} f_{\mathbf{1}}^{S/H} = f_{\mathbf{1}}^{S/H} \; ,$$

since $H\backslash S/H = S/H$ is the identity $(S/H, S/H)$-biset, and $f_{\mathbf{1}}^{S/H}$ is an idempotent.

Similarly, for Assertion 3, if H is expansive, then

$$(H\backslash SxS/H) \times_{S/H} f_{\mathbf{1}}^{S/H} = 0$$

unless $x \in S = N_G(H)$. Thus

$$\mathrm{Defres}_{N_G(H)/H}^{G} \circ \mathcal{I}_H = (H\backslash S/H) \times_{S/H} f_{\mathbf{1}}^{S/H} = (S/H) \times_{S/H} f_{\mathbf{1}}^{S/H} = f_{\mathbf{1}}^{S/H} \; ,$$

as was to be shown. □

6.4.5. Corollary : *Let G be a finite group. Let $\mathcal{G}$ be a set of weakly expansive subgroups of G, such that $H \neq_G K$ for any distinct elements H and K of $\mathcal{G}$. Then for any biset functor F, the map*

$$\mathop{\oplus}_{H\in\mathcal{G}} \mathrm{Indinf}_{N_G(H)/H}^G : \mathop{\oplus}_{H\in\mathcal{G}} \partial F\big(N_G(H)/H\big) \to F(G)$$

is split injective. A left inverse is the map

$$\mathop{\oplus}_{H\in\mathcal{G}} f_{\mathbf{1}}^{N_G(H)/H} \circ \mathrm{Defres}_{N_G(H)/H}^G : F(G) \to \mathop{\oplus}_{H\in\mathcal{G}} \partial F\big(N_G(H)/H\big) .$$

In particular, the elements

$$\gamma_H^G = \mathrm{Indinf}_{N_G(H)/H}^G \circ f_{\mathbf{1}}^{N_G(H)/H} \circ \mathrm{Defres}_{N_G(H)/H}^G ,$$

for $H \in \mathcal{G}$, are orthogonal idempotents of $RB(G,G)$.

Proof: This is straightforward, since the restriction of $\mathcal{I}_H$ to $\partial F\big(N_G(H)/H\big)$ is equal to the restriction of $\mathrm{Indinf}_{N_G(H)/H}^G$. For the last assertion, observe that

$$\mathrm{Indinf}_{N_G(H)/H}^G \circ f_{\mathbf{1}}^{N_G(H)/H} \circ \mathrm{Defres}_{N_G(H)/H}^G = \mathcal{I}_H \circ \mathcal{D}_H .$$

This is an idempotent since $\mathcal{D}_H \circ \mathcal{I}_H = f_{\mathbf{1}}^{N_G(H)/H}$. These idempotents are orthogonal, since $\mathcal{D}_K \circ \mathcal{I}_H = 0$ for $H \neq K$. □

6.4.6. Remark : The sum of the idempotents γ_H^G, for $H \in \mathcal{G}$, need not be equal to the identity of $RB(G,G)$.

6.5. The Case of Invertible Group Order

Let G be an object of $\mathcal{D}$. Recall that since $\mathcal{D}$ is a replete subcategory of $\mathcal{C}$, the algebra $\mathrm{End}_{R\mathcal{D}}(G)$ is equal to the whole Burnside biset algebra $RB(G,G)$, and that there is an algebra homomorphism, denoted by $u \mapsto \widetilde{u}$, from $RB(G)$ to $RB(G,G)$ (see Notation 2.5.9). In the case $|G|$ is invertible in R, the formula

$$e_H^G = \frac{1}{|N_G(H)|} \sum_{K \leq H} |K| \mu(K,H)\, [G/K]$$

makes sense in $RB(G)$, for any subgroup H of G. This yields a decomposition of the identity of $RB(G)$ as a sum of orthogonal idempotents, whose image by the above algebra homomorphism is a decomposition of the identity of $RB(G,G)$ as a sum of the orthogonal idempotents

$$\widetilde{e}_H^G = \frac{1}{|N_G(H)|} \sum_{K \leq H} |K| \mu(K,H)\, [(G \times G)/\Delta(K)] ,$$

when H runs through a set $[s_G]$ of representatives of conjugacy classes of subgroups of G.

6.5.1. Notation : *Let G be an object of $\mathcal{D}$, such that $|G|$ is invertible in R. If H is a subgroup of G, let $n_{G,H}$ be the element of* $\mathrm{End}_{R\mathcal{D}}(H)$ *defined by*

$$n_{G,H} = \frac{1}{|N_G(H):H|} \sum_{g\in[N_G(H)/H]} \mathrm{Iso}(\gamma_g) \ ,$$

where $[N_G(H)/H]$ is a set of representatives H-cosets in $N_G(H)$, and $\mathrm{Iso}(\gamma_g)$ is the (H,H)-biset associated to the automorphism $\gamma_g : h \mapsto {}^g h$ of H, for any $g \in [N_G(H)/H]$.

Let $\rho_{G,H}$ denote the element of $\mathrm{End}_{R\mathcal{D}}(H)$ *defined by*

$$\rho_{G,H} = n_{G,H} \circ \tilde{e}_H^H \ .$$

6.5.2. Remark : The element $n_{G,H}$ does not depend on the choice of the set of representatives $[N_G(H)/H]$, since for $h \in H$, the (H,H)-bisets $\mathrm{Iso}(\gamma_g)$ and $\mathrm{Iso}(\gamma_{gh})$ are isomorphic, for any $g \in N_G(H)$.

6.5.3. Lemma : *With the above notation:*

1. $n_{G,H} \circ n_{G,H} = n_{G,H}$.
2. $n_{G,H} \circ \tilde{e}_H^H = \tilde{e}_H^H \circ n_{G,H}$.
3. $\rho_{G,H} \circ \rho_{G,H} = \rho_{G,H}$.
4. $n_{G,H} \circ \mathrm{Res}_H^G = \mathrm{Res}_H^G$ *in* $\mathrm{Hom}_{R\mathcal{D}}(G,H)$, *and* $\mathrm{Ind}_H^G \circ n_{G,H} = \mathrm{Ind}_H^G$ *in* $\mathrm{Hom}_{R\mathcal{D}}(H,G)$.

Proof: Assertion 1 follows from the fact that $\mathrm{Iso}(\gamma_{gg'}) = \mathrm{Iso}(\gamma_g) \circ \mathrm{Iso}(\gamma_{g'})$ in $\mathrm{End}_{R\mathcal{D}}(H)$, for any $g, g' \in N_G(H)$. Assertion 2 follows from the fact that $e_H^H \in RB(H)$ is invariant under any automorphism of H, and that the (H,H)-bisets $\mathrm{Iso}(f) \times_H \widetilde{X}$ and $\widetilde{f(X)} \times_H \mathrm{Iso}(f)$ are isomorphic, for any $f \in \mathrm{Aut}(H)$ and any H-set X. Assertion 3 is an obvious consequence of Assertion 2. The first part of Assertion 4 follows from the fact that for any $g \in N_G(H)$, there are isomorphisms of (H,G)-bisets

$$\mathrm{Iso}(\gamma_g) \times_H \mathrm{Res}_H^G \cong \mathrm{Res}_H^G \times_G \mathrm{Iso}_G(\gamma_g) \cong \mathrm{Res}_H^G \ ,$$

where $\mathrm{Iso}_G(\gamma_g)$ is the (G,G)-biset associated to conjugation by g in G. Since this is an inner automorphism of G, this biset is isomorphic to the identity biset of G. The second part of Assertion 4 follows from the first one, by considering opposite bisets. □

6.5.4. Proposition : *Let G be an object of $\mathcal{D}$, such that $|G|$ is invertible in R, and let H be a subgroup of G. Then the maps*

$$a : u \in \mathrm{End}_{R\mathcal{D}}(G) \mapsto \frac{1}{|N_G(H):H|} \tilde{e}_H^H \circ \mathrm{Res}_H^G \circ u \circ \mathrm{Ind}_H^G \circ \tilde{e}_H^H \in \mathrm{End}_{R\mathcal{D}}(H)$$

and

$$b : v \in \mathrm{End}_{R\mathcal{D}}(H) \mapsto \frac{1}{|N_G(H):H|}\mathrm{Ind}_H^G \circ v \circ \mathrm{Res}_H^G \in \mathrm{End}_{R\mathcal{D}}(G)$$

restrict to mutual inverse isomorphisms of unital algebras between $\widetilde{e}_H^G \mathrm{End}_{R\mathcal{D}}(G)\widetilde{e}_H^G$ *and* $\rho_{G,H}\mathrm{End}_{R\mathcal{D}}(H)\rho_{G,H}$.

Proof: First observe that if $u \in \mathrm{End}_{R\mathcal{D}}(G)$, then by Lemma 6.5.3

$$a(u) = \frac{1}{|N_G(H):H|}\rho_{G,H} \circ \mathrm{Res}_H^G \circ u \circ \mathrm{Ind}_H^G \circ \rho_{G,H} ,$$

so $\mathrm{Im}\, a \subseteq \rho_{G,H}\mathrm{End}_{R\mathcal{D}}(H)\rho_{G,H}$.

Similarly, if $v \in \mathrm{End}_{R\mathcal{D}}(H)$, then by Corollary 2.5.12

$$\begin{aligned}\widetilde{e}_H^G \circ b(v) \circ \widetilde{e}_H^G &= \frac{1}{|N_G(H):H|}\widetilde{e}_H^G \circ \mathrm{Ind}_H^G \circ v \circ \mathrm{Res}_H^G \circ \widetilde{e}_H^G \\ &= \frac{1}{|N_G(H):H|}\mathrm{Ind}_H^G \circ \widetilde{\mathrm{Res}_H^G e_H^G} \circ v \circ \widetilde{\mathrm{Res}_H^G e_H^G} \circ \mathrm{Res}_H^G .\end{aligned}$$

Moreover $\mathrm{Res}_H^G e_H^G = e_H^H$ by Theorem 5.2.4. Hence if $v \in \widetilde{e}_H^H \mathrm{End}_{R\mathcal{D}}(H)\widetilde{e}_H^H$, then $\widetilde{e}_H^G \circ b(v) \circ \widetilde{e}_H^G = b(v)$, and in particular

$$b\big(\rho_{G,H}\mathrm{End}_{R\mathcal{D}}(H)\rho_{G,H}\big) \subseteq \widetilde{e}_H^G \mathrm{End}_{R\mathcal{D}}(G)\widetilde{e}_H^G .$$

Now let $u, u' \in \mathrm{End}_{R\mathcal{D}}(G)$. Then

$$a(u) \circ a(u') = \frac{1}{|N_G(H):H|^2}\widetilde{e}_H^H \circ \mathrm{Res}_H^G \circ u \circ \mathrm{Ind}_H^G \circ \widetilde{e}_H^H \circ \mathrm{Res}_H^G \circ u' \circ \mathrm{Ind}_H^G \circ \widetilde{e}_H^H ,$$

and by Corollary 2.5.12 and Theorem 5.2.4

$$\mathrm{Ind}_H^G \circ \widetilde{e}_H^H \circ \mathrm{Res}_H^G = \widetilde{\mathrm{Ind}_H^G e_H^H} = |N_G(H) : H|\widetilde{e}_H^G .$$

Now if $u \circ \widetilde{e}_H^G = u$, it follows that

$$a(u) \circ a(u') = \frac{1}{|N_G(H):H|}\widetilde{e}_H^H \circ \mathrm{Res}_H^G \circ u \circ u' \circ \mathrm{Ind}_H^G \circ \widetilde{e}_H^H = a(u \circ u') .$$

Moreover, for $u \in \widetilde{e}_H^G \mathrm{End}_{R\mathcal{D}}(G)\widetilde{e}_H^G$

$$\begin{aligned}b \circ a(u) &= \frac{1}{|N_G(H):H|^2}\mathrm{Ind}_H^G \circ \tilde{e}_H^H \circ \mathrm{Res}_H^G \circ u \circ \mathrm{Ind}_H^G \circ \tilde{e}_H^H \circ \mathrm{Res}_H^G \\ &= \widetilde{e}_H^G \circ u \circ \widetilde{e}_H^G \\ &= u .\end{aligned}$$

Now for $v \in \mathrm{End}_{R\mathcal{D}}(H)$

$$a \circ b(v) = \frac{1}{|N_G(H):H|^2} \widetilde{e}_H^H \circ \mathrm{Res}_H^G \circ \mathrm{Ind}_H^G \circ v \circ \mathrm{Res}_H^G \circ \mathrm{Ind}_H^G \circ \widetilde{e}_H^H \ .$$

But $\mathrm{Res}_H^G \circ \mathrm{Ind}_H^G$ is the endomorphism of H associated to the (H,H)-biset

$$G \times_G G \cong G \cong \bigsqcup_{g \in [H\backslash G/H]} HgH \ ,$$

where $[H\backslash G/H]$ is a set of representatives of double cosets of H in G. Now for $g \in G$, there is an isomorphism of (H,H)-bisets

$$HgH \cong (H \times H)/\{({}^g h, h) \mid h \in H \cap H^g\} \ .$$

It follows from Lemma 5.3.2 that $\widetilde{e}_H^H \times_H (HgH) = 0$ if ${}^gH \cap H \neq H$, i.e. if ${}^gH \neq H$. Similarly $(HgH) \times_H \widetilde{e}_H^H = 0$ if $H \cap H^g \neq H$, i.e. $H^g \neq H$. Moreover, if $g \in N_G(H)$, then there is an isomorphism of (H,H)-bisets $HgH \cong \mathrm{Iso}(\gamma_g)$. This gives finally

$$a \circ b(v) = \frac{1}{|N_G(H):H|^2} \sum_{g,g' \in [N_G(H)/H]} \widetilde{e}_H^H \circ \mathrm{Iso}(\gamma_g) \circ v \circ \mathrm{Iso}(\gamma_{g'}) \circ \widetilde{e}_H^H \ .$$

Now if $v \in \rho_{G,H}\mathrm{End}_{R\mathcal{D}}(H)\rho_{G,H}$, then in particular $v = n_{G,H} \circ v \circ n_{G,H}$, and since $\mathrm{Iso}(\gamma_g) \circ n_{G,H} = n_{G,H} = n_{G,H} \circ \mathrm{Iso}(\gamma_{g'})$ for any $g, g' \in N_G(H)$, this gives

$$a \circ b(v) = \widetilde{e}_H^H \circ n_{G,H} \circ v \circ n_{G,H} \circ \widetilde{e}_H^H = \rho_{G,H} \circ v \circ \rho_{G,H} = v \ .$$

It follows that a and b restrict to mutual inverse linear isomorphisms between $\widetilde{e}_H^G \mathrm{End}_{R\mathcal{D}}(G) \widetilde{e}_H^G$ and $\rho_{G,H}\mathrm{End}_{R\mathcal{D}}(H)\rho_{G,H}$. Since a is multiplicative, its inverse b is also multiplicative. Finally, by Lemma 6.5.3 and Corollary 2.5.12

$$\begin{aligned} b(\rho_{G,H}) &= \frac{1}{|N_G(H):H|} \mathrm{Ind}_H^G \circ n_{G,H} \circ \widetilde{e}_H^H \circ \mathrm{Res}_H^G \\ &= \frac{1}{|N_G(H):H|} \mathrm{Ind}_H^G \circ \widetilde{e}_H^H \circ \mathrm{Res}_H^G \\ &= \frac{1}{|N_G(H):H|} \widetilde{\mathrm{Ind}_H^G e_H^H} \\ &= \widetilde{e}_H^G \ , \end{aligned}$$

so a and b are isomorphisms of unital algebras. □

6.5.5. Proposition : *Let G be an object of $\mathcal{D}$, such that $|G|$ is invertible in R, and let F be a biset functor.*

If H is a subgroup of G, the map $\mathrm{Res}_H^G : F(G) \to F(H)$ induces an isomorphism of R-modules from $\widetilde{e}_H^G F(G)$ to $\big(\widetilde{e}_H^H F(H)\big)^{N_G(H)}$, whose inverse is induced by the map $\frac{1}{|N_G(H):H|}\mathrm{Ind}_H^G : F(H) \to F(G)$.

This yields an isomorphism of R-modules

$$F(G) \cong \mathop{\oplus}_{H\in[s_G]} \big(\widetilde{e}_H^H F(H)\big)^{N_G(H)} ,$$

where $[s_G]$ is a set of representatives of conjugacy classes of subgroups of G.

Proof: Let $u \in F(G)$ and $v = \widetilde{e}_H^G u$. Then by Corollary 2.5.12 and Theorem 5.2.4

$$\begin{aligned}\mathrm{Res}_H^G v &= \mathrm{Res}_H^G \circ \widetilde{e}_H^G u\\ &= \widetilde{\mathrm{Res}_H^G e_H^G} \circ \mathrm{Res}_H^G(u)\\ &= \widetilde{e}_H^H \mathrm{Res}_H^G u \ .\end{aligned}$$

This shows that $\mathrm{Res}_H^G\big(\widetilde{e}_H^G F(G)\big) \subseteq \widetilde{e}_H^H F(H)$. Moreover if $g \in N_G(H)$, then $g\mathrm{Res}_H^G v = \mathrm{Res}_H^G gv = \mathrm{Res}_H^G v$, so $\mathrm{Res}_H^G\big(\widetilde{e}_H^G F(G)\big) \subseteq \big(\widetilde{e}_H^H F(H)\big)^{N_G(H)}$.

Conversely, let $w \in \widetilde{e}_H^H F(H)$. Then by Corollary 2.5.12

$$\begin{aligned}\widetilde{e}_H^G \mathrm{Ind}_H^G w &= \mathrm{Ind}_H^G \widetilde{\mathrm{Res}_H^G e_H^G} w\\ &= \mathrm{Ind}_H^G \widetilde{e}_H^H w\\ &= \mathrm{Ind}_H^G w \ .\end{aligned}$$

This shows that

$$\mathrm{Ind}_H^G\big(\widetilde{e}_H^H F(H)\big) \subseteq \widetilde{e}_H^G F(G) \ .$$

Now for $u \in \widetilde{e}_H^G F(G)$

$$\begin{aligned}\mathrm{Ind}_H^G \mathrm{Res}_H^G u = \mathrm{Ind}_H^G \mathrm{Res}_H^G \widetilde{e}_H^G u &= \mathrm{Ind}_H^G \widetilde{\mathrm{Res}_H^G e_H^G} \mathrm{Res}_H^G u\\ &= \mathrm{Ind}_H^G \widetilde{e}_H^H \mathrm{Res}_H^G u\\ &= \widetilde{\mathrm{Ind}_H^G e_H^H} u\\ &= |N_G(H) : H| \widetilde{e}_H^G u\\ &= |N_G(H) : H| u \ .\end{aligned}$$

On the other hand, if $w \in \left(\widetilde{e}_H^H F(H)\right)^{N_G(H)}$, then

$$\mathrm{Res}_H^G \mathrm{Ind}_H^G w = \sum_{x \in [H\backslash G/H]} \mathrm{Ind}_{H \cap {}^xH}^H \mathrm{Iso}(\gamma_x) \mathrm{Res}_{H^x \cap H}^H \widetilde{e}_H^H w \;,$$

where $\gamma_x : H^x \cap H \to H \cap {}^xH$ is the isomorphism induced by conjugation by x. But

$$\mathrm{Res}_{H^x \cap H}^H \circ \widetilde{e}_H^H = \widetilde{\mathrm{Res}_{H^x \cap H}^H e_H^H} \circ \mathrm{Res}_{H^x \cap H}^H$$

by Corollary 2.5.12. Moreover $\mathrm{Res}_{H^x \cap H}^H e_H^H = 0$ if $H^x \neq H$. Thus

$$\begin{aligned} \mathrm{Res}_H^G \mathrm{Ind}_H^G w &= \sum_{x \in [N_G(H)/H]} \mathrm{Iso}(\gamma_x) w \\ &= |N_G(H) : H| w \;. \end{aligned}$$

It follows that the maps Res_H^G and $\frac{1}{|N_G(H):H|}\mathrm{Ind}_H^G$ induce mutual inverse isomorphisms of R-modules between $\widetilde{e}_H^G F(G)$ and $\left(\widetilde{e}_H^H F(H)\right)^{N_G(H)}$.

Since the idempotents $\widetilde{e}_H^G$, for $H \in [s_G]$, are orthogonal idempotents of $\mathrm{End}_{R\mathcal{D}}(G)$, whose sum is equal to Id_G, this yields isomorphisms of R-modules

$$F(G) \cong \mathop{\oplus}_{H \in [s_G]} \widetilde{e}_H^G F(G) \cong \mathop{\oplus}_{H \in [s_G]} \left(\widetilde{e}_H^H F(H)\right)^{N_G(H)} \;,$$

completing the proof of the proposition. □

Chapter 7
The Functor $\mathbb{C}R_{\mathbb{C}}$

Let $\mathbb{C}$ be any algebraically closed field of characteristic 0. This chapter is devoted to the study of the biset functor $\mathbb{C}R_{\mathbb{C}}$ sending a finite group G to the $\mathbb{C}$-vector space $\mathbb{C}R_{\mathbb{C}}(G) = \mathbb{C} \otimes_{\mathbb{Z}} R_{\mathbb{C}}(G)$. This functor $\mathbb{C}R_{\mathbb{C}}$ is defined on all finite groups, so it will be viewed as an object of $\mathcal{F}_{\mathcal{C},\mathbb{C}}$.

7.1. Definition

7.1.1. Notation : *If G is a finite group, denote by $R_{\mathbb{C}}(G)$ the Grothendieck group of the category of finite dimensional $\mathbb{C}G$-modules. If H is another finite group, and if U is a finite (H,G)-biset, denote by $R_{\mathbb{C}}(U) : R_{\mathbb{C}}(G) \to R_{\mathbb{Q}}(H)$ the group homomorphism defined by*

$$R_{\mathbb{C}}(U)([E]) = [\mathbb{C}U \otimes_{\mathbb{C}G} E] \; ,$$

where $[E] \in R_{\mathbb{C}}(G)$ denotes the isomorphism class of a finite dimensional $\mathbb{C}G$-module E, and $\mathbb{C}U$ is the $(\mathbb{C}H, \mathbb{C}G)$-permutation bimodule associated to U.

This construction can be extended by linearity: for any element α in the biset Burnside group $B(H,G)$, this gives a group homomorphism $R_{\mathbb{C}}(\alpha) : R_{\mathbb{C}}(G) \to R_{\mathbb{C}}(H)$, and this endows the correspondence $G \mapsto R_{\mathbb{C}}(G)$ with a structure of biset functor, simply denoted by $R_{\mathbb{C}}$. Let $\mathbb{C}R_{\mathbb{C}} = \mathbb{C} \otimes_{\mathbb{Z}} R_{\mathbb{C}}$ denote the object of $\mathcal{F}_{\mathcal{C},\mathbb{C}}$ obtained by $\mathbb{C}$-linear extension of this biset functor.

For any finite group G, the group $R_{\mathbb{C}}(G)$ identifies with the group of ordinary characters of G. Similarly, the vector space $\mathbb{C}R_{\mathbb{C}}(G)$ will be viewed as a set of central functions from G to $\mathbb{C}$. Before describing the action of bisets on characters, it is necessary to fix the notation for characters of bimodules:

7.1.2. Notation : *If G and H are finite groups, and if M is a finite dimensional $(\mathbb{C}H, \mathbb{C}G)$-bimodule, then* the character τ_M of M *is the function $H \times G \to \mathbb{C}$ sending $(h,g) \in H \times G$ to the trace of the endomorphism $m \mapsto hmg^{-1}$ of M.*

S. Bouc, *Biset Functors for Finite Groups*, Lecture Notes in Mathematics 1990, DOI 10.1007/978-3-642-11297-3_7,

In the following lemma, the field $\mathbb{C}$ need not be algebraically closed:

7.1.3. Lemma : *Let $\mathbb{C}$ be a field of characteristic 0.*

1. *Let G, H, and K be finite groups, let M be a finite dimensional $(\mathbb{C}H, \mathbb{C}G)$-bimodule, and N be a finite dimensional $(\mathbb{C}K, \mathbb{C}H)$-bimodule. Then the character of the $(\mathbb{C}K, \mathbb{C}G)$-bimodule $N \otimes_{\mathbb{C}H} M$ is given by*

$$\forall z \in K,\ \forall x \in G,\quad \tau_{N\otimes_{\mathbb{C}H}M}(z,x) = \frac{1}{|H|}\sum_{y\in H} \tau_N(z,y)\tau_M(y,x)\ .$$

2. *Let G and H be finite groups, and let U be a finite (H,G)-biset. If $\chi \in R_{\mathbb{C}}(G)$, then $R_{\mathbb{C}}(U)(\chi)$ is the character θ of H defined by*

$$\forall h \in H,\quad \theta(h) = \frac{1}{|G|}\sum_{\substack{u\in U,\, g\in G\\ hu=ug}} \chi(g)\ .$$

Proof: For Assertion 1, observe that the dual $(\mathbb{C}K, \mathbb{C}G)$-bimodule

$$(N \otimes_{\mathbb{C}H} M)^* = \mathrm{Hom}_{\mathbb{C}}(N \otimes_{\mathbb{C}H} M, \mathbb{C})$$

is isomorphic to the bimodule $P = \mathrm{Hom}_{\mathbb{C}H}(M, N^*)$, where N^* is the $\mathbb{C}$-dual of N. This isomorphism θ is obtained by sending $\alpha \in (N \otimes_{\mathbb{C}H} M)^*$ to the element of P defined by

$$\forall m \in M,\ \forall n \in N,\quad \big(\theta(\alpha)(m)\big)(n) = \alpha(n \otimes m)\ .$$

The bimodule P is the submodule of H-invariant elements in the vector space $Q = \mathrm{Hom}_{\mathbb{C}}(M, N^*)$. Now Q is a $(K \times H \times G)$-module for the following action

$$\big((z,y,x)\varphi)(m)\big)(n) = \varphi(y^{-1}mx)(z^{-1}ny)\ ,$$

where $(z,y,x) \in K \times H \times G$, $\varphi \in Q$, $m \in M$, and $n \in N$.

The character of the $\mathbb{C}(K \times G)$-module of fixed points by the normal subgroup $\mathbf{1} \times H \times \mathbf{1}$ of $K \times H \times G$ is given by

$$\forall (z,x) \in K \times G,\quad \psi_P(z,x) = \frac{1}{|H|}\sum_{y\in H} \psi_Q(z,y,x)\ ,$$

where ψ_Q is the character of the $\mathbb{C}(K \times H \times G)$-module Q. This character can be determined by choosing a $\mathbb{C}$-basis $(m_i)_{1\leq i\leq a}$ of M, and a $\mathbb{C}$-basis $(n_j)_{1\leq j\leq b}$ of N. Denote by $(n_j^*)_{1\leq j\leq b}$ the dual basis of N^*. Then Q has a basis $(u_{i,j})_{1\leq i\leq a,\, 1\leq j\leq b}$ defined by

$$\forall k \in \{1,\ldots,a\},\quad u_{i,j}(m_k) = \delta_{i,k}n_j^*\ ,$$

where $\delta_{i,k}$ is a Kronecker symbol. Let $(\rho_N(z,y)_{j,l})_{1\leq j,l\leq b}$ denote the matrix of the endomorphism $n \mapsto zny^{-1}$ of N, for $z \in K$ and $y \in H$, defined by

$$zn_l y^{-1} = \sum_j \rho_N(z,y)_{j,l}\, n_j \ .$$

It follows that

$$z^{-1}n_l^* y = \sum_j \rho_N(z,y)_{l,j}\, n_j^* \ .$$

Let $\big(\rho_M(y,x)_{i,p}\big)_{1\leq i,p\leq a}$ denote the matrix of the endomorphism $m \mapsto ymx^{-1}$ of M, for $y \in H$ and $x \in G$. It follows that

$$\begin{aligned}\big((z,y,x)u_{i,j}\big)(m_k)(n_l) &= u_{i,j}(y^{-1}m_k x)(z^{-1}n_l y)\\ &= \sum_p \sum_q \rho_M(y^{-1},x^{-1})_{p,k}\rho_N(z^{-1},y^{-1})_{q,l}u_{i,j}(m_p)(n_q)\\ &= \rho_M(y^{-1},x^{-1})_{i,k}\rho_N(z^{-1},y^{-1})_{j,l} \ .\end{aligned}$$

In other words $\big((z,y,x)u_{i,j}\big)(m_k) = \sum_l \rho_M(y^{-1},x^{-1})_{i,k}\rho_N(z^{-1},y^{-1})_{j,l}n_l^*$, i.e.

$$(z,y,x)u_{i,j} = \sum_{r,l} \rho_N(z^{-1},y^{-1})_{j,l}\rho_M(y^{-1},x^{-1})_{i,r}u_{r,l} \ ,$$

hence

$$\begin{aligned}\psi_Q(z,y,x) &= \sum_{i,j} \rho_N(z^{-1},y^{-1})_{j,j}\rho_M(y^{-1},x^{-1})_{i,i}\\ &= \tau_N(z^{-1},y^{-1})\tau_M(y^{-1},x^{-1}) \ .\end{aligned}$$

This proves Assertion 1, since moreover $\tau_{N\otimes_{\mathbb{C}H}M}(z,x) = \psi_P(z^{-1},x^{-1})$, as $N\otimes_{\mathbb{C}H} M$ is the dual of P.

Assertion 2 follows, by replacing G by $\mathbf{1}$, H by G, K by H, and N by $\mathbb{C}U$. The $(\mathbb{C}G,\mathbb{C}\mathbf{1})$-bimodule M is just an ordinary $\mathbb{C}G$-module, with character $\tau_M = \chi$. The character of the permutation bimodule $N = \mathbb{C}U$ is given by

$$\forall (z,y) \in H\times G,\ \ \tau_N(z,y) = |\{u \in U \mid zu = uy\}| \ .$$

It follows that

$$\begin{aligned}\forall z \in H,\ \ \tau_{\mathbb{C}U\otimes_{\mathbb{C}G}M}(z) &= \frac{1}{|G|}\sum_{y\in G}|\{u\in U \mid zu = uy\}|\,\chi(y)\\ &= \frac{1}{|G|}\sum_{\substack{y\in G,\, u\in U\\ zu=uy}} \chi(y) \ ,\end{aligned}$$

as was to be shown. □

7.2. Decomposition

7.2.1. Notation : *Let $\hat{\mathbb{Z}}^\times$ denote the inverse limit of the multiplicative groups $(\mathbb{Z}/m\mathbb{Z})^\times$, for $m \in \mathbb{N} - \{0\}$, and the system of morphisms $\pi_{m,n} : (\mathbb{Z}/m\mathbb{Z})^\times \to (\mathbb{Z}/n\mathbb{Z})^\times$, given by the natural projection maps, for $n|m$. If $m \in \mathbb{N} - \{0\}$, let $\pi_m : \hat{\mathbb{Z}}^\times \to (\mathbb{Z}/m\mathbb{Z})^\times$ denote the canonical morphism, and let U_m denote the kernel of π_m.*

In other words, the group $\hat{\mathbb{Z}}^\times$ is the subgroup of $\prod_{m\in\mathbb{N}-\{0\}} (\mathbb{Z}/m\mathbb{Z})^\times$ consisting of sequences $(u_m)_{m\in\mathbb{N}-\{0\}}$, such that $\pi_{m,n}(u_m) = u_n$ whenever $n \mid m$. A standard argument shows that

$$\hat{\mathbb{Z}}^\times \cong \prod_p \varprojlim_{n\geq 0} (\mathbb{Z}/p^n\mathbb{Z})^\times ,$$

where p runs through the set of prime numbers. In particular, the maps π_m are surjective, for any positive integer m.

7.2.2. Remark : Since $(\mathbb{Z}/m\mathbb{Z})^\times$ is the Galois group of the m-th cyclotomic extension $\mathbb{Q}(\zeta_m)/\mathbb{Q}$, where ζ_m is a primitive m-th root of unity in $\mathbb{C}$, the group $\hat{\mathbb{Z}}^\times$ is also the Galois group of the union of all these cyclotomic fields, i.e. the Galois group of the abelian closure of $\mathbb{Q}$, by the Kronecker-Weber Theorem.

7.2.3. Adams Operations. Recall that if G is a finite group, and $n \in \mathbb{Z}$, the Adams operation Ψ_G^n is the endomorphism of $R_{\mathbb{C}}(G)$ defined by

$$\forall \chi \in R_{\mathbb{C}}(G),\ \forall g \in G,\ \ \Psi_G^n(\chi)(g) = \chi(g^n) .$$

These Adams operations can be used to define an action of $\hat{\mathbb{Z}}^\times$ on $R_{\mathbb{C}}(G)$, in the following way: if $z \in \hat{\mathbb{Z}}^\times$, and $\chi \in R_{\mathbb{C}}(G)$, define ${}^z\chi$ by

$${}^z\chi = \Psi^{\widetilde{z_m}}(\chi) ,$$

where m is any integer multiple of the exponent of the group G, and $\widetilde{z_m}$ is a representative in $\mathbb{Z}$ of $z_m \in (\mathbb{Z}/m\mathbb{Z})^\times$. In other words, for any $g \in G$

$${}^z\chi(g) = \chi(g^{\widetilde{z_m}}) .$$

Since $z \in \hat{\mathbb{Z}}^\times$, this is well defined, i.e. it does not depend on the choice of such integers m and $\widetilde{z_m}$.

This clearly defines an action of the group $\hat{\mathbb{Z}}^\times$ on $R_{\mathbb{C}}(G)$, so $R_{\mathbb{C}}(G)$ is a $\mathbb{Z}\hat{\mathbb{Z}}^\times$-module. Moreover:

7.2.4. Proposition : *Let G and H be finite groups, and U be a finite (H,G)-biset. If $\chi \in R_{\mathbb{C}}(G)$ and $z \in \hat{\mathbb{Z}}^\times$, then*

$${}^{z}\big(R_{\mathbb{C}}(U)(\chi)\big) = R_{\mathbb{C}}(U)({}^{z}\chi) \,.$$

*In other words $R_{\mathbb{C}}$ is a biset functor with values in the category $\mathbb{Z}\hat{\mathbb{Z}}^\times$-*Mod *of $\mathbb{Z}\hat{\mathbb{Z}}^\times$-modules.*

Proof: Let m be a common multiple of the exponent of G and the exponent of H, and set $a = \widetilde{z_m}$. Then for any $h \in H$

$$\begin{aligned} {}^{z}\big(R_{\mathbb{C}}(U)(\chi)\big)(h) &= R_{\mathbb{C}}(U)(\chi)(h^a) \\ &= \frac{1}{|G|} \sum_{\substack{u\in U,\, g\in G\\ h^a u = ug}} \chi(g) \,. \end{aligned}$$

Since a is coprime to m, there exists integers b and c such that $ab + cm = 1$. Now the equality $h^a u = ug$ implies $h^{aj}u = ug^j$ for any integer j, so in particular $h^{ab}u = ug^b$, i.e. $hu = ug^b$ since $ab = 1 - cm$ is congruent to 1 modulo the exponent of H. Conversely, the equality $hu = ug^b$ implies $h^a u = ug^{ab} = ug$, since ab is congruent to 1 modulo the exponent of G. Moreover the map $g \mapsto g^b$ is a bijection of G, and the inverse bijection is the map $g \mapsto g^a$. It follows that

$$\begin{aligned} {}^{z}\big(R_{\mathbb{C}}(U)(\chi)\big)(h) &= \tfrac{1}{|G|} \sum_{\substack{u\in U,\, g\in G\\ hu = ug^b}} \chi(g) \\ &= \tfrac{1}{|G|} \sum_{\substack{u\in U,\, g\in G\\ hu = ug}} \chi(g^a) = R_{\mathbb{C}}(U)({}^{z}\chi)(h) \,, \end{aligned}$$

and this completes the proof. □

7.2.5. Locally Constant Functions.

7.2.6. Definition : *A function $f : \hat{\mathbb{Z}}^\times \to \mathbb{C}$ is called* locally constant *if there is an integer N such that f is invariant by U_N, i.e.*

$$\forall \gamma \in U_N,\ \forall z \in \hat{\mathbb{Z}}^\times,\ \ f(z\gamma) = f(z) \,.$$

Denote by $\mathcal{LC}(\hat{\mathbb{Z}}^\times, \mathbb{C})$ the set of locally constant functions from $\hat{\mathbb{Z}}^\times$ to $\mathbb{C}$. It is a subalgebra of the $\mathbb{C}$-algebra of all functions from $\hat{\mathbb{Z}}^\times$ to $\mathbb{C}$.

The rationale for this terminology is the following: the group $\hat{\mathbb{Z}}^\times$ is a profinite group, so it is compact and totally disconnected, and the subgroups U_N, for $N \in \mathbb{N} - \{0\}$, form a basis of neighbourhoods of the identity element.

For this topology, the above definition is equivalent to the usual one, for which a locally constant function is a function f such that any point admits a neighbourhood on which f is constant.

7.2.7. Example : Let G be a finite group, let $\chi \in R_{\mathbb{C}}(G)$, and $g \in G$. Then the function $z \in \hat{\mathbb{Z}}^\times \mapsto {}^z\chi(g) \in \mathbb{C}$ is locally constant: indeed, if N denotes the exponent of G, and if $\gamma \in U_N$, then ${}^{z\gamma}\chi(g) = \chi(g^{\widetilde{z_N}\widetilde{\gamma_N}}) = \chi(g^{\widetilde{z_N}}) = {}^z\chi(g)$.

7.2.8. Example (locally constant characters) : Set $\mathbb{C}^\times = \mathbb{C} - \{0\}$. Let $\theta : \hat{\mathbb{Z}}^\times \to \mathbb{C}^\times$ be a group homomorphism. Then θ is locally constant if and only if $\operatorname{Ker}\theta$ contains U_N, for some integer N, i.e. if θ factors as $\theta = \bar{\theta} \circ \pi_N$, where $\bar{\theta}$ is a group homomorphism $(\mathbb{Z}/N\mathbb{Z})^\times \to \mathbb{C}^\times$. This is equivalent to saying that $\operatorname{Ker}\theta$ is open in $\hat{\mathbb{Z}}^\times$.

7.2.9. Notation : *Let $\Gamma(\hat{\mathbb{Z}}^\times)$ denote the set of locally constant group homomorphisms $\hat{\mathbb{Z}}^\times \to \mathbb{C}^\times$. It is a multiplicative group, called* the group of locally constant characters *of* $\hat{\mathbb{Z}}^\times$.

7.2.10. Lemma : *Let $f : \hat{\mathbb{Z}}^\times \to \mathbb{C}$ be a locally constant function. Then the expression*

$$\frac{1}{\varphi(N)} \sum_{z \in [\hat{\mathbb{Z}}^\times/U_N]} f(z)$$

(where φ is the Euler totient function) does not depend on the choice of the integer N such that f is invariant by U_N, nor on the choice of the set $[\hat{\mathbb{Z}}^\times/U_N]$ of coset representatives of U_N in $\hat{\mathbb{Z}}^\times$.

Proof: Set $E_N = \frac{1}{\varphi(N)} \sum\limits_{z \in [\hat{\mathbb{Z}}^\times/U_N]} f(z)$. If f is invariant by U_N, then E_N does not depend on the choice of the coset representatives of $\hat{\mathbb{Z}}^\times/U_N$. Moreover if M is a multiple of N, then $U_M \leq U_N$, and

$$\begin{aligned} E_M &= \frac{1}{\varphi(M)} \sum_{z \in [\hat{\mathbb{Z}}^\times/U_M]} f(z) = \frac{1}{\varphi(M)} \sum_{\substack{z \in [\hat{\mathbb{Z}}^\times/U_N] \\ \gamma \in U_N/U_M}} f(z\gamma) \\ &= \frac{|U_N:U_M|}{\varphi(M)} \sum_{z \in [\hat{\mathbb{Z}}^\times/U_N]} f(z) = E_N \ , \end{aligned}$$

since $|U_N : U_M| = \dfrac{|\operatorname{Im}\pi_M|}{|\operatorname{Im}\pi_N|} = \dfrac{\varphi(M)}{\varphi(N)}$.

Now if N and N' are integers such that f is invariant by U_N and $U_{N'}$, then, denoting by M a common multiple of N and N', it follows that $E_N = E_M = E_{N'}$. □

7.2.11. Notation : *If $f : \hat{\mathbb{Z}}^\times \to \mathbb{C}$ is a locally constant function, set*

$$\int_{\hat{\mathbb{Z}}^\times} f(z)dz = \frac{1}{\varphi(N)} \sum_{z \in [\hat{\mathbb{Z}}^\times / U_N]} f(z) \; ,$$

where N is any integer such that f is invariant by U_N.

7.2.12. Remark : In the case where $\mathbb{C}$ is the field of complex numbers, the use of the Haar measure μ on the compact group $\hat{\mathbb{Z}}^\times$ allows for defining an integral $\int_{\hat{\mathbb{Z}}^\times} f(z)d\mu(z)$. If μ is normalized such that $\mu(\hat{\mathbb{Z}}^\times) = 1$, then $\mu(U_N) = \frac{1}{\varphi(N)}$, and it follows easily that $\int_{\hat{\mathbb{Z}}^\times} f(z)d\mu(z) = \int_{\hat{\mathbb{Z}}^\times} f(z)dz$, for any locally constant function $f : \hat{\mathbb{Z}}^\times \to \mathbb{C}$.

Example 7.2.7 allows for the following construction:

7.2.13. Notation : *Let $\theta \in \Gamma(\hat{\mathbb{Z}}^\times)$. When G is a finite group, and $\chi \in \mathbb{C}R_{\mathbb{C}}(G)$, let $p_{\theta,G}(\chi)$ denote the function from G to $\mathbb{C}$ defined by*

$$\forall g \in G, \;\; p_{\theta,G}(\chi)(g) = \int_{\hat{\mathbb{Z}}^\times} \theta(z^{-1}) \, {}^z\chi(g) \, dz \; ,$$

where the action $(z, \chi) \mapsto {}^z\chi$ from 7.2.3 is extended to $\mathbb{C}R_{\mathbb{C}}(G)$ by $\mathbb{C}$-linearity.

7.2.14. Theorem :

1. *If G is a finite group, and $\theta \in \Gamma(\hat{\mathbb{Z}}^\times)$, then the map $p_{\theta,G} : \chi \mapsto p_{\theta,G}(\chi)$ is an idempotent endomorphism of $\mathbb{C}R_{\mathbb{C}}(G)$, whose image is equal to*

$$\mathbb{C}R_{\mathbb{C}}^{\theta}(G) = \{f \in \mathbb{C}R_{\mathbb{C}}(G) \mid \forall z \in \hat{\mathbb{Z}}^\times, \;\; {}^z f = \theta(z)f\} \; .$$

2. *The maps $p_{\theta,G}$ define an idempotent endomorphism p_θ of the biset functor $\mathbb{C}R_{\mathbb{C}}$. In particular, the image $\mathbb{C}R_{\mathbb{C}}^{\theta}$ of p_θ is a direct summand of the biset functor $\mathbb{C}R_{\mathbb{C}}$.*
3. *The endomorphisms p_θ, for $\theta \in \Gamma(\hat{\mathbb{Z}}^\times)$, are orthogonal idempotent endomorphisms of $\mathbb{C}R_{\mathbb{C}}$, and*

$$\sum_{\theta \in \Gamma(\hat{\mathbb{Z}}^\times)} p_\theta = \mathrm{Id}_{\mathbb{C}R_{\mathbb{C}}} \; ,$$

where the summation in the left hand side is locally finite, i.e. for each finite group G, the set of $\theta \in \Gamma(\hat{\mathbb{Z}}^\times)$ such that $p_{\theta,G} \neq 0$ is finite.

Proof: For Assertion 1, the first thing to check is that $p_{\theta,G}(\chi) \in \mathbb{C}R_{\mathbb{C}}(G)$, for any $\chi \in R_{\mathbb{C}}(G)$. But this is clear, since by definition

$$p_{\theta,G}(\chi) = \frac{1}{\varphi(N)} \sum_{z \in [\hat{\mathbb{Z}}^\times/U_N]} \theta(z^{-1})\,{}^z\chi\ ,$$

for any integer N multiple of the exponent of G, and such that $U_N \leq \operatorname{Ker}\theta$. Since $p_{\theta,G}$ is clearly $\mathbb{C}$-linear, it is an endomorphism of $\mathbb{C}R_{\mathbb{C}}(G)$.

Now if $f \in \mathbb{C}R_{\mathbb{C}}^{\theta}(G)$

$$\begin{aligned} p_{\theta,G}(f) &= \frac{1}{\varphi(N)} \sum_{z \in [\hat{\mathbb{Z}}^\times/U_N]} \theta(z^{-1})\,{}^z f \\ &= \frac{1}{\varphi(N)} \sum_{z \in [\hat{\mathbb{Z}}^\times/U_N]} \theta(z^{-1})\,\theta(z) f \\ &= \frac{|\hat{\mathbb{Z}}^\times/U_N|}{\varphi(N)} f = f\ , \end{aligned}$$

so $f \in \operatorname{Im} p_{\theta,G}$, and the restriction of $p_{\theta,G}$ to $\mathbb{C}R_{\mathbb{C}}^{\theta}(G)$ is equal to the identity. Conversely, if $\chi \in R_{\mathbb{C}}(G)$ and $z_0 \in \hat{\mathbb{Z}}^\times$, then

$$\begin{aligned} {}^{z_0}p_{\theta,G}(\chi) &= \frac{1}{\varphi(N)} \sum_{z \in [\hat{\mathbb{Z}}^\times/U_N]} \theta(z^{-1})\,{}^{z_0 z}\chi \\ &= \frac{1}{\varphi(N)} \sum_{z \in [\hat{\mathbb{Z}}^\times/U_N]} \theta(z^{-1}z_0)\,{}^z\chi \\ &= \theta(z_0)p_{\theta,G}(\chi)\ , \end{aligned}$$

and it follows that $\operatorname{Im} p_{\theta,G} = \mathbb{C}R_{\mathbb{C}}^{\theta}(G)$. Thus $p_{\theta,G}$ is a projector on $\mathbb{C}R_{\mathbb{C}}^{\theta}(G)$.

For Assertion 2, let G and H be finite groups, and let $\alpha \in B(H,G)$. Let N be any integer multiple of the exponents of G and H, and such that $U_N \leq \operatorname{Ker}\theta$. Then, by Proposition 7.2.4, for $\chi \in R_{\mathbb{C}}(G)$

$$\begin{aligned} \mathbb{C}R_{\mathbb{C}}(\alpha)p_{\theta,G}(\chi) &= \frac{1}{\varphi(N)} \sum_{z \in [\hat{\mathbb{Z}}^\times/U_N]} \theta(z^{-1})\,R_{\mathbb{C}}(\alpha)({}^z\chi) \\ &= \frac{1}{\varphi(N)} \sum_{z \in [\hat{\mathbb{Z}}^\times/U_N]} \theta(z^{-1})\,{}^zR_{\mathbb{C}}(\alpha)(\chi) \\ &= p_{\theta,H}\mathbb{C}R_{\mathbb{C}}(\alpha)(\chi)\ , \end{aligned}$$

so $\mathbb{C}R_{\mathbb{C}}(\alpha)p_{\theta,G} = p_{\theta,H}\mathbb{C}R_{\mathbb{C}}(\alpha)$, as was to be shown.

For Assertion 3, if $\theta \neq \theta'$ are elements of $\Gamma(\hat{\mathbb{Z}}^\times)$, then if N is an integer multiple of the exponent of G, and such that $U_N \leq \operatorname{Ker}\theta \cap \operatorname{Ker}\theta'$, then for $\chi \in R_{\mathbb{C}}(G)$

$$\begin{aligned} p_{\theta,G}p_{\theta',G}(\chi) &= \frac{1}{\varphi(N)} \sum_{z\in[\hat{\mathbb{Z}}^\times/U_N]} \theta(z^{-1})\,{}^z p_{\theta',G}(\chi) \\ &= \left(\frac{1}{\varphi(N)} \sum_{z\in[\hat{\mathbb{Z}}^\times/U_N]} \theta(z^{-1})\,\theta'(z)\right) p_{\theta',G}(\chi)\,. \end{aligned}$$

Now the expression in parentheses is equal to

$$\frac{1}{\varphi(N)} \sum_{x\in(\mathbb{Z}/N\mathbb{Z})^\times} \bar{\theta}(x^{-1})\,\overline{\theta'}(x)\,,$$

where $\bar{\theta}$ and $\overline{\theta'}$ are the characters of $(\mathbb{Z}/N\mathbb{Z})^\times$ such that $\theta = \bar{\theta}\pi_N$ and $\theta' = \overline{\theta'}\pi_N$. This is equal to zero, since distinct characters of the finite group $(\mathbb{Z}/N\mathbb{Z})^\times$ are orthogonal (or, by a more elementary argument, since the sum of the values of its non trivial linear character $\bar{\theta}^{-1}\overline{\theta'}$ is equal to zero). Hence $p_{\theta,G}p_{\theta',G} = 0$.

It follows that the sum $\sum_{\theta\in\Gamma(\hat{\mathbb{Z}}^\times)} \mathrm{Im}\, p_{\theta,G}$ is a direct sum. Since $\mathbb{C}R_\mathbb{C}(G)$ is a finite dimensional $\mathbb{C}$-vector space, there is only a finite number of non zero terms in this summation.

So for each G, there is an integer N, multiple of the exponent of G, and such that $U_N \leq \mathrm{Ker}\,\theta$, for any θ such that $p_{\theta,G} \neq 0$. Hence for each $\chi \in R_\mathbb{C}(G)$

$$\begin{aligned} \sum_{\theta\in\Gamma(\hat{\mathbb{Z}}^\times)} p_{\theta,G}(\chi) &= \sum_{\theta\in\Gamma(\hat{\mathbb{Z}}^\times)} \frac{1}{\varphi(N)} \sum_{z\in[\hat{\mathbb{Z}}^\times/U_N]} \theta(z^{-1})\,{}^z\chi \\ &= \frac{1}{\varphi(N)} \sum_{\bar{\theta}\in\Gamma((\mathbb{Z}/N\mathbb{Z})^\times)} \sum_{z\in[\hat{\mathbb{Z}}^\times/U_N]} \bar{\theta}\pi_N(z^{-1})\,{}^z\chi \\ &= \frac{1}{\varphi(N)} \sum_{z\in[\hat{\mathbb{Z}}^\times/U_N]} \sum_{\bar{\theta}\in\Gamma((\mathbb{Z}/N\mathbb{Z})^\times)} \bar{\theta}\pi_N(z^{-1})\,{}^z\chi\,, \end{aligned}$$

where $\Gamma\big((\mathbb{Z}/N\mathbb{Z})^\times\big) = \mathrm{Hom}\big((\mathbb{Z}/N\mathbb{Z})^\times, \mathbb{C}^\times\big)$. Now for a given $z \in [\hat{\mathbb{Z}}^\times/U_N]$, the summation $\sum_{\bar{\theta}\in\Gamma((\mathbb{Z}/N\mathbb{Z})^\times)} \bar{\theta}\pi_N(z^{-1})$ is equal to 0 if $\pi_N(z) \neq 1$, i.e. if $z \notin U_N$, and to $\varphi(N)$ if $z \in U_N$. It follows that $\sum_{\theta\in\Gamma(\hat{\mathbb{Z}}^\times)} p_{\theta,G}(\chi) = \chi$, as was to be shown. □

7.2.15. Corollary : *The biset functor $\mathbb{C}R_\mathbb{C}$ splits as the direct sum of biset functors*

$$\mathbb{C}R_\mathbb{C} = \mathop{\oplus}_{\theta\in\Gamma(\hat{\mathbb{Z}}^\times)} \mathbb{C}R_\mathbb{C}^\theta\,.$$

7.3. Semisimplicity

7.3.1. Definition : *Let $m \in \mathbb{N} - \{0\}$. A character $\bar{\theta} : (Z/m\mathbb{Z})^\times \to \mathbb{C}^\times$ is called* primitive *if it cannot be factored through any proper quotient $(\mathbb{Z}/n\mathbb{Z})^\times$ of $(\mathbb{Z}/m\mathbb{Z})^\times$, i.e. if $n \mid m$ and $\operatorname{Ker} \pi_{m,n} \leq \operatorname{Ker} \bar{\theta}$ implies $n = m$.*

7.3.2. Lemma : *Let $\theta \in \Gamma(\hat{\mathbb{Z}}^\times)$. Then there exists a unique pair (m, ξ), where $m \in \mathbb{N} - \{0\}$ and $\xi : (Z/m\mathbb{Z})^\times \to \mathbb{C}^\times$ is a primitive character, such that $\theta = \xi \pi_m$.*

Proof: This follows from elementary properties of the subgroups U_m of $\hat{\mathbb{Z}}^\times$: if m and n are positive integers, then $U_m \cap U_n = U_{m \vee n}$, by the chinese remainder theorem, where $m \vee n$ is the l.c.m. of m and n. Moreover

$$U_{m \vee n} \leq U_m \cdot U_n \leq U_{(m,n)} \ .$$

The factor group $(U_m \cdot U_n)/U_{m \vee n}$ is isomorphic to the direct product of $U_m/U_{m \vee n}$ and $U_n/U_{m \vee n}$, hence it has order $\frac{\varphi(m \vee n)^2}{\varphi(m)\varphi(n)}$. The factor group $U_{(m,n)}/U_{m \vee n}$ has order $\frac{\varphi(m \vee n)}{\varphi((m,n))}$. Since $\varphi(m)\varphi(n) = \varphi(m \vee n)\varphi\big((m,n)\big)$, it follows that $U_m \cdot U_n = U_{(m,n)}$.

In other words there is a commutative diagram
(7.3.3)

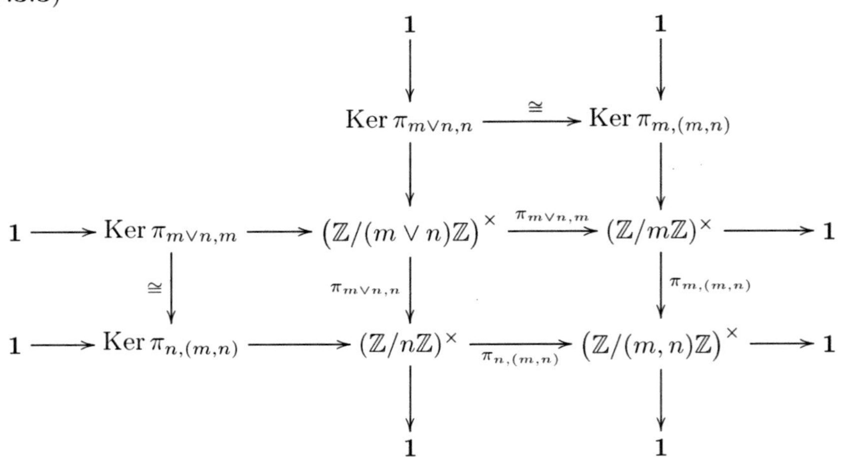

with exact rows and columns.

Now if $\theta \in \Gamma(\hat{\mathbb{Z}}^\times)$, the set of positive integers n such that $U_n \leq \operatorname{Ker} \theta$ is non empty, so it has a smallest element m, and there is a character $\xi : (\mathbb{Z}/m\mathbb{Z})^\times \to \mathbb{C}^\times$ such that $\theta = \xi \pi_m$. The minimality of m implies that ξ is primitive.

Conversely, suppose that $\theta = \psi\pi_n$, for some integer n, and some primitive character $\psi : (\mathbb{Z}/n\mathbb{Z})^\times \to \mathbb{C}^\times$. Then for any integer N

$$\begin{aligned}\theta(U_N) &= \psi\pi_n(U_N) = \psi\pi_n(U_N \cdot U_n)\\ &= \psi\pi_n(U_{(N,n)})\\ &= \psi(\mathrm{Ker}\,\pi_{n,(N,n)}) \ .\end{aligned}$$

Thus $U_N \leq \mathrm{Ker}\,\theta$ if and only if $\mathrm{Ker}\,\pi_{n,(N,n)} \leq \mathrm{Ker}\,\psi$, so if and only if $n = (N,n)$, i.e. $n \mid N$, since ψ is primitive. It follows that $n = m$, and $\psi = \xi$, as was to be shown. □

7.3.4. Theorem : *Let $m \in \mathbb{N} - \{0\}$, and $\xi : (\mathbb{Z}/m\mathbb{Z})^\times \to \mathbb{C}^\times$ be a primitive character. Set $\theta = \xi\pi_m$. Then the biset functor $\mathbb{C}R_{\mathbb{C}}^{\theta}$ is isomorphic to the simple functor $S_{\mathbb{Z}/m\mathbb{Z},\mathbb{C}_\xi}$, where $\mathbb{C}_\xi$ is the vector space $\mathbb{C}$ on which the group* $\mathrm{Out}(\mathbb{Z}/m\mathbb{Z}) \cong (\mathbb{Z}/m\mathbb{Z})^\times$ *acts via ξ.*

Proof: There are two steps in the proof:

• Denote by $\tilde{\xi}$ the function from $\mathbb{Z}/m\mathbb{Z}$ to $\mathbb{C}$ obtained by extending ξ by 0, i.e. the function defined by

$$\forall x \in \mathbb{Z}/m\mathbb{Z},\ \ \tilde{\xi}(x) = \begin{cases} \xi(x) \text{ if } x \in (\mathbb{Z}/m\mathbb{Z})^\times \\ \ \ 0 \quad \text{otherwise.}\end{cases}$$

Then $\tilde{\xi} \in \mathbb{C}R_{\mathbb{C}}(\mathbb{Z}/m\mathbb{Z})$, and actually $\tilde{\xi} \in \mathbb{C}R_{\mathbb{C}}^{\theta}(\mathbb{Z}/m\mathbb{Z})$, since obviously

$$\tilde{\xi}(i \cdot x) = \xi(i)\tilde{\xi}(x) \ ,$$

for any $x \in \mathbb{Z}/m\mathbb{Z}$, and any integer $i \in (\mathbb{Z}/m\mathbb{Z})^\times$ (here the additive notation $i \cdot x$ is used for x^i in the additive group $\mathbb{Z}/m\mathbb{Z}$): if $x \notin (\mathbb{Z}/m\mathbb{Z})^\times$, both sides are equal to 0, and if $x \in (\mathbb{Z}/m\mathbb{Z})^\times$, then equality holds because ξ is a character of $(\mathbb{Z}/m\mathbb{Z})^\times$. The first step in the proof consists in showing that $\mathbb{C}R_{\mathbb{C}}^{\theta}$ *is generated by* $\tilde{\xi}$.

Let Ξ denote the subfunctor of $\mathbb{C}R_{\mathbb{C}}^{\theta}$ generated by $\tilde{\xi}$. If $\Xi \neq \mathbb{C}R_{\mathbb{C}}^{\theta}$, let G be a minimal group for the quotient $\mathbb{C}R_{\mathbb{C}}^{\theta}/\Xi$, and let $f \in \mathbb{C}R_{\mathbb{C}}^{\theta}(G) - \Xi(G)$. By Proposition 6.5.5

$$f = \sum_{H \in [s_G]} \tfrac{1}{|N_G(H):H|} \mathrm{Ind}_H^G(\widetilde{e}_H^H \mathrm{Res}_H^G f) \ ,$$

where $[s_G]$ is a set of representatives of conjugacy classes of subgroups of G. By minimality of G, all the terms in the right hand side indexed by proper subgroups H of G are in $\Xi(G)$. It follows that the only remaining term $\widetilde{e}_G^G f$ lies in $\mathbb{C}R_{\mathbb{C}}^{\theta}(G) - \Xi(G)$. In other words, one can assume $f = \widetilde{e}_G^G f$. But

$$\begin{aligned}\widetilde{e}_G^G f &= \frac{1}{|G|}\sum_{H\leq G}|H|\mu(H,G)\mathrm{Ind}_H^G\mathrm{Res}_H^G f\\ &= \big(\frac{1}{|G|}\sum_{H\leq G}|H|\mu(H,G)\mathrm{Ind}_H^G 1\big)f\\ &= \chi_{e_G^G}\cdot f\ ,\end{aligned}$$

where $\chi_{e_G^G}$ is the character of e_G^G, i.e. the image of $e_G^G \in \mathbb{C}B(G)$ in $\mathbb{C}R_{\mathbb{C}}(G)$ by the linearization map $\mathbb{C}\chi_{\mathbb{C},G}$ (see Remark 1.2.3). Moreover

$$\forall g\in G,\ \ \chi_{e_G^G}(g) = |(e_G^G)^g| = \begin{cases}1 \text{ if } <g> = G\\ 0 \text{ otherwise.}\end{cases}$$

In other words $\widetilde{e}_G^G f(g)$ is equal to $f(g)$ if $<g> = G$, and equal to 0 otherwise. Since $f \neq 0$, it follows that G is cyclic, and that $f(g) = 0$ if g is not a generator of G. So if $G = \mathbb{Z}/n\mathbb{Z}$, then $f(g) = 0$ if $g \notin (\mathbb{Z}/n\mathbb{Z})^\times$.

Moreover since $f \in \mathbb{C}R_{\mathbb{C}}^{\theta}(\mathbb{Z}/n\mathbb{Z})$

$$f(i\cdot g) = \xi\pi_{m\vee n,m}(i)f(g)\ ,$$

for any $g \in \mathbb{Z}/n\mathbb{Z}$ and any $i \in \big(\mathbb{Z}/(m\vee n)\mathbb{Z}\big)^\times$. In particular, if $i \in \mathrm{Ker}\,\pi_{m\vee n,n}$, then $i\cdot g = g$ for any $g \in \mathbb{Z}/n\mathbb{Z}$, and since $f \neq 0$, this implies $\xi\pi_{m\vee n,m}(i) = 1$. But $\pi_{m\vee n,m}(\mathrm{Ker}\,\pi_{m\vee n,n}) = \mathrm{Ker}\,\pi_{m,(m,n)}$, as can be seen in Diagram 7.3.3. Thus $\mathrm{Ker}\,\pi_{m,(m,n)} \leq \mathrm{Ker}\,\xi$, and since ξ is primitive, this implies $m = (m,n)$, i.e. $m \mid n$.

Now $f(i\cdot g) = \xi\pi_{n,m}(i)f(g)$, for any $i \in (\mathbb{Z}/n\mathbb{Z})^\times$. Thus

$$f(i) = f(1)\xi\pi_{n,m}(i)\ ,$$

for $i \in (\mathbb{Z}/n\mathbb{Z})^\times$, and $f(i) = 0$ if $i \in (\mathbb{Z}/n\mathbb{Z}) - (\mathbb{Z}/n\mathbb{Z})^\times$. This shows that $f = f(1)\widetilde{e}_G^G\mathrm{Inf}_{\mathbb{Z}/m\mathbb{Z}}^{\mathbb{Z}/n\mathbb{Z}}\tilde{\xi}$, so $f \in \Xi(\mathbb{Z}/n\mathbb{Z})$ (there is a slight notational abuse here: the correct expression should be $f = f(1)\widetilde{e}_G^G\mathrm{Inf}_{(\mathbb{Z}/n\mathbb{Z})/(m\mathbb{Z}/n\mathbb{Z})}^{\mathbb{Z}/n\mathbb{Z}}\tilde{\xi})$. This contradiction shows that $\Xi = \mathbb{C}R_{\mathbb{C}}^{\theta}$, as claimed. The previous argument also shows that $\mathbb{Z}/m\mathbb{Z}$ is a minimal group for $\mathbb{C}R_{\mathbb{C}}^{\theta}$, and that $\mathbb{C}R_{\mathbb{C}}^{\theta}(Z/m\mathbb{Z})$ is one dimensional, generated by $\tilde{\xi}$.

• Let F be a non zero subfunctor of $\mathbb{C}R_{\mathbb{C}}^{\theta}$, and let G be a minimal group for F. If $f \in F(G) - \{0\}$, then again by Proposition 6.5.5

$$f = \sum_{H\in[s_G]}\tfrac{1}{|N_G(H):H|}\mathrm{Ind}_H^G(\widetilde{e}_H^H\mathrm{Res}_H^G f) = \widetilde{e}_G^G f\ ,$$

since all terms in the summation indexed by proper subgroups H of G vanish, by minimality of G. Since $f \neq 0$, the group G is cyclic, and $f(g) = 0$ for $g \in G$, if $<g> \neq G$. Moreover since $f \in \mathbb{C}R_{\mathbb{C}}^{\theta}(G)$, if $G = \mathbb{Z}/n\mathbb{Z}$, then

$$f(i \cdot g) = \xi\pi_{m\vee n,m}(i)f(g) ,$$

for any $i \in \big(\mathbb{Z}/(m \vee n)\mathbb{Z}\big)^{\times}$. It follows that

$$\pi_{m\vee n,m}(\mathrm{Ker}\,\pi_{m\vee n,n}) = \mathrm{Ker}\,\pi_{m,(m,n)} \leq \mathrm{Ker}\,\xi ,$$

thus $m \mid n$ since ξ is primitive. Then $f(i \cdot g) = \xi\pi_{n,m}(i)f(g)$, for any $i \in (\mathbb{Z}/n\mathbb{Z})^{\times}$, thus $f = f(1)\widetilde{e}_{\mathbb{Z}/n\mathbb{Z}}^{\mathbb{Z}/n\mathbb{Z}}\mathrm{Inf}_{\mathbb{Z}/m\mathbb{Z}}^{\mathbb{Z}/n\mathbb{Z}}\tilde{\xi}$, and $f(1) \neq 0$. Thus by Corollary 2.5.12, if $G = \mathbb{Z}/n\mathbb{Z}$ and $N = m\mathbb{Z}/n\mathbb{Z}$, so that $G/N = \mathbb{Z}/m\mathbb{Z}$,

$$\begin{aligned}
\mathrm{Def}^G_{G/N} f &= f(1)(\mathrm{Def}^G_{G/N}\widetilde{e}^G_G \mathrm{Inf}^G_{G/N})\tilde{\xi} \\
&= f(1)(\widetilde{\mathrm{Def}^G_{G/N} e^G_G})\tilde{\xi} \\
&= f(1)m_{G,N}\widetilde{e}^{G/N}_{G/N}\tilde{\xi} \\
&= f(1)m_{G,N}\tilde{\xi} .
\end{aligned}$$

Thus $f(1)m_{G,N}\tilde{\xi} \in F(G/N) = F(\mathbb{Z}/m\mathbb{Z})$. But $f(1) \neq 0$, and $m_{G,N} \neq 0$ since G is cyclic, by Propositions 5.3.1 and 5.6.1. Hence $\tilde{\xi} \in F(\mathbb{Z}/m\mathbb{Z})$. Since $\mathbb{C}R^{\theta}_{\mathbb{C}}$ is generated by $\tilde{\xi}$, it follows that $F = \mathbb{C}R^{\theta}_{\mathbb{C}}$.

Hence $\mathbb{C}R^{\theta}_{\mathbb{C}}$ is a simple biset functor. The group $\mathbb{Z}/m\mathbb{Z}$ is a minimal group for this functor, and $\mathbb{C}R^{\theta}_{\mathbb{C}}(\mathbb{Z}/m\mathbb{Z})$ is a one dimensional $\mathbb{C}$-vector space, generated by $\tilde{\xi}$. Clearly, if $x \in (\mathbb{Z}/m\mathbb{Z})^{\times}$, then $x\tilde{\xi} = \xi(x)\tilde{\xi}$, so $\mathbb{C}R^{\theta}_{\mathbb{C}}(\mathbb{Z}/m\mathbb{Z})$ is the $\mathbb{C}\mathrm{Out}(\mathbb{Z}/m\mathbb{Z})$-module $\mathbb{C}_{\xi}$, and $\mathbb{C}R^{\theta}_{\mathbb{C}}$ is isomorphic to the simple functor $S_{\mathbb{Z}/m\mathbb{Z},\mathbb{C}_{\xi}}$. □

7.3.5. Corollary : *The functor $\mathbb{C}R_{\mathbb{C}}$ is a semisimple object of $\mathcal{F}_{C,\mathbb{C}}$. More precisely*

$$\mathbb{C}R_{\mathbb{C}} \cong \bigoplus_{(m,\xi)} S_{\mathbb{Z}/m\mathbb{Z},\mathbb{C}_{\xi}} ,$$

where (m,ξ) runs through the set of pairs consisting of a positive integer m and a primitive character $\xi : (\mathbb{Z}/m\mathbb{Z})^{\times} \to \mathbb{C}^{\times}$.

7.4. The Simple Summands of $\mathbb{C}R_{\mathbb{C}}$

7.4.1. Notation : *Let m be a positive integer, and $\xi : (\mathbb{Z}/m\mathbb{Z})^{\times} \to \mathbb{C}^{\times}$ be a primitive character. If G is a finite group, let $C(G,m,\xi)$ denote the set of cyclic subgroups H of G, of order multiple of m, such that the image of the map*

$$N_G(H) \to \mathrm{Aut}(H) \to \mathrm{Aut}\big((\mathbb{Z}/m\mathbb{Z})^{\times}\big)$$

is contained in $\operatorname{Ker}\xi$. *Let* $[C(G,m,\xi)]$ *denote a set of representatives of conjugacy classes in* G *of subgroups in* $C(G,m,\xi)$.

If $H \in C(G,m,\xi)$, *set* $n = |H|$, *and choose a group isomorphism* $\phi_H : H \to \mathbb{Z}/n\mathbb{Z}$. *Define a function* $\zeta_H : H \to \mathbb{C}$ *by*

$$\forall h \in H,\;\; \zeta_H(h) = \begin{cases} \xi\pi_{n,m}\phi_H(h) & \textit{if} <h> = H \\ 0 & \textit{otherwise.} \end{cases}$$

Then ζ_H *does not depend on the choice of* ϕ_H *up to multiplication by a scalar in* $\mathbb{C}^\times$.

Indeed, choosing ϕ_H amounts to choosing a particular generator h of H. If h is replaced by h', then ζ_H is replaced by $\xi(h')\xi(h)^{-1}\zeta_H$.

7.4.2. Proposition : *Let* m *be a positive integer, and* $\xi : (\mathbb{Z}/m\mathbb{Z})^\times \to \mathbb{C}^\times$ *be a primitive character. Set* $\theta = \xi\pi_m \in \Gamma(\hat{\mathbb{Z}}^\times)$.

Then if G *is a finite group, the elements* $\operatorname{Ind}_H^G\zeta_H$, *for* $H \in [C(G,m,\xi)]$ *form a* $\mathbb{C}$*-basis of* $\mathbb{C}R_{\mathbb{C}}^\theta(G)$.

Proof: Let H be a subgroup of G, and let $f \in \tilde{e}_H^H\mathbb{C}R_{\mathbb{C}}^\theta(H)$. If $f \neq 0$, then as above, the group H is cyclic, of order n multiple of m, and $f(h) = 0$ for $h \in H$ if h is not a generator of H. Moreover, up to a non zero scalar, the function f is equal to

$$\forall h \in H,\;\; f(h) = \begin{cases} \xi\pi_{n,m}\phi_H(h) & \text{if} <h> = H \\ 0 & \text{otherwise.} \end{cases}$$

It follows that $\tilde{e}_H^H\mathbb{C}R_{\mathbb{C}}^\theta(H)$ is zero or one dimensional, generated by ζ_H. In the latter case, the group $N_G(H)$ has non zero invariants on $\tilde{e}_H^H\mathbb{C}R_{\mathbb{C}}^\theta(H)$ if and only if it fixes ζ_H. This is equivalent to $\xi\pi_{n,m}\phi_H({}^gh) = \xi\pi_{n,m}\phi_H(h)$, whenever h is a generator of H, and $g \in N_G(H)$. In other words $H \in C(G,m,\xi)$.

In this case $\big(\tilde{e}_H^H\mathbb{C}R_{\mathbb{C}}^\theta(H)\big)^{N_G(H)}$ is one dimensional, generated by ζ_H. By Proposition 6.5.5, the functions $\operatorname{Ind}_H^G\zeta_H$, for $H \in [C(G,m,\xi)]$ form a $\mathbb{C}$-basis of $\mathbb{C}R_{\mathbb{C}}^\theta(G)$. □

7.4.3. Corollary : *The dimension of* $S_{\mathbb{Z}/m\mathbb{Z},\mathbb{C}_\xi}(G)$ *is equal to the number of conjugacy classes of cyclic subgroups* H *of* G, *of order multiple of* m, *for which the natural image of* $N_G(H)$ *in* $(\mathbb{Z}/m\mathbb{Z})^\times$ *is contained in the kernel of* ξ.

Proof: This follows from the isomorphism $S_{\mathbb{Z}/m\mathbb{Z},\mathbb{C}_\xi} \cong \mathbb{C}R_{\mathbb{C}}^\theta$. □

Chapter 8
Tensor Product and Internal Hom

Hypothesis : *In this chapter $\mathcal{D}$ is a replete subcategory of $\mathcal{C}$, which is moreover* closed under direct product*: if G and H are objects of $\mathcal{D}$, then $G \times H$ is also an object of $\mathcal{D}$.*

8.1. Bisets and Direct Products

8.1.1. Notation : *Let G, G', H, H' be finite groups. If U is a finite (H, G)-biset and U' is a finite (H', G')-biset, then the cartesian product $U \times V$ has a natural structure of $(H \times H', G \times G')$-biset, given by*

$$(h, h') \cdot (u, u') \cdot (g, g') = (hug, h'u'g') .$$

By linearity, this construction yields a map

$$RB(H, G) \times RB(H', G') \to RB(H \times H', G \times G') ,$$

still denoted by $(\alpha, \beta) \mapsto \alpha \times \beta$.

The following lemma is similar to Assertion 2 of Proposition 2.5.14:

8.1.2. Lemma : *The correspondence sending a pair (H, G) of finite groups to $H \times G$, and a pair $(\alpha, \beta) \in RB(H, G) \times RB(H', G')$ to*

$$\alpha \times \beta \in RB(H \times H', G \times G') ,$$

is a bilinear functor from $R\mathcal{C} \times R\mathcal{C}$ to $R\mathcal{C}$.

Proof: This amounts to checking that if G, H, K, G', H', K' are finite groups, if U is an (H, G)-biset, if V is a (K, H)-biset, if U' is an (H', G')-biset, and V' is a (K', H')-biset, then there is an isomorphism of $(K \times K', G \times G')$-bisets

S. Bouc, *Biset Functors for Finite Groups*, Lecture Notes in Mathematics 1990,
DOI 10.1007/978-3-642-11297-3_8, © Springer-Verlag Berlin Heidelberg 2010

$$(V \times V') \times_{H \times H'} (U \times U') \cong (V \times_H U) \times (V' \times_{H'} U')\ ,$$

given by $\big((v,v')\,,_{_{H\times H'}}(u,u')\big) \mapsto \big((v,_{_H} u),(v',_{_{H'}} u')\big)$, which is straightforward, and that $\mathrm{Id}_G \times \mathrm{Id}_{G'} = \mathrm{Id}_{G\times G'}$, which is obvious. □

8.2. The Yoneda–Dress Construction

This section describes a construction which is connected both to the Yoneda functors on the biset category $\mathcal{C}$, and to the Dress construction for Mackey functors (see [34], or [7] 1.2).

8.2.1. Notation : *Let Y be a finite group. Let $\mathsf{p}_Y : \mathcal{C} \to \mathcal{C}$ denote the functor sending a finite group G to the direct product $G \times Y$, and a morphism $\alpha \in RB(H,G)$ to $\alpha \times \mathrm{Id}_Y \in RB(H \times Y, G \times Y)$.*

When Y is an object of $\mathcal{D}$, also denote by p_Y the restriction of the functor p_Y to the subcategory $\mathcal{D}$.

8.2.2. Lemma : *The correspondence p_Y is an endofunctor of $R\mathcal{D}$, which is R-linear on morphisms.*

Proof: This is straightforward. □

8.2.3. Definition : *If Y is an object of $\mathcal{D}$, then the* Yoneda–Dress construction *P_Y at Y is the endofunctor of $\mathcal{F}_{\mathcal{D},R}$ obtained by precomposition with p_Y, i.e. the functor sending the biset functor F to the biset functor $\mathsf{P}_Y(F) = F_Y = F \circ \mathsf{p}_Y$.*

In other words, for any object G of $\mathcal{D}$, one has that

$$F_Y(G) = F(G \times Y)\ ,$$

and for any $\varphi \in \mathrm{Hom}_{R\mathcal{D}}(G,H)$

$$F_Y(\varphi) = F(\varphi \times Y)\ .$$

8.2.4. Lemma : *Let X, Y, and Z be objects of $\mathcal{D}$.*

1. *The functors $\mathsf{P}_Y \circ \mathsf{P}_X$ and $\mathsf{P}_{Y\times X}$ are isomorphic.*
2. *Any (Y,X)-biset U induces a natural transformation $\mathsf{P}_U : \mathsf{P}_X \to \mathsf{P}_Y$, defined by*

$$\mathsf{P}_{U,F,G} = F(G \times U) : F(G \times X) \to F(G \times Y)$$

for any object F of $\mathcal{F}_{\mathcal{D},R}$ and any object G of $\mathcal{D}$. This construction extends linearly, giving a natural transformation $\mathsf{P}_f : \mathsf{P}_X \to \mathsf{P}_Y$ for any $f \in RB(Y,X)$.

3. *If $f \in RB(Y,X)$ and $g \in RB(Z,Y)$, then*

$$\mathsf{P}_g \circ \mathsf{P}_f = \mathsf{P}_{g \times_Y f}$$

as natural transformations $\mathsf{P}_X \to \mathsf{P}_Z$.

4. *The correspondence $\mathsf{P}_\bullet$ sending an object X of $\mathcal{D}$ to P_X, and a morphism $f \in RB(Y,X)$ in $\mathcal{D}$ to $\mathsf{P}_f : \mathsf{P}_X \to \mathsf{P}_Y$ is an R-linear functor from $R\mathcal{D}$ to the category $\mathsf{Fun}_R(\mathcal{F}_{\mathcal{D},R}, \mathcal{F}_{\mathcal{D},R})$ of R-linear endofunctors of $\mathcal{F}_{\mathcal{D},R}$.*

Proof: All the assertions are straightforward. □

8.2.5. Notation : *If $f \in RB(Y,X)$, and if F is a biset functor, the morphism of biset functors $\mathsf{P}_{f,F} : F_X \to F_Y$ will be denoted by F_f.*

8.2.6. Notation : *If G is a finite group, let $\overrightarrow{G}$ denote the set G, endowed with the $(G \times G, \mathbf{1})$-biset structure defined by the formula*

$$(x,y).h = xhy^{-1}$$

for x, y in G and h in $\overrightarrow{G}$. Denote by $\overleftarrow{G}$ the $(\mathbf{1}, G \times G)$-biset $(\overrightarrow{G})^{op}$

8.2.7. Proposition : *Let Y be an object of $\mathcal{D}$. Then the functor P_Y is a self-adjoint exact R-linear endofunctor of the category $\mathcal{F}_{\mathcal{D},R}$.*

Proof: The functor P_Y is obviously additive and exact. Proving that it is self-adjoint amounts to building the unit $\eta : \mathrm{Id} \to \mathsf{P}_Y \circ \mathsf{P}_Y$ and counit $\varepsilon : \mathsf{P}_Y \circ \mathsf{P}_Y \to \mathrm{Id}$ for this adjunction. But $\mathsf{P}_Y \circ \mathsf{P}_Y \cong \mathsf{P}_{Y \times Y}$, by Lemma 8.2.4, and $\mathsf{P}_\mathbf{1}$ is isomorphic to the identity functor of $\mathcal{F}_{\mathcal{D},R}$.

Define η to be the natural transformation

$$\mathsf{P}_{\overrightarrow{Y}} : \mathsf{P}_\mathbf{1} \cong \mathrm{Id} \to \mathsf{P}_{Y \times Y} \cong \mathsf{P}_Y \circ \mathsf{P}_Y \ .$$

Similarly, using the opposite biset $\overleftarrow{Y}$, define ε to be the natural transformation

$$\mathsf{P}_{\overleftarrow{Y}} : \mathsf{P}_{Y \times Y} \cong \mathsf{P}_Y \circ \mathsf{P}_Y \to \mathsf{P}_\mathbf{1} \cong \mathrm{Id} \ .$$

In other words, for any object F of $\mathcal{F}_{\mathcal{D},R}$ and any object G of $\mathcal{D}$, the map

$$\eta_{F,G} : F(G) \to \big(\mathsf{P}_Y \circ \mathsf{P}_Y(F)\big)(G) = F(G \times Y \times Y)$$

is equal to $F(G \times \overrightarrow{Y})$, and the map

$$\varepsilon_{F,G} : \big(\mathsf{P}_Y \circ \mathsf{P}_Y(F)\big)(G) = F(G \times Y \times Y) \to F(G)$$

is equal to $F(G \times \overleftarrow{Y})$.

It only remains to check that η and ε are the unit and counit of an adjunction. It means that for any object F of $\mathcal{F}_{\mathcal{D},R}$, the following equalities hold:

$$\mathsf{P}_Y(\varepsilon_F) \circ \eta_{\mathsf{P}_Y(F)} = \mathrm{Id}_{\mathsf{P}_Y(F)}\ , \qquad \varepsilon_{\mathsf{P}_Y(F)} \circ \mathsf{P}_Y(\eta_F) = \mathrm{Id}_{\mathsf{P}_Y(F)}\ .$$

Equivalently, for any object G of $\mathcal{D}$, the compositions
(**8.2.8**)

$$F(G \times Y) \xrightarrow{F(G\times Y\times\overrightarrow{Y})} F(G \times Y \times Y \times Y) \xrightarrow{F(G\times\overleftarrow{Y}\times Y)} F(G \times Y)$$

and
(**8.2.9**)

$$F(G \times Y) \xrightarrow{F(G\times\overrightarrow{Y}\times Y)} F(G \times Y \times Y \times Y) \xrightarrow{F(G\times Y\times\overleftarrow{Y})} F(G \times Y)$$

should be equal to the identity of $F(G \times Y)$.

Composition 8.2.8 is also equal to $F(W)$, where W is the $(G \times Y, G \times Y)$-biset

$$W = (G \times \overleftarrow{Y} \times Y) \times_{G\times Y^3} (G \times Y \times \overrightarrow{Y})\ .$$

In other words $W = G \times W'$, where W' is the (Y, Y)-biset

$$W' = (\overleftarrow{Y} \times Y) \times_{Y^3} (Y \times \overrightarrow{Y})\ .$$

Here the (Y^3, Y)-biset structure on $Y \times \overrightarrow{Y}$ is given by

$$(x, y, z) \cdot (h, u) \cdot t = (xht, yuz^{-1})\ ,$$

for x, y, z, h, t, u in Y, whereas the (Y, Y^3)-biset structure on $(\overleftarrow{Y} \times Y)$ is given by

$$t.(u, h).(x, y, z) = (x^{-1}uy, thz)\ .$$

Now W' is the quotient of Y^4 by the right action of Y^3 given by

$$(u, h, h', u')(x, y, z) = (x^{-1}uy, hz, x^{-1}h', y^{-1}u'z)\ ,$$

for x, y, z, h, h', u, u' in Y. The map

$$(u, h, h', u') \in Y^4 \mapsto hu'^{-1}u^{-1}h' \in Y$$

passes down to the quotient, giving a bijection $a : W' \to Y$, with inverse bijection b given by

$$b : h \in Y \mapsto (1, h, 1, 1)Y^3 \in W' .$$

With these bijections, the set Y inherits an (Y, Y)-biset structure given by

$$x \cdot h \cdot y = a\big(x \cdot b(h) \cdot y\big) = a\big(x \cdot (1, h, 1, 1)Y^3 \cdot y\big) = a\big((1, xh, y, 1)Y^3\big) = xhy ,$$

so this biset structure is given by left and right multiplication. It follows that $F(W)$ is the identity map of $F(G \times Y)$. A similar proof shows that Composition 8.2.9 is equal to the identity map, and this completes the proof of Proposition 8.2.7. □

8.3. Internal Hom for Biset Functors

8.3.1. Definition : *Let M and N be objects of $\mathcal{F}_{\mathcal{D},R}$. Denote by $\mathcal{H}(M, N)$ (or by $\mathcal{H}_{R\mathcal{D}}(M, N)$ in case of ambiguity on R or $\mathcal{D}$) the object of $\mathcal{F}_{\mathcal{D},R}$ defined by*

$$\mathcal{H}(M, N) = \mathrm{Hom}_{\mathcal{F}_{\mathcal{D},R}}\big(M, P_{\bullet}(N)\big) .$$

In other words, if G is an object of $\mathcal{D}$, then

$$\mathcal{H}(M, N)(G) = \mathrm{Hom}_{\mathcal{F}_{\mathcal{D},R}}(M, N_G) .$$

If G and H are objects of $\mathcal{D}$, and if $f \in RB(H, G)$, then $\mathcal{H}(M, N)(f)$ is the map from $\mathcal{H}(M, N)(G)$ to $\mathcal{H}(M, N)(H)$ defined by

$$\varphi \in \mathrm{Hom}_{\mathcal{F}_{\mathcal{D},R}}(M, N_G) \mapsto N_f \circ \varphi \in \mathrm{Hom}_{\mathcal{F}_{\mathcal{D},R}}(M, N_H) .$$

8.3.2. Proposition and Definition : *The correspondence*

$$\mathcal{H}(-,-) : (M, N) \mapsto \mathcal{H}(M, N)$$

is a bilinear functor from $\mathcal{F}_{\mathcal{D},R}^{op} \times \mathcal{F}_{\mathcal{D},R} \to \mathcal{F}_{\mathcal{D},R}$, called the internal hom functor.

Proof: This is a straightforward consequence of Lemma 8.2.4. □

8.3.3. Proposition : *Let G be an object of $\mathcal{D}$. There is an isomorphism of bifunctors*

$$\mathsf{P}_G \circ \mathcal{H}(-,-) \cong \mathcal{H}(-, \mathsf{P}_G(-)) .$$

Proof: Indeed for any object H of $\mathcal{D}$, and any objects M and N of $\mathcal{F}_{\mathcal{D},R}$

$$\begin{aligned}\mathcal{H}(M,N)_G(H) &= \mathcal{H}(M,N)(H\times G) = \mathrm{Hom}_{\mathcal{F}_{\mathcal{D},\mathcal{R}}}(M, N_{H\times G})\\ &\cong \mathrm{Hom}_{\mathcal{F}_{\mathcal{D},\mathcal{R}}}\big(M,(N_G)_H\big)\\ &= \mathcal{H}(M,N_G)(H)\ ,\end{aligned}$$

and this isomorphism is functorial in H. Thus $\mathcal{H}(M,N)_G \cong \mathcal{H}(M,N_G)$, and this isomorphism is functorial in M and N. □

8.3.4. Proposition : *Let RB be the restriction of the Burnside functor to $R\mathcal{D}$.*

1. *If M is an object of $\mathcal{F}_{\mathcal{D},R}$, then the map*

$$f\in \mathrm{Hom}_{\mathcal{F}_{\mathcal{D},R}}(RB,M)\mapsto f_{\mathbf{1}}(\bullet)\in M(\mathbf{1})$$

   is an isomorphism of abelian groups, where $\bullet$ is the image in $RB(\mathbf{1})$ of a set of cardinality 1.
2. *The functor $M\mapsto \mathcal{H}(RB,M)$ is isomorphic to the identity functor of $\mathcal{F}_{\mathcal{D},R}$.*

Proof: Assertion 1 is a special case of Yoneda Lemma, since RB is isomorphic to $\mathrm{Hom}_{\mathcal{F}_{\mathcal{D},R}}(\mathbf{1},-)$. In other words, for any object M of $\mathcal{F}_{\mathcal{D},R}$, the map sending $f\in \mathrm{Hom}_{\mathcal{F}_{\mathcal{D},R}}(B,M)$ to $f_{\mathbf{1}}(\bullet)\in M(\mathbf{1})$ is an isomorphism.

Now for any object M of $\mathcal{F}_{\mathcal{D},R}$ and any object H of $\mathcal{D}$

$$\mathcal{H}(RB,M)(H) = \mathrm{Hom}_{\mathcal{F}_{\mathcal{D},R}}(RB,M_H)\cong M_H(\mathbf{1})\cong M(H)\ ,$$

and one checks easily that this isomorphism is functorial in M. □

8.4. Tensor Product of Biset Functors

The construction of tensor product of biset functors is left adjoint to the internal hom functor. More precisely (see Chap. IX of [39] for details on ends and coends):

8.4.1. Definition : *The tensor product $M\otimes N$ (also denoted by $M\otimes_{R\mathcal{D}} N$ in case of ambiguity on R or $\mathcal{D}$) of the objects M and N of $\mathcal{F}_{\mathcal{D},R}$ is the object of $\mathcal{F}_{\mathcal{D},R}$ defined as the following coend on $R\mathcal{D}\times R\mathcal{D}$*

$$M\otimes N = \int^{D,D'} M(D)\otimes_R N(D')\otimes_R \mathrm{Hom}_{R\mathcal{D}}(D\times D',-)\ .$$

In more concrete words, this means that for any object G of $\mathcal{D}$, the module $(M\otimes N)(G)$ can be computed as follows: let $T_{M,N}(G)$ denote the direct sum

$$T_{M,N}(G) = \bigoplus_{D,D',f} \big(M(D)\otimes_R N(D')\big) ,$$

where D and D' run through a given set $\mathcal{S}$ of equivalence classes of objects of $\mathcal{D}$ up to group isomorphism, and $f : D\times D' \to G$ is a morphism in $R\mathcal{D}$. For such a triple (D,D',f), and for $m\in M(D)$ and $n\in N(D')$, denote by $[m\otimes n]_{D,D',f}$ the element $m\otimes n$ of the summand of $T_{M,N}(G)$ indexed by (D,D',f). Then $(M\otimes N)(G)$ is equal to the quotient of $T_{M,N}(G)$ by the R-submodule generated by the elements of the form

$$\begin{aligned}
&[m\otimes n]_{D,D',rf+r'f'} - \big(r[m\otimes n]_{D,D',f} + r'[m\otimes n]_{D,D',f'}\big), \;\; \text{for } r,r'\in R ,\\
&[m\otimes n]_{D,D',f\circ(\alpha\times\beta)} - [M(\alpha)(m)\otimes N(\beta)(n)]_{D_1,D_1',f} ,
\end{aligned}$$

where $\alpha : D\to D_1$ and $\beta : D'\to D_1'$ are morphisms in $R\mathcal{D}$. If $\varphi : G\to H$ is a morphism in $R\mathcal{D}$, then $(M\otimes N)(f) : (M\otimes N)(G) \to (M\otimes N)(H)$ is the map induced by the map $T_{M,N}(G)\to T_{M,N}(H)$ sending $[m\otimes n]_{D,D',f}$ to $[m\otimes n]_{D,D',\varphi\circ f}$.

8.4.2. Proposition and definition :

1. *The correspondence*

$$-\otimes - : (M,N) \to M\otimes N$$

is a bilinear functor from $\mathcal{F}_{\mathcal{D},R}\times\mathcal{F}_{\mathcal{D},R}$ to $\mathcal{F}_{\mathcal{D},R}$, called the tensor product of biset functors.

2. *There are functorial isomorphisms*

$$\mathrm{Hom}_{\mathcal{F}_{\mathcal{D},R}}(M\otimes N,P) \cong \mathrm{Hom}_{\mathcal{F}_{\mathcal{D},R}}\big(N,\mathcal{H}(M,P)\big)$$

for any objects M, N and P of $\mathcal{F}_{\mathcal{D},R}$. In other words, for any object M of $\mathcal{F}_{\mathcal{D},R}$, the functor $M\otimes -$ is left adjoint to the functor $\mathcal{H}(M,-)$.

Proof: The proof is classical, and not specific to the particular monoidal category $R\mathcal{D}$: set $L=\mathrm{Hom}_{\mathcal{F}_{\mathcal{D},R}}(M\otimes N,P)$. Then:

$$\begin{aligned}
L &= \int_G \mathrm{Hom}_R\big((M\otimes N)(G),P(G)\big)\\
&\cong \int_G \mathrm{Hom}_R\Big(\int^{D,D'} M(D)\otimes_R N(D')\otimes_R \mathrm{Hom}_R(D\times D',G),P(G)\Big)\\
&\cong \int_G\int_{D,D'} \mathrm{Hom}_R\big(M(D)\otimes_R N(D')\otimes_R \mathrm{Hom}_R(D\times D',G),P(G)\big)
\end{aligned}$$

$$\begin{aligned}
L &\cong \int_{D,D'} \int_G \mathrm{Hom}_R\big(M(D) \otimes_R N(D') \otimes_R \mathrm{Hom}_R(D \times D', G), P(G)\big) \\
&\quad \text{(by Fubini's theorem see [39] IX.8)} \\
&\cong \int_{D,D'} \mathrm{Hom}_R\big(M(D) \otimes_R N(D'), P(D \times D')\big) \\
&\quad \text{(by Yoneda's lemma)} \\
&\cong \int_{D'} \mathrm{Hom}_R\Big(N(D'), \int_D \mathrm{Hom}_R\big(M(D), P(D \times D')\big)\Big) \\
&\cong \int_{D'} \mathrm{Hom}_R\big(N(D'), \mathcal{H}(M,P)(D')\big) \\
&\cong \mathrm{Hom}_{\mathcal{F}_{\mathcal{D},R}}\big(N, \mathcal{H}(M,P)\big) \ ,
\end{aligned}$$

as was to be shown. □

8.4.3. Remark (universal property of tensor products) : The above proof shows in particular that

$$\mathrm{Hom}_{\mathcal{F}_{\mathcal{D},R}}(M \otimes N, P) = \int_{D,D'} \mathrm{Hom}_R\big(M(D) \otimes_R N(D'), P(D \times D')\big) \ .$$

This means that $M \otimes N$ has the following universal property: define a *bilinear pairing* $\Phi : (M,N) \to P$ as a natural transformation from the functor $(G,H) \mapsto M(G) \otimes_R N(H)$, from $R\mathcal{D} \times R\mathcal{D}$ to R-Mod, to the functor $(G,H) \mapsto P(G \times H)$. In other words Φ is a collection of bilinear maps $\Phi_{G,H} : M(G) \otimes_R N(H) \to P(G \times H)$, for objects G, H of $\mathcal{D}$, satisfying obvious functoriality conditions. Then the set of bilinear pairings from (M,N) to P is in one to one correspondence with the set $\mathrm{Hom}_{\mathcal{F}_{\mathcal{D},R}}(M \otimes N, P)$.

8.4.4. Corollary : *Let M be an object of $\mathcal{F}_{\mathcal{D},R}$. Then the functor $M \otimes -$ is right exact, and the functor $\mathcal{H}(M,-)$ is left exact.*

Proof: Indeed an additive functor admitting a right (resp. left) adjoint is right (resp. left) exact. □

8.4.5. Proposition : *Let M, N, and P be object of $\mathcal{F}_{\mathcal{D},R}$. There is an isomorphism of biset functors*

$$\mathcal{H}(M \otimes N, P) \cong \mathcal{H}\big(N, \mathcal{H}(M,P)\big)$$

which is functorial in M, N, and P.

Proof: Let G be an object of $\mathcal{D}$. Then by Propositions 8.4.2 and 8.3.3

$$\begin{aligned}\mathcal{H}(M \otimes N, P)(G) &= \mathrm{Hom}_{\mathcal{F}_{\mathcal{D},R}}(M \otimes N, P_G) \cong \mathrm{Hom}_{\mathcal{F}_{\mathcal{D},R}}\big(N, \mathcal{H}(M, P_G)\big)\\ &\cong \mathrm{Hom}_{\mathcal{F}_{\mathcal{D},R}}\big(N, \mathcal{H}(M,P)_G\big) = \mathcal{H}\big(N, \mathcal{H}(M,P)\big)(G)\ ,\end{aligned}$$

and this isomorphism is functorial in G. □

8.4.6. Proposition :

1. *The functor $M \mapsto RB \otimes M$ is isomorphic to the identity functor on $\mathcal{F}_{\mathcal{D},R}$.*
2. *The functors $(M, N) \mapsto M \otimes N$ and $(M, N) \mapsto N \otimes M$ from $\mathcal{F}_{\mathcal{D},R} \times \mathcal{F}_{\mathcal{D},R}$ to $\mathcal{F}_{\mathcal{D},R}$ are isomorphic.*
3. *The functors $(M, N, P) \mapsto (M \otimes N) \otimes P$ and $(M, N, P) \mapsto M \otimes (N \otimes P)$ from $\mathcal{F}_{\mathcal{D},R} \times \mathcal{F}_{\mathcal{D},R} \times \mathcal{F}_{\mathcal{D},R}$ to $\mathcal{F}_{\mathcal{D},R}$ are isomorphic.*

In other words, the category $(\mathcal{F}_{\mathcal{D},R}, \otimes, RB)$ is a symmetric monoidal category.

Proof: For Assertion 1, by Proposition 8.4.2, the functor $E : M \mapsto RB \otimes M$ is left adjoint to the functor $N \mapsto \mathcal{H}(RB, N)$, which is isomorphic to the identity functor by Proposition 8.3.4. Hence E is isomorphic to the identity functor.

For Assertion 2, let M and N be objects of $\mathcal{F}_{\mathcal{D},R}$. If G, H, and K are objects of $\mathcal{D}$, and if $\alpha : H \times K \to G$ is a morphism in $R\mathcal{D}$, denote by $\hat{\alpha} : K \times H \to G$ the morphism obtained by precomposition of α with the (isomorphism in the biset category associated to the) canonical group isomorphism $(H \times K) \to (K \times H)$. Then it is clear that the correspondence $[m \otimes n]_\alpha \mapsto [n \otimes m]_{\hat{\alpha}}$, where $m \in M(H)$ and $n \in N(K)$, extends to a linear isomorphism $(M \otimes N)(G) \to (N \otimes M)(G)$, which gives the required functorial isomorphism in Assertion 2.

For Assertion 3, if M, N, P, and Q are objects of $\mathcal{F}_{\mathcal{D},R}$, then by Propositions 8.4.2 and 8.4.5

$$\begin{aligned}\mathrm{Hom}_{\mathcal{F}_{\mathcal{D},R}}\big(M \otimes (N \otimes P), Q\big) &\cong \mathrm{Hom}_{\mathcal{F}_{\mathcal{D},R}}\big(N \otimes P, \mathcal{H}(M, Q)\big)\\ &\cong \mathrm{Hom}_{\mathcal{F}_{\mathcal{D},R}}\Big(P, \mathcal{H}\big(N, \mathcal{H}(M.Q)\big)\Big)\\ &\cong \mathrm{Hom}_{\mathcal{F}_{\mathcal{D},R}}\big(P, \mathcal{H}(M \otimes N, Q)\big)\\ &\cong \mathrm{Hom}_{\mathcal{F}_{\mathcal{D},R}}\big((M \otimes N) \otimes P, Q\big)\ .\end{aligned}$$

Assertion 3 follows, since all these isomorphisms are functorial in M, N, P, and Q. □

8.4.7. Remark : The isomorphism $\iota : RB \otimes M \to M$ of Assertion 1 can be described explicitly as follows: if G, H, and K are objects of $\mathcal{D}$, if $\alpha : H \times K \to G$ is a morphism in $R\mathcal{D}$, then

$$\iota_G([x \otimes m]_{H,K,\alpha}) = M(\alpha)M(x \times \mathrm{Id}_K)M(\gamma_K)(m)\ ,$$

for $x \in RB(H)$ and $m \in M(K)$, where $\gamma_K : K \to \mathbf{1} \times K$ is the canonical group isomorphism, and $x \times \mathrm{Id}_K : \mathbf{1} \times K \to H \times K$ is the morphism in $R\mathcal{D}$ associated to $x \in RB(H) \cong \mathrm{Hom}_{R\mathcal{C}}(\mathbf{1}, H)$. This map ι_G is an isomorphism $(RB \otimes M)(G) \to M(G)$, the inverse isomorphism sending $m \in M(G)$ to the element $[\bullet \otimes m]_{\mathbf{1},K,\gamma_G^{-1}} \in (RB \otimes M)(G)$.

8.4.8. Another Expression for $M \otimes N$. Let $\tau : R\mathcal{D} \to R\mathcal{D}^{op}$ denote the functor equal to the identity on objects, and sending a morphism $\varphi : G \to H$ in $R\mathcal{D}$ to $\varphi^{op} : H \to G$. If D, D', and G are objects of $\mathcal{D}$, there are isomorphisms

$$\mathrm{Hom}_{R\mathcal{D}}(D \times D', G) \cong \mathrm{Hom}_{R\mathcal{D}}\big(D', \tau(D) \times G\big)$$

and these isomorphisms are functorial in (D, D'). It follows that if M and N are objects of $\mathcal{F}_{\mathcal{D},R}$, then

$$\begin{aligned}
(M \otimes N)(G) &= \int^{D,D'} M(D) \otimes_R N(D') \otimes_R \mathrm{Hom}_{R\mathcal{D}}(D \times D', G)\\
&\cong \int^{D,D'} M(D) \otimes_R N(D') \otimes_R \mathrm{Hom}_{R\mathcal{D}}\big(D', \tau(D) \times G\big)\\
&\cong \int^{D} \int^{D'} M(D) \otimes_R N(D') \otimes_R \mathrm{Hom}_{R\mathcal{D}}\big(D', \tau(D) \times G\big)\\
&\cong \int^{D} M(D) \otimes_R N\big(\tau(D) \times G\big)\\
&\cong \int^{D} M(D) \otimes_R N_G\big(\tau(D)\big) \cong M \odot N_G\ ,
\end{aligned}$$

where the notation is as follows:

8.4.9. Notation : *If M and N are objects of $\mathcal{F}_{\mathcal{D},R}$, let $M \odot N$ denote the R-module defined by*

$$M \odot N = \int^{D} M(D) \otimes_R N\big(\tau(D)\big)\ .$$

In other words

$$M \odot N \cong \Big(\mathop{\oplus}_{H \in \mathcal{S}} \big(M(H) \otimes_R N(H)\big) \Big) / \mathcal{R}\ ,$$

where $\mathcal{S}$ is as before a fixed set of representatives of equivalence classes of objects of $\mathcal{D}$ up to group isomorphism. If $H \in \mathcal{S}$, if $m \in M(H)$ and if

$n \in N(H)$, denote by $\langle m \otimes n \rangle_H$ the element $m \otimes n$ of the summand indexed by H in the direct sum $\Sigma = \bigoplus_{H \in \mathcal{S}} \big(M(H) \otimes_R N(H)\big)$. Then $\mathcal{R}$ is the R-submodule of Σ generated by the elements of the form

$$\langle \varphi(m) \otimes n \rangle_H - \langle m \otimes \varphi^{op}(n) \rangle_G \ ,$$

for any $\varphi \in \mathrm{Hom}_{R\mathcal{D}}(G, H)$, any $m \in M(G)$, and any $n \in N(H)$.

The construction $(M, N) \mapsto M \odot N$ has the following duality property:

8.4.10. Lemma : *Let M and N be objects of $\mathcal{F}_{\mathcal{D},R}$, and V be an R-module. Then then is an isomorphism of R-modules*

$$\mathrm{Hom}_R(M \odot N, V) \cong \mathrm{Hom}_{\mathcal{F}_{\mathcal{D},R}}\big(M, \mathrm{Hom}_R(N, V)\big) \ .$$

Proof: By definition:

$$\begin{aligned}
\mathrm{Hom}_R(M \odot N, V) &= \mathrm{Hom}_R\Big(\int^D M(D) \otimes N\big(\tau(D)\big), V\Big) \\
&\cong \int_D \mathrm{Hom}_R\Big(M(D), \mathrm{Hom}_R\big(N(D), V\big)\Big) \\
&\cong \mathrm{Hom}_{\mathcal{F}_{\mathcal{D},R}}\big(M, \mathrm{Hom}_R(N, V)\big) \ .
\end{aligned}$$

The lemma follows. □

This yields the following alternative description of the tensor product of biset functors:

8.4.11. Proposition :

1. *The correspondence $(M, N) \mapsto M \odot N$ is a bilinear functor from $\mathcal{F}_{\mathcal{D},R} \times \mathcal{F}_{\mathcal{D},R}$ to R-*Mod.
2. *There are isomorphisms of biset functors*

$$M \otimes N \cong M \odot \mathsf{P}_\bullet(N) \ ,$$

which are functorial in M and N.

Proof: Assertion 1 is straightforward, and Assertion 2 follows from the above discussion. □

8.4.12. Corollary : *Let M and N be objects of $\mathcal{F}_{\mathcal{D},R}$, and let Y be an object of $\mathcal{D}$. Then there are isomorphisms of functors:*

1. $(M \otimes N)_Y \cong M \otimes (N_Y) \cong M_Y \otimes N$.
2. $\mathcal{H}(M, N)_Y \cong \mathcal{H}(M, N_Y) \cong \mathcal{H}(M_Y, N)$.

Proof: For Assertion 1, let X be any object of $\mathcal{D}$. Then:

$$\begin{aligned}(M \otimes N_Y)(X) &\cong M \odot P_X(N_Y)\\ &\cong M \odot P_X P_Y(N)\\ &\cong M \odot P_{X \times Y}(N)\\ &\cong (M \otimes N)(X \times Y)\\ &\cong (M \otimes N)_Y(X)\ ,\end{aligned}$$

and one checks easily that this isomorphism is functorial in X.

Thus $(M \otimes N)_Y \cong M \otimes (N_Y)$. By Proposition 8.4.6, it follows that

$$(M \otimes N)_Y \cong (N \otimes M)_Y \cong N \otimes M_Y \cong M_Y \otimes N\ .$$

Assertion 2 follows from a direct argument, or by adjunction, from Assertion 1, since for any object P of $\mathcal{D}$

$$\begin{aligned}\mathrm{Hom}_{\mathcal{F}_{\mathcal{D},R}}(M \otimes N_Y, P) &\cong \mathrm{Hom}_{\mathcal{F}_{\mathcal{D},R}}\big(N_Y, \mathcal{H}(M, P)\big)\\ &\cong \mathrm{Hom}_{\mathcal{F}_{\mathcal{D},R}}\big(N, \mathcal{H}(M, P)_Y\big)\\ &\cong \mathrm{Hom}_{\mathcal{F}_{\mathcal{D},R}}\big((M \otimes N)_Y, P\big)\\ &\cong \mathrm{Hom}_{\mathcal{F}_{\mathcal{D},R}}(M \otimes N, P_Y)\\ &\cong \mathrm{Hom}_{\mathcal{F}_{\mathcal{D},R}}\big(N, \mathcal{H}(M, P_Y)\big)\ .\end{aligned}$$

The isomorphism between the second and the last line gives the first isomorphism in Assertion 2. The additional following isomorphisms:

$$\begin{aligned}\mathrm{Hom}_{\mathcal{F}_{\mathcal{D},R}}(M \otimes N_Y, P) &\cong \mathrm{Hom}_{\mathcal{F}_{\mathcal{D},R}}(M_Y \otimes N, P)\\ &= \mathrm{Hom}_{\mathcal{F}_{\mathcal{D},R}}\big(N, \mathcal{H}(M_Y, P)\big)\end{aligned}$$

show that $\mathcal{H}(M_Y, P) \cong \mathcal{H}(M, P_Y)$, completing the proof. □

8.4.13. Corollary : *Let M and N be objects of $\mathcal{F}_{\mathcal{D},R}$, and V be an R-module. There are isomorphisms of biset functors*

$$\mathrm{Hom}_R(M \otimes N, V) \cong \mathcal{H}\big(N, \mathrm{Hom}_R(M, V)\big) \cong \mathcal{H}\big(M, \mathrm{Hom}_R(N, V)\big)\ .$$

Proof: For any object Y of $\mathcal{D}$

$$\begin{aligned}\mathrm{Hom}_R(M\otimes N,V)(Y) &= \mathrm{Hom}_R\big((M\otimes N)(Y),V\big)\\ &\cong \mathrm{Hom}_R(M\odot N_Y,V)\\ &\cong \mathrm{Hom}_{\mathcal{F}_{\mathcal{D},R}}\big(N_Y,\mathrm{Hom}_R(M,V)\big)\\ &\cong \mathrm{Hom}_{\mathcal{F}_{\mathcal{D},R}}\big(N,\mathrm{Hom}_R(M,V)_Y\big)\\ &= \mathcal{H}\big(N,\mathrm{Hom}_R(M,V)\big)(Y)\ ,\end{aligned}$$

and these isomorphisms are functorial in Y. This shows the first isomorphism in Corollary 8.4.13. The second one follows by exchanging M and N, since the tensor product of biset functors is commutative, by Proposition 8.4.6. □

8.5. Green Biset Functors

A Green biset functor is a "biset functor with a ring structure", just like a Green functor is a Mackey functor with a ring structure (see [7]). More precisely:

8.5.1. Definition : *A* Green biset functor *(on $\mathcal{D}$, with values in R-*Mod*) is a monoid in the monoidal category $\mathcal{F}_{\mathcal{D},R}$.*

This means (see [39] VII.3) that a Green biset functor is an object A of $\mathcal{F}_{\mathcal{D},R}$, together with maps of bisets functors

$$\mu : A\otimes A \to A,\ \ e : RB \to A\ ,$$

such that the following diagrams commute:

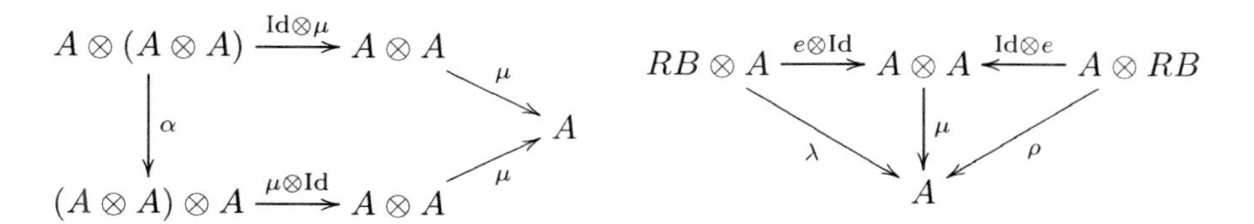

where α, λ, and ρ are the isomorphisms afforded by Proposition 8.4.6.

Equivalently, by Remark 8.4.3 and Proposition 5.1.2, a Green biset functor is an object A of $\mathcal{F}_{\mathcal{D},R}$, together with *bilinear products* $A(G)\times A(H)\to A(G\times H)$, denoted by $(a,b)\mapsto a\times b$, for objects G,H of $\mathcal{D}$, and an element $\varepsilon_A\in A(\mathbf{1})$, satisfying the following conditions:

- (Associativity) Let G,H and K be objects of $\mathcal{D}$. If

$$\alpha_{G,H,K} : G\times (H\times K)\to (G\times H)\times K$$

is the canonical group isomorphism, then for any $a\in A(G)$, $b\in A(H)$, and $c\in A(K)$

$$(a \times b) \times c = \mathrm{Iso}(\alpha_{G,H,K})\big(a \times (b \times c)\big) \ .$$

- (Identity element) Let G be an object of $\mathcal{D}$. Let $\lambda_G : \mathbf{1} \times G \to G$ and $\rho_G : G \times \mathbf{1} \to G$ denote the canonical group isomorphisms. Then for any $a \in A(G)$
$$a = \mathrm{Iso}(\lambda_G)(\varepsilon_A \times a) = \mathrm{Iso}(\rho_G)(a \times \varepsilon_A) \ .$$
- (Functoriality) If $\varphi : G \to G'$ and $\psi : H \to H'$ are morphisms in $R\mathcal{D}$, then for any $a \in A(G)$ and $b \in A(H)$
$$A(\varphi \times \psi)(a \times b) = A(\varphi)(a) \times A(\psi)(b) \ .$$

8.5.2. Definition : *If (A, μ, e) and (A', μ', e') are Green biset functors on $\mathcal{D}$, with values in R-Mod, a morphism of Green biset functors from A to A' is a morphism $f : A \to A'$ in $\mathcal{F}_{\mathcal{D},R}$ such that the diagrams*

$$\begin{array}{ccc} A \otimes A & \xrightarrow{\mu} & A \\ {\scriptstyle f\otimes f}\downarrow & & \downarrow{\scriptstyle f} \\ A' \otimes A' & \xrightarrow{\mu'} & A' \end{array} \qquad \begin{array}{ccc} & \overset{e}{\nearrow} & A \\ RB & & \downarrow{\scriptstyle f} \\ & \underset{e'}{\searrow} & A' \end{array}$$

are commutative.

Morphisms of Green biset functors can be composed, and Green biset functors on $\mathcal{D}$, with values in R-Mod, form a category $\mathsf{Green}_{\mathcal{D},R}$.

In other words $f_{G\times H}(a \times b) = f_G(a) \times f_H(b)$, for any objects G and H of $\mathcal{D}$, and for any $a \in A(G)$ and $b \in A(H)$.

8.5.3. Examples : The Burnside functor B is a Green biset functor on $\mathcal{C}$, with values in $\mathbb{Z}$-mod, for the bilinear product $B(G) \times B(H) \to B(G \times H)$, for finite groups G and H, defined by sending (X, Y), where X is (the class of) a G-set and Y is (the class of) an H-set, to (the class of) the $(G \times H)$-set $X \times Y$. The identity element is $1 \in B(\mathbf{1}) = \mathbb{Z}$.

Similarly, if $\mathbb{F}$ is a field of characteristic 0, then the functor $R_{\mathbb{F}}$ sending a finite group G to the group $R_{\mathbb{F}}(G)$ of its representations over $\mathbb{F}$ (see Chap. 1), is a Green biset functor: the product of (the class of the $\mathbb{F}G$-module) $V \in R_{\mathbb{F}}(G)$ and (the class of the $\mathbb{F}H$-module) $W \in R_{\mathbb{F}}(H)$ is defined as (the class of) the external tensor product $V \boxtimes_{\mathbb{F}} W \in R_{\mathbb{F}}(G \times H)$, and the identity element is (the class of the trivial module) $\mathbb{F} \in R_{\mathbb{F}}(\mathbf{1})$.

The linearization morphism $\chi_{\mathbb{F}} : B \to R_{\mathbb{F}}$ (see Remark 1.2.3) is a morphism of Green biset functors.

8.5.4. Modules over a Green Biset Functor. For a Green biset functor A, there is a natural notion of A-module:

8.5.5. Definition : *Let A be a Green biset functor over $\mathcal{D}$, with values in R-Mod. A left A-module M is an object of $\mathcal{F}_{\mathcal{D},R}$, equipped with a morphism of biset functors $\mu_M : A \otimes M \to M$ such that the following diagrams commute:*

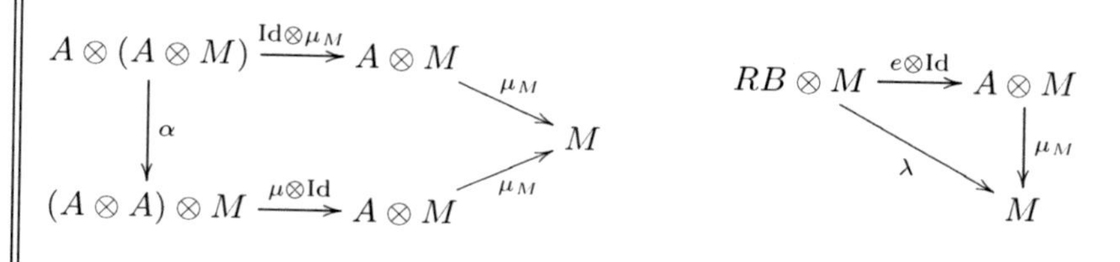

Equivalently, for any object G, H of $\mathcal{D}$, there are product maps

$$A(G) \times M(H) \to M(G \times H) \ ,$$

denoted by $(a, m) \mapsto a \times m$, fulfilling the following conditions:

- (Associativity) Let G, H and K be objects of $\mathcal{D}$. If

$$\alpha_{G,H,K} : G \times (H \times K) \to (G \times H) \times K$$

  is the canonical group isomorphism, then for any $a \in A(G)$, $b \in A(H)$, and $m \in M(K)$

$$(a \times b) \times m = \mathrm{Iso}(\alpha_{G,H,K})\big(a \times (b \times m)\big) \ .$$

- (Identity element) Let G be an object of $\mathcal{D}$. Let $\lambda_G : \mathbf{1} \times G \to G$ denote the canonical group isomorphism. Then for any $m \in M(G)$

$$m = \mathrm{Iso}(\lambda_G)(\varepsilon_A \times m) \ .$$

- (Functoriality) If $\varphi : G \to G'$ and $\psi : H \to H'$ are morphisms in $R\mathcal{D}$, then for any $a \in A(G)$ and $m \in M(H)$

$$M(\varphi \times \psi)(a \times m) = A(\varphi)(a) \times M(\psi)(m) \ .$$

8.5.6. Example : Fix an object X of $\mathcal{D}$. Then the functor A_X obtained by the Yoneda–Dress construction is a left A-module, for the product map induced by the product in A, namely, since $A_X(H) = A(H \times X)$

$$(a, m) \in A(G) \times A(H \times X) \mapsto a \times m \in A(G \times H \times X) = A_X(G \times H) \ .$$

In particular $A_{\mathbf{1}} = A$ is a left A-module.

8.5.7. Definition : *If M and N are A-modules, then a morphism of A-modules from M to N is a morphism of biset functors $f : M \to N$ such that the diagram*

$$\begin{array}{ccc} A \otimes M & \xrightarrow{\mu_M} & M \\ \downarrow{\scriptstyle \mathrm{Id}\otimes f} & & \downarrow{\scriptstyle f} \\ A \otimes N & \xrightarrow{\mu_N} & N \end{array}$$

is commutative.

*Morphisms of A-modules can be composed, and A-modules form a category, denoted by A-*Mod.

In other words $f_{G\times H}(a \times m) = a \times f_H(m)$, for any objects G, H of $\mathcal{D}$, any $a \in A(G)$ and any $m \in M(H)$.

8.5.8. Remark : The category A-Mod is an abelian category: if M and N are A-modules, then the direct sum $M \oplus N$ (as a biset functor) has an obvious structure of A-module, and this yields a direct sum in the category A-Mod. Similarly, if M and N are A-modules, and $f : M \to N$ is a morphism of A-modules, then the kernel and cokernel of f (as a morphism of biset functors) have natural structures of A-modules, and this yields a kernel and a cokernel of f in the category A-Mod.

8.5.9. Remark : The definition of a right A-module is similar, as a biset functor equipped with a morphism of biset functors $M \otimes A \to M$, fulfilling suitable conditions of associativity, identity element, and functoriality.

8.5.10. Ideals of a Green Biset Functor.

8.5.11. Definition : *Let A be a Green biset functor on $\mathcal{D}$, with value in R-*Mod. *A* left ideal *of A is an A-submodule of the left A-module A. In other words is it a biset subfunctor I of A such that*

$$A(G) \times I(H) \subseteq I(G \times H)$$

for any objects G and H of $\mathcal{D}$.

One defines similarly a right ideal *of A. A* two sided *ideal of A is a left ideal which is also a right ideal.*

In other words, a two sided ideal I of A is a biset subfunctor such that

$$A(G) \times I(H) \times A(K) \subseteq I(G \times H \times K) ,$$

for any objects G, H, K of $\mathcal{D}$, with a slight abuse in notation, a more correct formulation being rather

$$\big(A(G) \times I(H)\big) \times A(K) \subseteq I\big((G \times H) \times K\big) .$$

When I is a two sided ideal of A, then the quotient biset functor A/I is a Green biset functor in the obvious way, and the projection morphism from A to A/I is a morphism of Green functors. A Green biset functor A is called *simple* if its only two sided ideals are $\{0\}$ and A.

8.6. More on A-Modules

The following proposition lists without proof additional properties of Green biset functors and modules over them, inspired by Chaps. 3 and 6 of [7]:

8.6.1. Proposition : *Let A be an Green biset functor over $\mathcal{D}$, with values in R-*Mod*.*

1. *There is a unique homomorphism of Green biset functors $RB \rightarrow A$. In other words RB is an initial object of the category* $\mathsf{Green}_{\mathcal{D},R}$.
2. *Let X be an object of $\mathcal{D}$. Then A_X is a projective object in A-*Mod*. More precisely, for any A-module M, there is an isomorphism of R-modules*
$$\mathrm{Hom}_{A\text{-}\mathsf{Mod}}(A_X, M) \cong M(X) .$$
3. *If M is an A-module, then there exists a set S of objects of $\mathcal{D}$, and for each $X \in S$, a set I_X, such that M is isomorphic to a quotient of the module $\bigoplus_{X\in S}(A_X)^{(I_X)}$. In particular A-*Mod* has enough projective objects.*
4. *Let I be an injective cogenerator of the category R-*Mod*. If M is a left A-module, then $\mathrm{Hom}_R(M, I)$ is a right A-module. Moreover, if $f : P \rightarrow \mathrm{Hom}_R(M, I)$ is a surjective morphism of right A-modules, where P is a projective right A-module, then the map $M \rightarrow \mathrm{Hom}_R(P, I)$ obtained by duality is injective, and $\mathrm{Hom}_R(P, I)$ is an injective left A-module. In particular A-*Mod* has enough injective objects.*
5. *Let $\mathcal{P}_A$ be the following category:*
   - *The objects of $\mathcal{P}_A$ are the objects of $\mathcal{D}$.*
   - *If G and H are objects of $\mathcal{D}$, then $\mathrm{Hom}_{\mathcal{P}_A}(G, H) = A(H \times G)$.*
   - *If G, H and K are objects of $\mathcal{D}$, then the composition of $\alpha \in A(H\times G)$ and $\beta \in A(K \times H)$ is equal to*
   $$\beta \circ \alpha = A(G \times \overleftarrow{H} \times K)(\beta \times \alpha) .$$
   - *If G is an object of $\mathcal{D}$, then the identity morphism of G in $\mathcal{P}_A$ is equal to $A(\overrightarrow{G})(\varepsilon_A) \in A(G \times G)$.*

   *Then $\mathcal{P}_A$ is an R-linear category, and A-*Mod* is equivalent to the category of R-linear functors from $\mathcal{P}_A$ to R-*Mod*.*
6. *Let M, N and P be a objects of $\mathcal{F}_{\mathcal{D},R}$. If G and H are objects of $\mathcal{D}$, if $a \in \mathcal{H}(M, N)(X) = \mathrm{Hom}_{\mathcal{F}_{\mathcal{D},R}}(M, N_X)$ and $b \in \mathcal{H}(N, P)(Y) =$*

$\mathrm{Hom}_{\mathcal{F}_{\mathcal{D},R}}(N, P_Y)$, *let* $a \times b \in \mathcal{H}(M,P)(X \times Y)$ *be the morphism* $M \to P_{X\times Y} \cong (P_Y)_X$ *defined as the composition*

$$a \times b : M \xrightarrow{a} N_X \xrightarrow{P_X(b)} (P_Y)_X \cong P_{X\times Y} \ .$$

For this product, the functor $\mathcal{H}(M,M)$ *becomes a Green functor, the identity element being* $\mathrm{Id}_M \in \mathcal{H}(M,M)(\mathbf{1}) = \mathrm{End}_{\mathcal{F}_{\mathcal{D},R}}(M)$.
Moreover giving M *the structure of an* A*-module is equivalent to giving a morphism of Green biset functors* $A \to \mathcal{H}(M,M)$.

Part III
p-Biset Functors

Chapter 9
Rational Representations of p-Groups

Notation : *Throughout the present part of this book, the symbol p denotes a fixed prime number.*

The present chapter is preliminary to the study of *p-biset functors*, i.e. biset functors defined on finite p-groups. It collects tools about rational representations of p-groups, which will motivate the definition of the important class of *rational p-biset functors*.

9.1. The Functor of Rational Representations

9.1.1. Notation : *Recall that if G is a finite group, then $R_{\mathbb{Q}}(G)$ is* the group of rational representations *of G, i.e. the Grothendieck group of the category of finite dimensional $\mathbb{Q}G$-modules. This category is semisimple, so $R_{\mathbb{Q}}(G)$ is a free abelian group on the set* $\mathrm{Irr}_{\mathbb{Q}}(G)$ *of isomorphism classes of simple $\mathbb{Q}G$-modules. If V is a finite dimensional $\mathbb{Q}G$-module, its isomorphism class is denoted by $[V]$, and viewed as an element of $R_{\mathbb{Q}}(G)$.*

If V is a simple $\mathbb{Q}G$-module and W is a finite dimensional $\mathbb{Q}G$-module, then the multiplicity of V as a summand of W is denoted by $m(V,W)$. This extends by linearity to a linear form $m(V,-) : R_{\mathbb{Q}}(G) \to \mathbb{Z}$..

There is a unique bilinear form $\langle\ ,\ \rangle_G$ on $R_{\mathbb{Q}}(G)$, with values in $\mathbb{Z}$, such that

$$\langle [V],[W]\rangle_G = \dim_{\mathbb{Q}} \mathrm{Hom}_{\mathbb{Q}G}(V,W) \ ,$$

for any finite dimensional $\mathbb{Q}G$-modules V and W.

This bilinear form has the following crucial property:

9.1.2. Lemma (Frobenius reciprocity) : *Let G and H be finite groups, and U be a finite (H,G)-biset. If V is a finite dimensional $\mathbb{Q}G$-module, and W is a finite dimensional $\mathbb{Q}H$-module, then*

DOI 10.1007/978-3-642-11297-3_9,

$$\langle W, \mathbb{Q}U \otimes_{\mathbb{Q}G} V\rangle_H = \langle \mathbb{Q}U^{op} \otimes_{\mathbb{Q}H} W, V\rangle_G \; .$$

Proof: Let θ and φ denote the characters of V and W, respectively. Then the character ψ of $\mathbb{Q}U \otimes_{\mathbb{Q}G} V$ can be computed thanks to Lemma 7.1.3, and

$$\begin{aligned}
\langle W, \mathbb{Q}U \otimes_{\mathbb{Q}G} V\rangle_H &= \frac{1}{|H|}\sum_{h\in H}\varphi(h^{-1})\psi(h)\\
&= \frac{1}{|H||G|}\sum_{h\in H}\sum_{\substack{u\in U,\, g\in G\\ hu=ug}}\varphi(h^{-1})\theta(g)\\
&= \frac{1}{|H||G|}\sum_{g\in G}\sum_{\substack{u^{op}\in U^{op},\, h\in H\\ g^{-1}u^{op}=u^{op}h^{-1}}}\varphi(h^{-1})\theta(g)\\
&= \frac{1}{|H||G|}\sum_{g\in G}\chi(g^{-1})\theta(g) = \langle \mathbb{Q}U^{op}\otimes_{\mathbb{Q}H} W, V\rangle_G \; ,
\end{aligned}$$

where χ is the character of $\mathbb{Q}U^{op} \otimes_{\mathbb{Q}H} W$. □

9.1.3. Proposition : *Let G be a finite group. Then the $\mathbb{Z}$-module $\partial R_{\mathbb{Q}}(G)$ has a basis consisting of the elements $[V]$, where V runs through a set of representatives of isomorphism classes of faithful rational irreducible representations of G.*

Proof: Let V be a simple $\mathbb{Q}G$-module. If V is not faithful, then there exists a non trivial normal subgroup N of G such that $V = \mathrm{Inf}_{G/N}^G\mathrm{Def}_{G/N}^G V$, so $f_1^G V = 0$ by Proposition 6.2.6. Conversely, if N is a non trivial normal subgroup of G, since the unit map $V \to \mathrm{Inf}_{G/N}^G\mathrm{Def}_{G/N}^G V$ is surjective, it follows that $\mathrm{Def}_{G/N}^G V = \{0\}$ if V is faithful. Thus $V = f_1^G V$ in this case. This completes the proof. □

9.2. The Ritter–Segal Theorem Revisited

The following theorem has been proved independently by J. Ritter [41] and G. Segal [43]:

9.2.1. Theorem : [Ritter–Segal] *Let P be a finite p-group. Then the linearization morphism*

$$\chi_P : B(P) \to R_{\mathbb{Q}}(P)$$

is surjective.

The following more precise form was stated in [11]. It shows that any irreducible $\mathbb{Q}P$-module different from $\mathbb{Q}$ can be written as $\chi_P(P/Q)-\chi_P(P/Q')$, where Q and Q' are subgroups of P such that $Q < Q'$ and $|Q' : Q| = p$:

9.2.2. Theorem : *Let P be a finite p-group. If V is a non trivial simple $\mathbb{Q}P$-module, then there exist subgroups $Q < Q'$ of P, with $|Q' : Q| = p$, and a short exact sequence of $\mathbb{Q}P$-modules*

$$0 \longrightarrow V \longrightarrow \mathbb{Q}P/Q \xrightarrow{\mathbb{Q}\pi_Q^{Q'}} \mathbb{Q}P/Q' \longrightarrow 0\,,$$

where $\pi_Q^{Q'} : P/Q \to P/Q'$ is defined by $\pi_Q^{Q'}(xQ) = xQ'$, for $x \in P$.

This raises the question of knowing when the kernel of $\mathbb{Q}\pi_Q^{Q'}$ is a simple $\mathbb{Q}P$-module. The answer is as follows ([11] Proposition 4):

9.2.3. Proposition : *Let P be a finite p-group, and $Q < Q'$ be subgroups of P such that $|Q' : Q| = p$. Then the following conditions are equivalent:*

1. *The kernel of the map $\mathbb{Q}\pi_Q^{Q'} : \mathbb{Q}P/Q \to \mathbb{Q}P/Q'$ is a simple $\mathbb{Q}P$-module.*
2. *If S is any subgroup of P such that $S \cap Q' \leq Q$, then $|S| \leq |Q|$.*
3. *The group $N_P(Q)/Q$ is cyclic or generalized quaternion, the group Q'/Q is its unique subgroup of order p, and for any subgroup S of P*

$$S \cap N_P(Q) \leq Q \Rightarrow |S| \leq |Q|\,.$$

This proposition shows in particular that Q' is determined by Q. This motivates the following definition:

9.2.4. Definition : *Let P be a finite p-group.* A basic subgroup *of P is a subgroup Q such that the following two conditions hold:*

1. *The group $N_P(Q)/Q$ is cyclic or generalized quaternion.*
2. *If S is any subgroup of Q such that $S \cap N_P(Q) \leq Q$, then $|S| \leq |Q|$.*

With this definition, the group P is a basic subgroup of itself. If Q is a proper basic subgroup of P, define $\tilde{Q}$ as the unique subgroup of P containing Q, and such that $|\tilde{Q} : Q| = p$. Then $\operatorname{Ker} \pi_Q^{\tilde{Q}}$ is a simple $\mathbb{Q}P$-module, by Proposition 9.2.3.

9.2.5. Notation : *Let P be a finite p-group, and Q be a basic subgroup of P. Denote by V_Q the simple $\mathbb{Q}P$-module defined by*

$$V_Q = \begin{cases} \mathbb{Q} & \text{if } Q = P \\ \operatorname{Ker} \pi_Q^{\tilde{Q}} & \text{if } Q < P\,. \end{cases}$$

With this notation, any simple $\mathbb{Q}P$-module is isomorphic to V_Q, for some basic subgroup Q of P. This subgroup Q is not unique in general, but this ambiguity can be precisely described using the following definition:

9.2.6. Definition : *Let P be a finite p-group, and $\mathfrak{B}_P$ denote the set of basic subgroups of P. Define a relation $\div_P$ on $\mathfrak{B}_P$ by*

$$Q \div_P Q' \Leftrightarrow |Q| = |Q'| \text{ and } \exists x \in P,\ \ Q'^x \cap N_P(Q) \leq Q\ .$$

If Q and Q' are basic subgroups of P, then the simple $\mathbb{Q}P$-modules V_Q and $V_{Q'}$ are isomorphic if and only if $Q \div_P Q'$:

9.2.7. Proposition : [12, Theorem 2.8] *Let P be a finite p-group.*

1. *The relation $\div_P$ is an equivalence relation on $\mathfrak{B}_P$.*
2. *The correspondence $Q \mapsto V_Q$ induces a one to one correspondence between the set of equivalence classes of basic subgroups of P for the relation $\div_P$, and the set of isomorphism classes of simple $\mathbb{Q}P$-modules.*

9.2.8. Example (faithful simple $\mathbb{Q}P$-modules) : Let Q be a basic subgroup of P. Then it is easy to see that V_Q is faithful if and only if Q does not contain any non trivial normal subgroup of P, or equivalently, if $Q \cap Z(P) = \mathbf{1}$, where $Z(P)$ is the centre $Z(P)$. In this case $Z(P)$ maps injectively in the centre of $N_P(Q)/Q$, so $Z(P)$ is cyclic.

Conversely, if P is non trivial, and $Z(P)$ is cyclic, denote by Z the only subgroup of order p of $Z(P)$. Let Q be a subgroup of P of maximal order such that $Q \cap Z(P) = \mathbf{1}$. Then any subgroup R of P properly containing Q contains QZ.

This shows that $N_P(Q)/Q$ is cyclic or generalized quaternion. Moreover if $S \leq P$ is such that $S \cap N_P(Q) \leq Q$, then in particular $S \cap Z(P) = \mathbf{1}$, thus $|S| \leq |Q|$. It follows that Q is basic, and it is easy to see that V_Q is a faithful simple $\mathbb{Q}P$-module in this case.

9.3. Groups of Normal p-Rank 1

The following class of p-groups plays a crucial rôle for representation theory of p-groups:

9.3.1. Definition : *A p-group P is said to* have normal p-rank 1 *if it does not have any normal subgroup isomorphic to $(C_p)^2$.*

Denote by $\mathcal{N}$ the class of finite p-groups of normal p-rank 1, and by $[\mathcal{N}]$ a set of isomorphism classes of elements of $\mathcal{N}$.

If $P \in \mathcal{N}$, then by Theorem 4.10 of Chap. 5 of [36], the group P is cyclic if p is odd, and if $p = 2$, the group P is isomorphic to one of the groups C_{p^n}, for $n \geq 0$, or Q_{2^n}, for $n \geq 3$, or D_{2^n}, for $n \geq 4$, or SD_{2^n}, for $n \geq 4$, where:

- C_{p^n} is a cyclic group of order p^n.
- Q_{2^n} is a generalized quaternion group of order 2^n. It has the following presentation:

$$Q_{2^n} = <x, y \mid x^{2^{n-1}} = 1,\ {}^y x = x^{-1},\ x^{2^{n-2}} = y^2> .$$

- D_{2^n} is a dihedral group of order 2^n. It has the following presentation:

$$D_{2^n} = <x, y \mid x^{2^{n-1}} = y^2 = 1,\ {}^y x = x^{-1}> .$$

- SD_{2^n} is a semi-dihedral group of order 2^n. It has the following presentation :

$$SD_{2^n} = <x, y \mid x^{2^{n-1}} = y^2 = 1,\ {}^y x = x^{2^{n-2}-1}> .$$

9.3.2. Example : The following schematic diagrams represent the lattices of subgroups of the dihedral group D_{16}, the quaternion group Q_{16}, and the semi-dihedral group SD_{16}. An horizontal dotted link between two vertices means that the corresponding subgroups are conjugate. The vertex marked with a $\circ$ is the centre of the group. It has order 2 in each case.

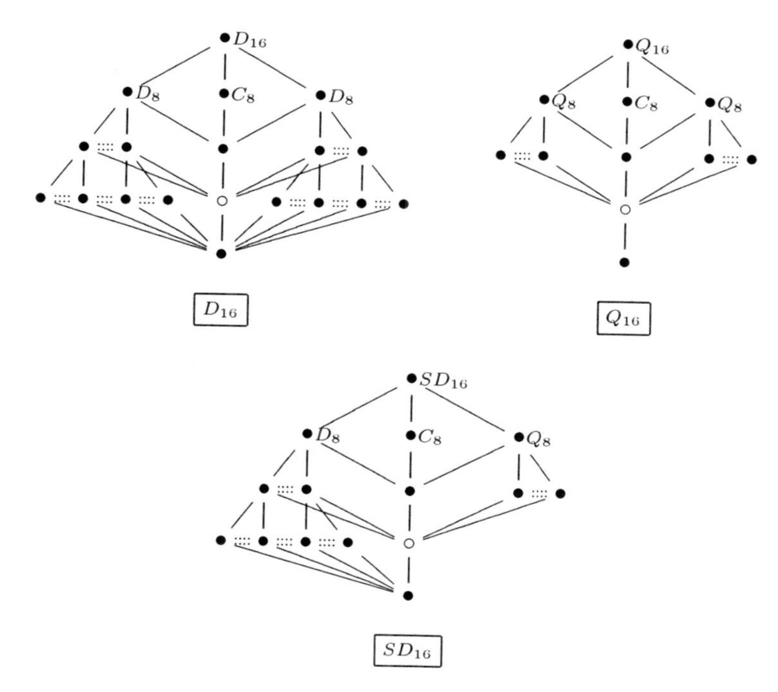

This diagram gives a good idea of the general case:

9.3.3. Lemma : *Let P be a non-cyclic group in $\mathcal{N}$. Then $p = 2$, and P has exactly 3 maximal subgroups A, B, and C. The group C is cyclic, and moreover:*

1. *If P is dihedral, then A and B are dihedral.*
2. *If $P \cong Q_8$, then A and B are cyclic.*
3. *If P is generalized quaternion of order at least 16, then A and B are generalized quaternion.*
4. *If P is semi-dihedral, then A is dihedral and B is generalized quaternion.*

There are 0,1, or 2 conjugacy classes of non-central involutions in P, according to P being generalized quaternion, semi-dihedral, or dihedral.

Proof: Let $P = <x, y>$, according to the presentations following Definition 9.3.1. Since P is non cyclic, the quotient of P by its Frattini subgroup is elementary abelian of rank 2, and P has exactly 3 maximal subgroups. One of them is the subgroup C generated by x. The other two are the subgroups $A = <x^2, y>$ and $B = <x^2, xy>$. It is easy to check that A and B are as described in the lemma.

When P is generalized quaternion, its unique involution is central. When P is dihedral, the non central involutions of P are all the elements $x^\alpha y$, where $\alpha \in \mathbb{N}$. They form two conjugacy classes in P, depending on the parity of α. When P is semi-dihedral, the non central involutions of P are the elements $x^\alpha y$, where alpha is even. They are all conjugate in P. □

9.3.4. Remark : An easy counting argument shows that when P is dihedral or semi-dihedral, the centralizer of a non central involution τ of P is the elementary abelian group of rank 2 generated by τ and the centre of P.

9.3.5. Proposition : *Let P be a p-group of normal p-rank 1.*

1. *If $Q \leq P$ and $Q \cap Z(P) = \mathbf{1}$, then $Q = \mathbf{1}$ if P is cyclic or generalized quaternion, and $|Q| \leq 2$ if P is dihedral or semi-dihedral.*
2. *The basic subgroups Q of P such that $Q \cap Z(P) = \mathbf{1}$ are the trivial subgroup if P is cyclic or generalized quaternion, or the non central subgroups of order 2 if P is dihedral or semi-dihedral.*
3. *The group P has a unique faithful rational irreducible representation, up to isomorphism. In other words $\partial R_{\mathbb{Q}}(P) \cong \mathbb{Z}$.*

Proof: Assertion 1 is clear if P is cyclic or generalized quaternion, since in this case if P is non trivial, then P has a unique subgroup of order p. If P is dihedral or semi-dihedral, then $Q \neq P$, so Q is contained in some maximal subgroup H of P. Moreover H contains the centre Z of P, of order 2, and Z is also the unique central subgroup of order 2 of H, by Lemma 9.3.3. Assertion 1 follows from an easy induction argument, on the order of P.

For Assertion 2, if $Q \cap Z(P) = \mathbf{1}$, and if P is cyclic or generalized quaternion, then $Q = \mathbf{1}$, and $\mathbf{1}$ is indeed a basic subgroup of P in this case (since $N_P(\mathbf{1})/\mathbf{1} \cong P$ is cyclic or generalized quaternion). If P is dihedral or semi-dihedral, then $|Q| \leq 2$ by Assertion 1. Moreover $\mathbf{1}$ is not a basic subgroup of P, since $N_P(\mathbf{1})/\mathbf{1} \cong P$ is neither cyclic nor generalized quaternion in this case. And if Q is non central of order 2, then Q is a subgroup of P of maximal order such that $Q \cap Z(P) = \mathbf{1}$. Hence Q is basic, by Example 9.2.8.

For Assertion 3, observe that there is a single conjugacy class of basic subgroups Q of P such that $Q \cap Z(P) = \mathbf{1}$, except in the case where P is dihedral. Now obviously, if Q and Q' are conjugate basic subgroups of P, then $Q \doteqdot_P Q'$. And when P is dihedral, there are two conjugacy classes of non central subgroups of order 2 in P. If Q and Q' are representatives of these classes, then $N_P(Q) = QZ(P)$, by Remark 9.3.4, so $Q' \cap N_P(Q) = \mathbf{1} \leq Q$, and $Q' \doteqdot_P Q$. Hence in this case also, there is a single equivalence class of basic subgroups Q of P such that $Q \cap Z(P) = \mathbf{1}$, for the relation $\doteqdot_P$. Assertion 3 follows. □

9.3.6. Remark : The argument used for Assertion 2 also applies to the dihedral group D_8 (though D_8 is not of normal p-rank 1). So D_8 also has a unique faithful rational irreducible representation, denoted by Φ_{D_8}.

9.3.7. Notation : *If P is a finite p-group of normal p-rank 1, or if $P \cong D_8$, let Φ_P denote the unique faithful rational irreducible $\mathbb{Q}P$-module.*

9.3.8. Corollary : *Let P be a non trivial finite p-group of normal p-rank 1, and order p^n. Then*

$$\dim_{\mathbb{Q}} \Phi_P = \begin{cases} (p-1)p^{n-1} & \text{if } P \cong C_{p^n} \text{ or } P \cong Q_{2^n} \\ 2^{n-2} & \text{if } P \cong D_{2^n} \text{ or } P \cong SD_{2^n}\,. \end{cases}$$

Proof: Indeed, for any proper basic subgroup Q of P

$$\dim_{\mathbb{Q}} V_Q = |P:Q| - |P:\tilde{Q}| = \tfrac{p-1}{p}|P:Q|\,.$$

Moreover $V_Q = \Phi_P$ if and only if $Q = \mathbf{1}$ if P is cyclic or generalized quaternion, or $|Q| = 2$ and $Q \neq Z(P)$ if P is dihedral or semi-dihedral. □

9.3.9. Lemma : *Let P be a non trivial finite p-group of normal p-rank 1, and H be a maximal subgroup of P. Then:*

1. $\mathrm{Res}^P_H \Phi_P = (p-1)\Phi_H$ *if* $|P| = p$.
2. $\mathrm{Res}^P_H \Phi_P = p\Phi_H$ *if P is cyclic or generalized quaternion.*

3. $\mathrm{Res}^P_H \Phi_P = 2\Phi_H$ *if P is dihedral or semi-dihedral, and H is dihedral.*
4. $\mathrm{Res}^P_H \Phi_P = \Phi_H$ *if P is dihedral or semi-dihedral, and H is cyclic or generalized quaternion.*

Proof: If $|P| = p$, then $\Phi_P = \mathbb{Q}P/1 - \mathbb{Q}P/P$ in $R_\mathbb{Q}(P)$, and $H = \mathbf{1}$. So $\mathrm{Res}^P_H \Phi_P = (p-1)\Phi_\mathbf{1}$, and Assertion 1 holds.

Assume now that $|P| \geq p^2$. Let Q be a basic subgroup of P, such that $Q \cap Z(P) = \mathbf{1}$. Then $Q = \mathbf{1}$ if P is cyclic or generalized quaternion, and $|Q| = 2$ if P is dihedral or semi-dihedral. Let Z denote the unique central subgroup of order p in P. Then $Z \leq H$, since H is a non trivial normal subgroup of P (for $|P| \geq p^2$). Moreover $\Phi_P = \mathbb{Q}P/Q - \mathbb{Q}P/QZ$ in $R_\mathbb{Q}(P)$. There are two cases:

• Either $Q \leq H$. In this case by the Mackey formula

$$\mathrm{Res}^P_H \Phi_P = p(\mathbb{Q}H/Q - \mathbb{Q}H/QZ) .$$

If P is cyclic or generalized quaternion, then so is H. Moreover $Q = \mathbf{1}$ in this case, and $\mathbb{Q}H/Q - \mathbb{Q}H/QZ = \Phi_H$. If P is dihedral or semi-dihedral, and if $H \geq Q$, then H is also dihedral (possibly of order 8), the group Z is equal to the centre of H, and Q is a basic subgroup of H. So $\Phi_H = \mathbb{Q}H/Q - \mathbb{Q}H/QZ$ also in this case. It follows that $\mathrm{Res}^P_H \Phi_P = p\Phi_H$ if $Q \leq H$. This proves Assertion 2.

• Or $Q \not\leq H$, then $p = 2$, and the group P is dihedral or semi-dihedral. Moreover $Q \cap H = \mathbf{1}$, and $QZ \cap H = Z$, and by the Mackey formula

$$\mathrm{Res}^P_H \Phi_P = \mathbb{Q}H/\mathbf{1} - \mathbb{Q}H/Z .$$

If the subgroup H is dihedral, then H contains a basic subgroup Q' of P such that $Q' \cap Z(P) = \mathbf{1}$. Thus, replacing Q by Q' yields $\mathrm{Res}^P_H \Phi_P = 2\Phi_H$ as shown in the previous case. Together with the previous case, this proves Assertion 3.

If H is cyclic or generalized quaternion, then $\Phi_H = \mathbb{Q}H/\mathbf{1} - \mathbb{Q}H/Z$, so $\mathrm{Res}^P_H \Phi_P = \Phi_H$, as was to be shown for Assertion 4. □

9.4. The Roquette Theorem Revisited

The following theorem is a reformulation of a result of P. Roquette [42]:

9.4.1. Theorem : [Roquette] *Let P be a finite p-group, and V be a simple $\mathbb{Q}P$-module. Then there exist a section (T, S) of P and a faithful simple $\mathbb{Q}(T/S)$-module W such that*

1. *The module V is isomorphic to* $\mathrm{Indinf}^P_{T/S} W$.

2. *This isomorphism induces an isomorphism of* $\mathbb{Q}$*-algebras*

$$\mathrm{End}_{\mathbb{Q}P}V \cong \mathrm{End}_{\mathbb{Q}(T/S)}W \ .$$

3. *The group* T/S *has normal* p*-rank 1.*

Proof: By induction on the order of P, one can suppose V faithful, since otherwise V is inflated from a proper factor group of P, for which the theorem holds. Moreover inflation preserves the required properties.

Now if P has normal p-rank 1, the result holds trivially, with $T = P$ and $S = \mathbf{1}$. Otherwise P has an elementary abelian normal subgroup $E \cong (C_p)^2$. Let L be a direct summand of $\mathrm{Res}_E^P V$. Let I denote the inertial subgroup of L in P, and let $\tilde{L}$ denote the isotypic component of L in $\mathrm{Res}_E^P V$. Then $V \cong \mathrm{Ind}_I^P \tilde{L}$ by Clifford theory. Moreover I contains the centralizer $C_P(E)$ of E, which has index at most p in P, since $P/C_P(E)$ is isomorphic to a subgroup of the automorphism group of E. Thus $|P : I| \leq p$, and in particular $I \trianglelefteq P$. If $I = P$, then $\tilde{L} \cong V$ is faithful, thus L is faithful. But E has no faithful irreducible rational representations, by Example 9.2.8. Hence I has index p in P. Moreover $\tilde{L}$ is an irreducible representation of I, and if $x \in P$ is such that ${}^x\tilde{L} = \tilde{L}$, then ${}^xL \cong L$, thus $x \in I$. It follows that $\mathrm{End}_{\mathbb{Q}P}(V) \cong \mathrm{End}_{\mathbb{Q}I}(\tilde{L})$. Now the induction hypothesis applies to I and $\tilde{L}$, completing the proof. □

9.4.2. Remark : Since T/S has normal p-rank 1, the $\mathbb{Q}(T/S)$-module W is isomorphic to $\Phi_{T/S}$, by Proposition 9.3.5. Moreover the $\mathbb{Q}$-algebras of Assertion 2 are actually skew fields, since V and W are simple modules.

Conversely, let (T, S) be any section of P, let W be any $\mathbb{Q}(T/S)$-module, and set $V = \mathrm{Indinf}_{T/S}^P W$. The functor $\mathrm{Indinf}_{T/S}^P$ always induces an injective algebra homomorphism from $\mathrm{End}_{\mathbb{Q}(T/S)}W$ to $\mathrm{End}_{\mathbb{Q}P}V$. It follows that these algebras are isomorphic if and only if they have the same dimension, i.e. if

$$\langle V, V\rangle_P = \langle W, W\rangle_{T/S} \ .$$

In particular, if this condition holds and if W is a simple $\mathbb{Q}(T/S)$-module, then $\mathrm{End}_{\mathbb{Q}P}V$ is a skew field, so V is an indecomposable $\mathbb{Q}P$-module, hence it is simple.

9.4.3. Lemma : *Let* P *be a finite* p*-group, and* (T, S) *be a section of* P *fulfilling the following two conditions:*

1. *The group* T/S *has normal* p*-rank 1.*
2. *Let* $V = \mathrm{Indinf}_{T/S}^P \Phi_{T/S}$*. Then* $\langle V, V\rangle_P = \langle \Phi_{T/S}, \Phi_{T/S}\rangle_{T/S}$*.*

Then $T = N_P(S)$*.*

Proof: Indeed by Frobenius reciprocity

$$\begin{aligned}
\langle V, V\rangle_P &= \langle \mathrm{Inf}_{T/S}^{T}\Phi_{T/S}, \mathrm{Res}_T^P\mathrm{Ind}_T^P\mathrm{Inf}_{T/S}^T\Phi_{T/S}\rangle_T \\
&= \sum_{x\in[T\backslash P/T]} \langle \mathrm{Inf}_{T/S}^{T}\Phi_{T/S}, \mathrm{Ind}_{T\cap {}^xT}^{T}\gamma_x\mathrm{Res}_{T^x\cap T}^{T}\mathrm{Inf}_{T/S}^T\Phi_{T/S}\rangle_T \\
&\geq \sum_{x\in[N_P(T,S)/T]} \langle \mathrm{Inf}_{T/S}^{T}\Phi_{T/S}, \gamma_x\mathrm{Inf}_{T/S}^T\Phi_{T/S}\rangle_T \; .
\end{aligned}$$

Now each $x \in N_P(T,S)$ induces an automorphism $\bar{x}$ of T/S, and

$$\gamma_x\mathrm{Inf}_{T/S}^T\Phi_{T/S} = \mathrm{Inf}_{T/S}^T\gamma_{\bar{x}}\Phi_{T/S} = \mathrm{Inf}_{T/S}^T\Phi_{T/S} \; ,$$

since $\gamma_{\bar{x}}\Phi_{T/S} = \Phi_{T/S}$, as $\Phi_{T/S}$ is the only faithful simple $\mathbb{Q}(T/S)$-module. It follows that

$$\begin{aligned}
\langle V, V\rangle_P &\geq |N_P(T,S) : T|\langle \mathrm{Inf}_{T/S}^T\Phi_{T/S}, \mathrm{Inf}_{T/S}^T\Phi_{T/S}\rangle_T \\
&= |N_P(T,S) : T|\langle \Phi_{T/S}, \Phi_{T/S}\rangle_{T/S} \; .
\end{aligned}$$

Now $\langle V, V\rangle_P = \langle \Phi_{T/S}, \Phi_{T/S}\rangle_{T/S}$ by Condition 2, thus $|N_P(T,S) : T| = 1$. Hence T is equal to its normalizer in $N_P(S)$, so $T = N_P(S)$. □

9.4.4. Definition and Notation : *Let P be a finite p-group. A subgroup S of P is called* genetic *if the following two conditions are satisfied:*

1. *The group $N_P(S)/S$ has normal p-rank 1.*
2. *Let $V(S) = \mathrm{Indinf}_{N_P(S)/S}^P\Phi_{N_P(S)/S}$. Then*

$$\langle V(S), V(S)\rangle_P = \langle \Phi_{N_P(S)/S}, \Phi_{N_P(S)/S}\rangle_{N_P(S)/S} \; .$$

If S is a genetic subgroup of P, define an integer d_S by

$$\mathsf{d}_S = \begin{cases} 1 \text{ if } N_P(S)/S \text{ is cyclic or generalized quaternion} \\ 2 \text{ if } N_P(S)/S \text{ is dihedral or semi-dihedral.} \end{cases}$$

9.4.5. Corollary : *Let P be a finite p-group, and V be a simple $\mathbb{Q}P$-module. Then there exists a genetic subgroup S of P such that $V \cong V(S)$.*

9.4.6. Lemma : *Let P be a finite p-group. If S is a genetic subgroup of P, then the kernel of the irreducible representation $V(S)$ is equal to the intersection of the conjugates of S in P. In particular $V(S) \cong \mathbb{Q}$ if and only if $S = P$.*

Proof: Indeed $V(S) = \mathrm{Indinf}_{N_P(S)/S}^{P} \Phi_{N_P(S)/S}$, so an element $g \in P$ acts trivially on $V(S)$ if and only if $gxN_P(S) = xN_P(S)$, i.e. $g^x \in N_P(S)$, for any $x \in P$, and if moreover g^x acts trivially on $\Phi_{N_P(S)/S}$, i.e. $g^x \in S$, since $\Phi_{N_P(S)/S}$ is faithful. The first assertion follows, and the second one is a straightforward consequence. □

9.5. Characterization of Genetic Subgroups

The main result of this section is Theorem 9.5.6, giving a combinatorial characterization of genetic subgroups of a p-group P, in which rational representations of P no longer appear.

9.5.1. Notation : *Let P be a finite p-group. If S is a subgroup of P, denote by $Z_P(S)$ the preimage in $N_P(S)$ of the centre of $N_P(S)/S$, i.e. the subgroup defined by $S \leq Z_P(S) \leq N_P(S)$ and*

$$Z_P(S)/S = Z\big(N_P(S)/S\big) .$$

Denote by $\hat{S}$ the preimage in $N_P(S)$ of the largest elementary abelian central subgroup of $N_P(S)/S$, i.e. the subgroup of $Z_P(S)$, containing S, defined by

$$\hat{S}/S = \Omega_1 Z\big(N_P(S)/S\big) .$$

9.5.2. Lemma : *Let P be a finite group.*

1. *Let S and T be subgroups of P. Then*

$$S \overset{P}{\uparrow} T = T \text{ if and only if } S \cap Z_P(T) \leq T .$$

In particular

$$S \frown_P T \;\Leftrightarrow\; \exists x \in P,\; S^x \cap Z_P(T) \leq T \text{ and } {}^xT \cap Z_P(S) \leq S .$$

2. *Let S be a subgroup of P. Then the following conditions are equivalent:*
   a. *The subgroup S is an expansive subgroup of P.*
   b. *If $x \in P$ is such that $S^x \cap Z_P(S) \leq S$, then $S^x = S$.*

   Similarly, the following conditions are equivalent:
   a. *The subgroup S is a weakly expansive subgroup of P.*
   b. *If $x \in P$ is such that ${}^xS \cap Z_P(S) \leq S$ and $S^x \cap Z_P(S) \leq S$, then $S^x = S$.*

Proof: Indeed

$$(S \overset{P}{\uparrow} T)/T = \bigcap_{g \in N_P(T)} (S^g \cap N_P(T))T/T$$

is a normal subgroup of the p-group $N_P(T)/T$. It is non trivial if and only if it intersects the centre of $N_P(T)/T$ non trivially, i.e. equivalently if

$$\Big(\bigcap_{g \in N_P(T)} (S^g \cap N_P(T))T\Big) \cap Z_P(T) \neq T \ .$$

Equivalently again, this means that

$$\bigcap_{g \in N_P(T)} (S^g \cap N_P(T) \cap Z_P(T))T \neq T \ .$$

The left hand side is equal to

$$\begin{aligned} \bigcap_{g \in N_P(T)} (S^g \cap Z_P(T))T &= \bigcap_{g \in N_P(T)} \Big((S \cap Z_P(T))T\Big)^g \\ &= \big(S \cap Z_P(T)\big)T \ , \end{aligned}$$

since $N_P(T)$ acts trivially by conjugation on $Z_P(T)/T$.

So T is a proper subgroup of $S \overset{P}{\uparrow} T$ if and only if $S \cap Z_P(T) \nleq T$, and this completes the proof of Assertion 1. Assertion 2 is a straightforward consequence of Assertion 1, by definition of a (weakly) expansive subgroup. □

9.5.3. Lemma : *Let P be a finite p-group, and S be a genetic subgroup of P. If Y is a subgroup of P, define integers a_Y and b_Y by*

$$\begin{aligned} a_Y &= |\{x \in [N_P(S)\backslash P/Y] \mid N_P(S) \cap {}^xY \leq S\}| \\ b_Y &= |\{x \in [N_P(S)\backslash P/Y] \mid |I_x(S,Y)| = p,\ I_x(S,Y) \nleq Z\big(N_P(S)/S\big)\}| \ , \end{aligned}$$

where $I_x(S,Y) = \big(N_P(S) \cap {}^xY\big)S/S$. Then

$$m\big(V(S), \mathbb{Q}(P/Y)\big) = \mathsf{d}_S a_Y + b_Y \ .$$

In particular, the following conditions are equivalent:

1. *The simple module $V(S)$ is a summand of $\mathbb{Q}P/Y$.*
2. *There exists $x \in P$ such that ${}^xY \cap Z_P(S) \leq S$.*

Proof: If $S = P$, then $V(S) = \mathbb{Q}$ is always a summand of $\mathbb{Q}P/Y$, with multiplicity 1. In this case $\mathsf{d}_S = 1$, $a_Y = 1$ and $b_Y = 0$, so $m\big(V(S), \mathbb{Q}(P/Y)\big) = \mathsf{d}_S a_Y + b_Y$. Moreover $Z_P(S) = P$, and Condition 2 is always satisfied. So one can suppose $S < P$.

Set $T = N_P(S)$. By Frobenius reciprocity

$$\begin{aligned} m(V(S), \mathbb{Q}P/Y)\langle V(S), V(S)\rangle_P &= \langle V(S), \mathbb{Q}P/Y\rangle_P \\ &= \langle \Phi_{T/S}, \mathrm{Defres}^P_{T/S}\mathbb{Q}P/Y\rangle_{T/S} \\ &= \langle \Phi_{T/S}, \mathbb{Q}\mathrm{Defres}^P_{T/S}P/Y\rangle_{T/S} \\ &= m(\Phi_{T/S}, \mathbb{Q}\mathrm{Defres}^P_{T/S}P/Y)\langle \Phi_{T/S}, \Phi_{T/S}\rangle_{T/S} \; . \end{aligned}$$

Since $\langle V(S), V(S)\rangle_P = \langle \Phi_{T/S}, \Phi_{T/S}\rangle_{T/S}$, it follows that $m\big(V(S), \mathbb{Q}P/Y\big) = m(\Phi_{T/S}, \mathbb{Q}\mathrm{Defres}^P_{T/S}P/Y)$. Now $\mathbb{Q}\mathrm{Defres}^P_{T/S}P/Y$ is the permutation module associated to the (T/S)-set $S\backslash P/Y$, and

$$S\backslash P/Y = \bigsqcup_{x\in[T\backslash P/Y]} S\backslash TxY/Y \; .$$

As a (T/S)-set, the set $S\backslash TxY/Y$ is transitive, and the stabilizer of its point SxY in (T/S) is equal to

$$\{tS \in T/S \mid tx \in SxY\} = \{tS \mid t \in (T \cap {}^xY)S\} = I_x(S,Y) \; .$$

It follows that

$$m\big(V(S), \mathbb{Q}P/Y\big) = \sum_{x\in[T\backslash P/Y]} m\big(\Phi_{T/S}, \mathbb{Q}(T/S)/I_x(S,Y)\big) \; .$$

Suppose that the group $I_x(S,Y)$ intersect $Z(T/S)$ non trivially. Then the module $\mathbb{Q}(T/S)/I_x(S,Y)$ is inflated from some proper quotient of T/S, and so are all its simple summands as $\mathbb{Q}(T/S)$-module. Since $\Phi_{T/S}$ is faithful, it follows that $m\big(\Phi_{T/S}, \mathbb{Q}(T/S)/I_x(S,Y)\big) = 0$ in this case.

And if $I_x(S,Y) \cap Z(T/S) = \mathbf{1}$, by Proposition 9.3.5, there are two cases:

• Either $I_x(S,Y) = \mathbf{1}$, i.e. $T \cap {}^xY \le S$. Then $m\big(\Phi_{T/S}, \mathbb{Q}(T/S)/I_x(S,Y)\big)$ is the multiplicity of $\Phi_{T/S}$ in the regular representation of T/S. Since the group $L = T/S$ has a unique minimal normal subgroup Z, the kernel of the projection map $\mathbb{Q}L/\mathbf{1} \to \mathbb{Q}L/Z$ is a sum of simple faithful $\mathbb{Q}L$-modules, hence a sum of copies of Φ_L. For dimension reasons, by Corollary 9.3.8, the number of copies is equal to

$$\frac{\frac{p-1}{p}|L|}{\dim_{\mathbb{Q}} \Phi_L} = \begin{cases} 1 \text{ if } L \text{ is cyclic or generalized quaternion} \\ 2 \text{ if } L \text{ is dihedral or semi-dihedral.} \end{cases}$$

In other words it is equal to d_S. So

$$m\big(\Phi_{T/S}, \mathbb{Q}(T/S)/I_x(S,Y)\big) = \mathsf{d}_S$$

in this case, and there are a_Y such double cosets TxY in P.

• Or $p = 2$ and the group $M = I_x(S,Y)$ is a non central subgroup of order 2 in the group $L = T/S$, which is dihedral or semi-dihedral. Moreover M is a basic subgroup of L, and there is an exact sequence of $\mathbb{Q}L$-modules

$$0 \to \Phi_L \to \mathbb{Q}L/M \to \mathbb{Q}L/MZ \to 0 ,$$

where Z is the unique central subgroup of order 2 of L. Since $\mathbb{Q}L/MZ$ is inflated from L/Z, the multiplicity of the faithful simple module Φ_L in $\mathbb{Q}L/MZ$ is equal to 0. It follows that the multiplicity of Φ_L in $\mathbb{Q}L/M$ is equal to 1. Hence

$$m\big(\Phi_{T/S}, \mathbb{Q}(T/S)/I_x(S,Y)\big) = 1$$

in this case, and there are b_Y such double cosets TxY in P. It follows that

$$m\big(V(S), \mathbb{Q}(P/Y)\big) = \mathsf{d}_S a_Y + b_Y .$$

In particular, the multiplicity of $V(S)$ in $\mathbb{Q}P/Y$ is non zero if and only if $a_Y + b_Y > 0$. But

$$a_Y + b_Y = |\{x \in [N_P(S)\backslash P/Y] \mid {}^xY \cap Z_P(S) \leq S\} ,$$

and this completes the proof of the lemma. □

9.5.4. Lemma : *Let P be a finite p-group, and S be a proper genetic subgroup of P. Then $|\hat{S} : S| = p$ and*

1. *The kernel of the projection map $\mathbb{Q}P/S \to \mathbb{Q}P/\hat{S}$ is isomorphic to a direct sum of d_S copies of $V(S)$.*
2. *The module $V(S)$ is not a direct summand of $\mathbb{Q}P/\hat{S}$.*

Proof: Since the centre of the group $N = N_P(S)/S$ is cyclic, and non trivial, the group $Z = \hat{S}/S$ has order p. Now by Lemma 9.5.3, the multiplicity of Φ_N as a summand of $\mathbb{Q}N/\mathbf{1}$ is equal to d_S, and the multiplicity of Φ_N as a summand of $\mathbb{Q}N/Z$ is equal to 0. So there is an exact sequence of $\mathbb{Q}N$-modules

$$0 \to \mathsf{d}_S\Phi_N \to \mathbb{Q}N/\mathbf{1} \xrightarrow{f} \mathbb{Q}N/Z \to 0 ,$$

where f is the projection map, and $\mathsf{d}_S\Phi_N$ is a direct sum of d_S copies of Φ_N. Applying $\mathrm{Indinf}_{N_P(S)/S}^P$ to this sequence gives the exact sequence of $\mathbb{Q}P$-modules

$$0 \to \mathsf{d}_S V(S) \to \mathbb{Q}P/S \xrightarrow{\varphi} \mathbb{Q}P/\hat{S} \to 0 ,$$

where φ is the projection map, and $\mathsf{d}_S V(S)$ is a direct sum of d_S copies of $V(S)$. This shows Assertion 1.

For Assertion 2, observe that by Corollary 9.3.8 and Definition 9.4.4

$$\dim_{\mathbb{Q}} V(S) = \frac{(p-1)|P:S|}{p\,\mathsf{d}_S} = \frac{(p-1)|P:\hat{S}|}{\mathsf{d}_S} \; .$$

If $V(S)$ is a direct summand of $\mathbb{Q}P/\hat{S}$, since $\mathbb{Q} \not\cong V(S)$ is another direct summand, it follows that $\dim_{\mathbb{Q}} V(S) < \dim_{\mathbb{Q}} \mathbb{Q}P/\hat{S} = |P:\hat{S}|$, thus

$$\frac{(p-1)|P:\hat{S}|}{\mathsf{d}_S} < |P:\hat{S}| \; ,$$

so $\mathsf{d}_S > p-1 \geq 1$, i.e. $\mathsf{d}_S \geq 2$. This implies that $p=2$, and that $N_P(S)/S$ is dihedral or semi-dihedral. Let S' be a subgroup of P containing $\hat{S}$, with $|S':\hat{S}| = 2$. In particular $S' \neq P$, since $|P:S| \geq 16$.

Then $\dim_{\mathbb{Q}} V(S) = \frac{|P:S|}{4} = |P:S'|$, so $V(S)$ is not a direct summand of $\mathbb{Q}P/S'$, since otherwise $V(S) \cong \mathbb{Q}P/S'$, which implies that $\mathbb{Q}$ is a direct summand of $V(S)$, so $V(S) = \mathbb{Q}$. It follows that $V(S)$ is a direct summand of the kernel W of the projection map

$$\mathbb{Q}P/\hat{S} \to \mathbb{Q}P/S' \; ,$$

so $V(S) \cong W$ since $\dim_{\mathbb{Q}} W = \frac{|P:S|}{4} = \dim_{\mathbb{Q}} V(S)$. Hence W is a simple $\mathbb{Q}P$-module, and by Proposition 9.2.3, this implies that $\hat{S}$ is a basic subgroup of P. In particular the group $N_P(\hat{S})/\hat{S}$ is cyclic or generalized quaternion. But this group contains the dihedral group $N_P(S)/\hat{S}$, and this contradiction completes the proof of Assertion 2. □

9.5.5. Corollary : *Let P be a finite p-group, and S be a genetic subgroup of P.*

1. *If $x \in P$ and $N_P(S) \cap {}^xS \leq S$, then ${}^xS = S$.*
2. *If $x \in P$, then $(N_P(S) \cap {}^xS)S/S$ cannot be a non central subgroup of order p of $N_P(S)/S$.*
3. *If $x \in P$ and $Z_P(S) \cap {}^xS \leq S$, then ${}^xS = S$. In other words S is an expansive subgroup of P.*

Proof: If $S = P$, all the assertions are trivial. Otherwise by Lemma 9.5.4, the multiplicity of $V(S)$ as a summand of $\mathbb{Q}P/S$ is equal to d_S. With the notation of Lemma 9.5.3, this gives

$$\mathsf{d}_S a_S + b_S = \mathsf{d}_S \; ,$$

i.e. $\mathsf{d}_S(a_S - 1) + b_S = 0$. But $a_S \geq 1$ and $b_S \geq 0$, so $a_S = 1$ and $b_S = 0$. Assertion 2 follows from $b_S = 0$, and Assertion 1 from $a_S = 1$. Finally $a_S + b_S$

is equal to the number of double cosets $N_P(S)xS$, for $x \in P$, such that $Z_P(S) \cap {}^xS \leq S$. This number is equal to 1, and together with Lemma 9.5.2, this proves Assertion 3. □

9.5.6. Theorem : *Let P be a finite p-group, and S be a subgroup of P, such that $N_P(S)/S$ has normal p-rank 1. Then the following conditions are equivalent:*

1. *The subgroup S is a genetic subgroup of P.*
2. *If $x \in P$ is such that ${}^xS \cap Z_P(S) \leq S$, then ${}^xS = S$. In other words S is an expansive subgroup of P.*
3. *If $x \in P$ is such that ${}^xS \cap Z_P(S) \leq S$ and $S^x \cap Z_P(S) \leq S$, then ${}^xS = S$. In other words S is a weakly expansive subgroup of P.*

Proof: If S is a genetic subgroup of P, then it fulfills Condition 2 by Corollary 9.5.5.

Condition 2 obviously implies Condition 3.

Suppose now that Condition 3 holds. Set $V(S) = \mathrm{Indinf}_{N_P(S)/S}^P \Phi_{N_P(S)/S}$. Now $\Phi_{N_P(S)/S} = f_{\mathbf{1}}^{N_P(S)/S}\Phi_{N_P(S)/S}$ since $\Phi_{N_P(S)/S}$ is faithful, so

$$V(S) = \mathcal{I}_S \Phi_{N_P(S)/S} \ , \tag{9.5.7}$$

with the notation of Proposition 6.4.4. By Frobenius reciprocity, and since $(f_{\mathbf{1}}^{N_P(S)/S})^{op} = f_{\mathbf{1}}^{N_P(S)/S}$

$$\begin{aligned}
\langle V(S), V(S)\rangle_P &= \langle \Phi_{N_P(S)/S}, \mathrm{Defres}_{N_P(S)/S}^P V(S)\rangle_{N_P(S)/S}\\
&= \langle \Phi_{N_P(S)/S}, \mathrm{Defres}_{N_P(S)/S}^P \mathcal{I}_S\Phi_{N_P(S)/S}\rangle_{N_P(S)/S}\\
&= \langle f_{\mathbf{1}}^{N_P(S)/S}\Phi_{N_P(S)/S}, \mathrm{Defres}_{N_P(S)/S}^P \mathcal{I}_S\Phi_{N_P(S)/S}\rangle_{N_P(S)/S}\\
&= \langle \Phi_{N_P(S)/S}, f_{\mathbf{1}}^{N_P(S)/S}\mathrm{Defres}_{N_P(S)/S}^P \mathcal{I}_S\Phi_{N_P(S)/S}\rangle_{N_P(S)/S}\\
&= \langle \Phi_{N_P(S)/S}, \mathcal{D}_S\mathcal{I}_S\Phi_{N_P(S)/S}\rangle_{N_P(S)/S}\\
&= \langle \Phi_{N_P(S)/S}, \Phi_{N_P(S)/S}\rangle_{N_P(S)/S} \ ,
\end{aligned}$$

since $\mathcal{D}_S\mathcal{I}_S\Phi_{N_P(S)/S} = f_{\mathbf{1}}^{N_P(S)/S}\Phi_{N_P(S)/S} = \Phi_{N_P(S)/S}$ by Proposition 6.4.4. Thus $\langle V(S), V(S)\rangle_P = \langle \Phi_{N_P(S)/S}, \Phi_{N_P(S)/S}\rangle_{N_P(S)/S}$, so S is a genetic subgroup of P. □

9.6. Genetic Bases

The following theorem answers the question of knowing when two genetic subgroups of a p-group P yield isomorphic irreducible representations of P:

9.6.1. Theorem : *Let P be a finite p-group. If S and T are genetic subgroups of P, the following conditions are equivalent:*

1. *The $\mathbb{Q}P$-modules $V(S)$ and $V(T)$ are isomorphic.*
2. *There exist $x, y \in P$ such that ${}^xT \cap Z_P(S) \leq S$ and ${}^yS \cap Z_P(T) \leq T$.*
3. *There exists $x \in P$ such that ${}^xT \cap Z_P(S) \leq S$ and $S^x \cap Z_P(T) \leq T$. In other words $S \mathrel{\widehat{=}_P} T$.*
4. *The sections $\big(N_P(S), S\big)$ and $\big(N_P(T), T\big)$ of P are linked modulo P.*

If these conditions hold, then in particular the groups $N_P(S)/S$ and $N_P(T)/T$ are isomorphic.

Proof: There are four steps:

• Suppose that Condition 4 holds. Then there exists $x \in P$ such that

$$\big(N_P(S), S\big) \;\text{—}\; \big(N_P({}^xT), {}^xT\big) ,$$

so in particular $N_P(S) \cap {}^xT = S \cap N_P({}^xT)$. Since $Z_P(S) \leq N_P(S)$ and $Z_P({}^xT) = {}^xZ_P(T)$, it follows that $Z_P(S) \cap {}^xT \leq S$ and $S^x \cap Z_P(T) \leq T$, so Condition 3 holds.

• Condition 3 obviously implies Condition 2.

• Suppose that Condition 2 holds. If $S = P$, then $Z_P(T) = T$, so $T = P$, and $V(S) \cong V(T) \cong \mathbb{Q}$. So one can suppose that S and T are proper subgroups of P. By Lemma 9.5.3, the module $V(S)$ is a direct summand of $\mathbb{Q}P/T$, and the module $V(T)$ is a direct summand of $\mathbb{Q}P/S$. Suppose that $V(S)$ and $V(T)$ are not isomorphic. By Lemma 9.4.6, none of these two modules is isomorphic to $\mathbb{Q}$, since S and T are proper subgroups of P. Then by Lemma 9.5.4, the module $V(S)$ is a direct summand of $\mathbb{Q}P/\hat{T}$. Since $\mathbb{Q}$ is also a direct summand of $\mathbb{Q}P/\hat{T}$

$$\dim_{\mathbb{Q}} V(S) = \frac{(p-1)|P:S|}{p\,\mathsf{d}_S} < |P:\hat{T}| = \frac{|P:T|}{p} .$$

Thus $(p-1)|P:S| < \mathsf{d}_S|P:T|$, and by symmetry $(p-1)|P:T| < \mathsf{d}_T|P:S|$. Multiplying these two inequalities yields $(p-1)^2 < d_S d_T$. Hence $d_S d_T > 1$, so one of d_S or d_T is greater than 1, and $p = 2$. In this case $|P:S| < \mathsf{d}_S|P:T|$, so $|P:S| \leq |P:T|$ since $\mathsf{d}_S \in \{1,2\}$, and both $|P:S|$ and $|P:T|$ are powers of 2. By symmetry $|P:T| \leq |P:S|$, so $|S| = |T|$.

Now $|P:S| < \mathsf{d}_S|P:T|$ implies $\mathsf{d}_S > 1$, so $\mathsf{d}_S = 2$, and $\mathsf{d}_T = 2$ by symmetry. Moreover $N_P(S)/S$ and $N_P(T)/T$ are dihedral or semi-dihedral, and in particular $|P:S| = |P:T| \geq 16$.

Let T' be a subgroup of P, containing $\hat{T}$, such that $|T':\hat{T}| = 2$. Then $\dim_{\mathbb{Q}} V(S) = \frac{|P:S|}{4} = |P:T'|$, so $V(S)$ cannot be a direct summand of $\mathbb{Q}P/T'$, since $\mathbb{Q}$ is another direct summand of QP/T'. It follows that $V(S)$ is a direct summand of the kernel W of the projection map $\mathbb{Q}P/\hat{T} \to \mathbb{Q}P/T'$.

Since $\dim_{\mathbb{Q}} W = \frac{|P:S|}{4} = \dim_{\mathbb{Q}} V(S)$, it follows that $W \cong V(S)$ is irreducible. By Proposition 9.2.3, it follows that $\hat{T}$ is a basic subgroup of P, so in particular $N_P(\hat{T})/\hat{T}$ is cyclic or generalized quaternion. But this group contains the dihedral group $N_P(T)/\hat{T}$, and this contradiction shows that $V(S) \cong V(T)$, so Condition 1 holds.

• Finally, suppose that Condition 1 holds, i.e. that $V(S) \cong V(T)$. If $V(S) \cong \mathbb{Q}$, then both S and T are equal to P, by Lemma 9.4.6, and Condition 4 holds trivially. Assume then that $V(S) \not\cong \mathbb{Q} \not\cong V(T)$, i.e. that S and T are proper subgroups of P. Set $N = N_P(S)$ and $M = N_P(T)$. Since $V(S) = \mathcal{I}_S \Phi_{N/S}$, as observed in 9.5.7, and since S is an expansive subgroup of P, Proposition 6.4.4 shows that

$$\mathrm{Defres}^P_{N/S} V(S) = f_1^{N/S} \Phi_{N/S} = \Phi_{N/S} \ .$$

It follows that $\mathrm{Defres}^P_{N/S} V(T) = \Phi_{N/S}$, i.e.

$$(S\backslash P/T)\Phi_{M/T} = \Phi_{N/S} \ ,$$

where $S\backslash P/T \cong (S\backslash P) \times_P (P/T)$ is the $\big(N/S, M/T\big)$-biset whose effect is $\mathrm{Defres}^P_{N/S} \circ \mathrm{Indinf}^P_{M/T}$. By symmetry of S and T

$$(T\backslash P/S)\Phi_{N/S} = \Phi_{M/T} \ ,$$

where $(T\backslash P/S) = (S\backslash P/T)^{op}$. Now the biset $S\backslash P/T$ splits as a disjoint union of transitive ones

$$S\backslash P/T = \bigsqcup_{x \in [N\backslash P/M]} S\backslash NxM/T \ .$$

The stabilizer L_x in $\big(N/S \times M/T\big)$ of the element SxT of the biset $S\backslash NxM/T$ is equal to

$$L_x = \{(aS, bT) \in N/S \times M/T \mid axb^{-1} \in SxT\} \ .$$

Thus, setting $p_{i,x} = p_i(L_x)$ and $k_{i,x} = k_i(L_x)$, for $i = 1, 2$, one has that

(9.6.2) $$p_{1,x} = S(N \cap {}^xM)/S, \qquad k_{1,x} = S\big(N \cap {}^xT\big)/S \ ,$$

(9.6.3) $$p_{2,x} = T(N^x \cap M)/T, \qquad k_{2,x} = T\big(S^x \cap M\big)/T \ .$$

It follows that

$$\Phi_{N/S} = \sum_{x \in [N \backslash P/M]} \mathrm{Indinf}_{p_{1,x}/k_{1,x}}^{N/S} \mathrm{Iso}(f_x) \mathrm{Defres}_{p_{2,x}/k_{2,x}}^{M/T} \Phi_{M/T} ,$$

where $f_x : p_{2,x}/k_{2,x} \to p_{1,x}/k_{1,x}$ is the canonical group isomorphism. This equality in $R_{\mathbb{Q}}(N/S)$ implies an isomorphism of $\mathbb{Q}N/S$-modules

$$\Phi_{N/S} \cong \bigoplus_{x \in [N \backslash P/M]} \mathrm{Indinf}_{p_{1,x}/k_{1,x}}^{N/S} \mathrm{Iso}(f_x) \mathrm{Defres}_{p_{2,x}/k_{2,x}}^{M/T} \Phi_{M/T} ,$$

and since $\Phi_{N/S}$ is a simple $\mathbb{Q}N/S$-module, it follows that there exists a unique element $x \in [N \backslash P/M]$ such that the corresponding summand in the right hand side is non zero, and

$$(\mathbf{9.6.4}) \qquad \Phi_{N/S} \cong \mathrm{Indinf}_{p_{1,x}/k_{1,x}}^{N/S} \mathrm{Iso}(f_x) \mathrm{Defres}_{p_{2,x}/k_{2,x}}^{M/T} \Phi_{M/T} .$$

In particular, for any $y \in [N \backslash P/M] - \{x\}$

$$\langle \Phi_{N/S}, \mathrm{Indinf}_{p_{1,y}/k_{1,y}}^{N/S} \mathrm{Iso}(f_y) \mathrm{Defres}_{p_{2,y}/k_{2,y}}^{M/T} \Phi_{M/T} \rangle_{N/S} = 0 ,$$

i.e.

$$(\mathbf{9.6.5}) \qquad \langle \mathrm{Defres}_{p_{1,y}/k_{1,y}}^{N/S} \Phi_{N/S}, \mathrm{Iso}(f_y) \mathrm{Defres}_{p_{2,y}/k_{2,y}}^{M/T} \Phi_{M/T} \rangle_{p_{1,y}/k_{1,y}} = 0 .$$

By symmetry of S and T there exists a unique element $x' \in [N \backslash P/M]$ such that

$$\Phi_{M/T} \cong \mathrm{Indinf}_{p_{2,x'}/k_{2,x'}}^{M/T} \mathrm{Iso}(f_{x'}^{-1}) \mathrm{Defres}_{p_{1,x'}/k_{1,x'}}^{N/S} \Phi_{N/S} ,$$

and for any $y \in [N \backslash P/M] - \{x'\}$

$$(\mathbf{9.6.6}) \qquad \langle \mathrm{Iso}(f_y^{-1}) \mathrm{Defres}_{p_{1,y}/k_{1,y}}^{N/S} \Phi_{N/S}, \mathrm{Defres}_{p_{2,y}/k_{2,y}}^{M/T} \Phi_{M/T} \rangle_{p_{1,y}/k_{1,y}} = 0 .$$

Since Equations 9.6.5 and 9.6.6 are equivalent, for any y, it follows that $x' = x$.

Since the functor $\mathrm{Indinf}_{p_{1,x}/k_{1,x}}^{N/S}$ is faithful, it follows from 9.6.4 that the module

$$W = \mathrm{Iso}(f_x) \mathrm{Defres}_{p_{2,x}/k_{2,x}}^{M/T} \Phi_{M/T}$$

is a simple $\mathbb{Q}(p_{1,x}/k_{1,x})$-module, and that

$$\langle W, W \rangle_{p_{1,x}/k_{1,x}} \leq \langle \Phi_{N/S}, \Phi_{N/S} \rangle_{N/S} .$$

Set $W' = \mathrm{Defres}_{p_{2,x}/k_{2,x}}^{M/T} \Phi_{M/T}$. Then W' is a simple $\mathbb{Q}(p_{2,x}/k_{2,x})$-module, and one has also that

$$\begin{aligned}\langle W, W\rangle_{p_{1,x}/k_{1,x}} &= \langle W', W'\rangle_{p_{2,x}/k_{2,x}}\\ &= \langle \Phi_{M/T}, \mathrm{Indinf}_{p_{2,x}/k_{2,x}}^{M/T} W'\rangle_{M/T}\end{aligned}$$

Since this is non zero, the module $\Phi_{M/T}$ is a summand of $\mathrm{Indinf}_{p_{2,x}/k_{2,x}}^{M/T} W'$, so the right hand side is at least equal to $\langle \Phi_{M/T}, \Phi_{M/T}\rangle_{M/T}$. Thus

$$\begin{aligned}\langle V(S), V(S)\rangle_P &= \langle \Phi_{N/S}, \Phi_{N/S}\rangle_{N/S}\\ &\geq \langle W, W\rangle_{p_{1,x}/k_{1,x}} \geq \langle \Phi_{M/T}, \Phi_{M/T}\rangle_{M/T}\\ &= \langle V(T), V(T)\rangle_P\ ,\end{aligned}$$

so

$$\begin{aligned}\langle V(S), V(S)\rangle_P &= \langle \Phi_{N/S}, \Phi_{N/S}\rangle_{N/S} = \langle W, W\rangle_{p_{1,x}/k_{1,x}}\\ &= \langle W', W'\rangle_{p_{2,x}/k_{2,x}} = \langle \Phi_{M/T}, \Phi_{M/T}\rangle_{M/T}\\ &= \langle V(T), V(T)\rangle_P\ .\end{aligned} \tag{9.6.7}$$

Now suppose that $p_{2,x} \neq (M/T)$, and choose a maximal subgroup H of M/T containing $p_{2,x}$. By Lemma 9.3.9, the restriction of $\Phi_{M/T}$ to H is equal to $m\Phi_H$, for some integer m. Since $W' = \mathrm{Defres}_{p_{2,x}/k_{2,x}}^H m\Phi_H$, it cannot be simple if $m > 1$. Thus $m = 1$, and this happens only if $p = 2$ and $|M/T| = 2$, or if M/T is dihedral or semi-dihedral, and H is cyclic or generalized quaternion.

If $|M/T| = 2$ and $p_{2,x} \neq (M/T)$, then $p_{2,x} = \mathbf{1} = k_{2,x}$, so $W' = \mathrm{Res}_{\mathbf{1}}^{M/T}\Phi_{M/T} \cong \mathbb{Q}$. It follows that $W \cong \mathbb{Q}$, and that

$$\Phi_{N/S} = \mathrm{Indinf}_{p_{1,x}/k_{1,x}}^{N/S}\mathbb{Q} \cong \mathbb{Q}(N/S)/k_{1,x}\ .$$

So $\mathbb{Q}$ is a direct summand of $\Phi_{N/S}$, and this is implies $\Phi_{N/S} = \mathbb{Q}$, so $N = N_P(S) = S$, hence $S = P$. It follows that $|M/T| \neq 2$.

If M/T is dihedral or semi-dihedral, and H is cyclic or generalized quaternion, then $\mathrm{Res}_H^{M/T}\Phi_{M/T} = \Phi_H$. Since any restriction of Φ_H to a proper subgroup of H is a multiple of 2, by Lemma 9.3.9, it follows that $W' = \mathrm{Defres}_{p_{2,x}/k_{2,x}}\Phi_H$ is also a multiple of 2 if $p_{2,x} \neq H$. Hence $p_{2,x} = H$ since W' is irreducible. Moreover any deflation of Φ_H to a proper quotient of H is equal to 0, so $k_{2,x} = \mathbf{1}$. In this case $W' = \Phi_H$. Thus

$$\langle W', W'\rangle_{p_{2,x}/k_{2,x}} = \langle \Phi_H, \Phi_H\rangle_H = \langle \Phi_{M/T}, \mathrm{Ind}_H^{M/T}\Phi_H\rangle_{M/T}\ .$$

Let Z denote the unique subgroup of order 2 of H. Then $\Phi_H = \mathbb{Q}H/\mathbf{1} - \mathbb{Q}H/Z$, so $\mathrm{Ind}_H^{M/T}\Phi_H = \mathbb{Q}(M/T)/\mathbf{1} - \mathbb{Q}(M/T)/Z$. By Lemma 9.5.4, this module is isomorphic to the direct sum of $\mathsf{d}_{M/T} = 2$ copies of $\Phi_{M/T}$. It follows that

$$\langle W', W' \rangle_{p_{2,x}/k_{2,x}} = 2\langle \Phi_{M/T}, \Phi_{M/T} \rangle_{M/T} \ ,$$

which contradicts 9.6.7. This contradiction shows that $p_{2,x} = (M/T)$.

Thus $W' = \mathrm{Def}^{M/T}_{(M/T)/k_{2,x}} \Phi_{M/T}$. Since any proper deflation of $\Phi_{M/T}$ is equal to 0, it follows that $k_{2,x} = \mathbf{1}$, and $W' = \Phi_{M/T}$.

By symmetry of S and T, and since $x' = x$, it follows that $p_{1,x} = (N/S)$ and $k_{1,x} = \mathbf{1}$. Now coming back to the expressions 9.6.2 and 9.6.3 shows that

$$(N, S) \longrightarrow ({}^xM, {}^xT) \ ,$$

so Condition 4 holds.

In particular, by Lemma 4.3.15, the groups $N_P(S)/S$ and $N_P(T)/T$ are isomorphic. □

9.6.8. Definition : *Let P be a finite p-group. If V is a rational irreducible representation of P, let S be a genetic subgroup of P such that $V \cong V(S)$. Then* the type of V *is the isomorphism class of the group $N_P(S)/S$.*

The type of V is well defined by Theorem 9.6.1. It is (the isomorphism class of) a finite p-group of normal p-rank 1.

9.6.9. Proposition : *Let P be a finite p-group, and let S and T be genetic subgroups of P fulfilling the equivalent conditions of Theorem 9.6.1. Then:*

1. *For $x \in P$, the following conditions are equivalent:*
   a. $\big(N_P(S), S\big) \longrightarrow \big(N_P({}^xT), {}^xT\big)$.
   b. $S^x \cap Z_P(T) \leq T$.
   c. ${}^xT \cap Z_P(S) \leq S$.
2. *The set $\mathcal{L}_{S,T}$ of elements $x \in P$ such that the conditions of Assertion 1 hold is a single $\big(N_P(S), N_P(T)\big)$-double coset in P. In particular, the $\big(N_P(S)/S, N_P(T)/T\big)$-biset $S\backslash N_P(S)xN_P(T)/T$ does not depend on the choice of $x \in \mathcal{L}_{S,T}$, up to isomorphism. Up to composition with an inner automorphism of $N_P(S)/S$ or $N_P(T)/T$, there is a unique group isomorphism $\lambda_T^S : N_P(T)/T \to N_P(S)/S$ such that this biset is isomorphic to* $\mathrm{Iso}(\lambda_T^S)$.
3. *In $B\big(N_P(S)/S, N_P(T)/T\big)$*

$$\mathcal{D}_S \circ \mathcal{I}_T = f_{\mathbf{1}}^{N_P(S)/S} \circ \mathrm{Iso}(\lambda_T^S) = \mathrm{Iso}(\lambda_T^S) \circ f_{\mathbf{1}}^{N_P(T)/T} \ .$$

Proof: For Assertion 1, observe that

(9.6.10)

$$\{x \in P \mid \big(N_P(S), S\big) \longrightarrow \big(N_P({}^xT), {}^xT\big)\} \subseteq \{x \in P \mid N_P(S) \cap {}^xT \leq S\} \ .$$

If Condition (a) holds, then in particular $N_P(S) \cap {}^xT = S \cap N_P({}^xT)$, so

$$S \cap Z_P({}^xT) \leq S \cap N_P({}^xT) \leq {}^xT ,$$

and Condition (b) holds. Similarly ${}^xT \cap Z_P(S) \leq S$, so Condition (c) holds.

Moreover by assumption $V(S) \cong V(T)$, and the groups $N_P(S)/S$ and $N_P(T)/T$ are isomorphic. In particular $\mathsf{d}_S = \mathsf{d}_T$. By Lemma 9.5.3,

$$m\big(V(S), \mathbb{Q}P/T\big) = \mathsf{d}_S a_T + b_T ,$$

where a_T and b_T are the integers defined by

$$\begin{aligned} a_T &= |\{x \in [N_P(S)\backslash P/T] \mid N_P(S) \cap {}^xT \leq S\}| \\ b_T &= |\{x \in [N_P(S)\backslash P/T] \mid |I_x(S,T)| = p,\ I_x(S,T) \nleq Z\big(N_P(S)/S\big)\}| , \end{aligned}$$

where $I_x(S,T) = \big(N_P(S) \cap {}^xT\big)S/S$. Since $V(S) \cong V(T)$, this multiplicity is also equal to $m\big(V(T), \mathbb{Q}P/T\big)$, which is equal to d_T by Lemma 9.5.4 if $T \neq P$, and also to $\mathsf{d}_T = 1$ if $T = P$, since $V(T) = \mathbb{Q}$ in this case. Thus

$$\mathsf{d}_S a_T + b_T = \mathsf{d}_T = \mathsf{d}_S ,$$

i.e. $\mathsf{d}_S(a_T - 1) + b_T = 0$. Now 9.6.10 shows that a_T is non zero, thus $a_T \geq 1$. It follows that $a_T = 1$ and $b_T = 0$. Moreover, as already observed,

$$a_T + b_T = |\{x \in [N_P(S)\backslash P/T] \mid Z_P(S) \cap {}^xT \leq S\}| ,$$

since any subgroup of $N_P(S)/S$ which intersect trivially the centre is either trivial of non central and of order p.

Since $b_T = 0$, the condition $Z_P(S) \cap {}^xT \leq S$ implies $N_P(S) \cap {}^xT \leq S$, and since $a_T = 1$, there is a single $\big(N_P(S), T\big)$-double coset of elements $x \in P$ such that this condition holds. Since $N_P(T)$ acts on this set on the right by multiplication, these elements form also a single $\big(N_P(S), N_P(T)\big)$-double coset.

Now Inclusion 9.6.10 becomes

$$\{x \in P \mid \big(N_P(S), S)\big) \text{---} \big(N_P({}^xT), {}^xT\big)\} \subseteq \{x \in P \mid Z_P(S) \cap {}^xT \leq S\} ,$$

and both sides form a single $\big(N_P(S), N_P(T)\big)$-double coset, so this inclusion is an equality. This shows that Condition (c) implies Condition (a), and by symmetry of the rôles of S and T, Condition (b) also implies Condition (a). This completes the proof of Assertion 1.

Assertion 2 follows easily from Assertion 1 and Lemma 4.3.15.

The proof of Assertion 3 is similar to the proof of Proposition 6.4.4: the composition $\mathcal{D}_S \circ \mathcal{I}_T$ is equal to

$$\begin{aligned}\mathcal{D}_S \circ \mathcal{I}_T &= f_1^{N_P(S)/S}(S\backslash P/T)f_1^{N_P(T)/T} \\ &= \sum_{x\in[N_P(S)\backslash P/N_P(T)]} f_1^{N_P(S)/S}\big(S\backslash N_P(S)xN_P(T)/T\big)f_1^{N_P(T)/T} .\end{aligned}$$

Now the group $M = (S^x \cap Z_P(T))T/T$ is a central subgroup of the group $N = N_P(T)/T$, and since

$$SxT \cdot (S^x \cap Z_P(T)) = Sx(S^x \cap Z_P(T))T = S(S \cap Z_P({}^xT))xT = SxT ,$$

it acts trivially on $\big(S\backslash N_P(S)xN_P(T)/T\big)$, so

$$\big(S\backslash N_P(S)xN_P(T)/T\big) = \big(S\backslash N_P(S)xN_P(T)/T\big)\mathrm{Inf}_{N/M}^N\mathrm{Def}_{N/M}^N .$$

By Proposition 6.2.6, it follows that

$$\big(S\backslash N_P(S)xN_P(T)/T\big)f_1^{N_P(T)/T} = 0$$

if $M \neq \mathbf{1}$, i.e. if $S^x \cap Z_P(T) \nleq T$. By Assertion 1, there is a single $\big(N_P(S), N_P(T)\big)$-double coset of elements $x \in P$ such that $S^x \cap Z_P(T) \leq T$. It follows that

$$\mathcal{D}_S \circ \mathcal{I}_T = f_1^{N_P(S)/S}\big(S\backslash N_P(S)xN_P(T)/T\big)f_1^{N_P(T)/T} ,$$

where $x \in P$ is such that $S^x \cap Z_P(T) \leq T$, i.e. equivalently

$$\big(N_P(S), S\big) \text{ --- } \big(N_P({}^xT), {}^xT\big) .$$

By Lemma 4.3.15, the $\big(N_P(S)/S, N_P(T)/T\big)$-biset $\big(S\backslash N_P(S)xN_P(T)/T\big)$ is isomorphic to $\mathrm{Iso}(\lambda_T^S)$. Assertion 3 follows, since $\mathrm{Iso}(\lambda_T^S)f_1^G = f_1^H\mathrm{Iso}(\lambda_T^S)$ for any group isomorphism $f : G \to H$. □

9.6.11. Definition : *By Theorem 9.6.1, the relation* $\,\widehat{=}_P\,$ *is an equivalence relation on the set of genetic subgroups of P. A* genetic basis *of P is by definition a set of representatives of these equivalence classes of genetic subgroups of P.*

It follows that if $\mathcal{G}$ is a genetic basis of P, then the $\mathbb{Q}P$-modules $V(S)$, for $S \in \mathcal{G}$, form a set of representatives of $\mathrm{Irr}_{\mathbb{Q}}(P)$.

9.6.12. Proposition : *Let P be a finite p-group, and $\mathcal{G}$ be a genetic basis of P. Then the map*

$$\mathfrak{I}_{\mathcal{G}} = \mathop{\oplus}_{S\in\mathcal{G}} \mathrm{Indinf}_{N_P(S)/S}^P : \mathop{\oplus}_{S\in\mathcal{G}} \partial R_{\mathbb{Q}}\big(N_P(S)/S\big) \to R_{\mathbb{Q}}(P)$$

is an isomorphism. The inverse isomorphism is the map

$$\mathfrak{D}_{\mathcal{G}} = \bigoplus_{S\in\mathcal{G}} \mathcal{D}_S : R_{\mathbb{Q}}(P) \to \bigoplus_{S\in\mathcal{G}} \partial R_{\mathbb{Q}}\big(N_P(S)/S\big) ,$$

where $\mathcal{D}_S = f_{\mathbf{1}}^{N_P(S)/S} \circ \mathrm{Defres}^P_{N_P(S)/S}$, *for* $S \in \mathcal{G}$.

9.6.13. Remark : The map $\mathfrak{I}_{\mathcal{G}}$ is actually an isometry from the (orthogonal) direct sum of the modules $\partial R_{\mathbb{Q}}\big(N_P(S)/S\big)$ (each of which is free of rank 1 over $\mathbb{Z}$, with basis $\Phi_{N_P(S)/S}$), to $R_Q(P)$.

9.7. Genetic Bases and Subgroups

The definition of genetic subgroups of a finite p-group originates in the reformulation of Roquette's Theorem 9.4.1, which reduces in some sense the study of rational representations of p-groups to the case of p-groups of normal p-rank 1. So if P is a finite p-group, which is not of normal p-rank 1, there should be a relation between the genetic subgroups of P, which correspond to rational irreducible representations of P, and genetic subgroups of some groups of order smaller than $|P|$. This section is devoted to a precise statement of such a result.

The following is a more precise version of Lemma 5.2 of [15], and the proof is entirely different:

9.7.1. Proposition : *Let P be a finite p-group, and E be a normal subgroup of P, which is elementary abelian of rank 2, and not contained in $Z(P)$. Set $Z = E \cap Z(P)$, and $H = C_P(E)$.*

1. *If S is a genetic subgroup of H, such that $S \not\geq Z$, then S is a genetic subgroup of P. Moreover, if $x, y \in P$, then $S^x \frown_H S^y$ if and only if $xy^{-1} \in H$. Conversely, if T is a genetic subgroup of P such that $T \not\geq Z$ and $T \leq H$, then $N_P(T) \leq H$, and T is a genetic subgroup of H.*
2. *If T is a genetic subgroup of P, such that $T \not\geq Z$ and $T \not\leq H$, then set $S = (T \cap H)F$, where F is some subgroup of order p of E, with $F \neq Z$. Then S is a genetic subgroup of H and of P, such that $S \not\geq Z$, and $S \frown_P T$.*

Proof: Since E is non central in P, the group $H = C_P(E)$ has index p in P, and the group Z has order p. If $L \leq P$ and $L \not\leq H$, then the commutator group $[L, E]$ has order p. It is normalized, hence centralized by L. Thus $[L, E] = Z$, since $L \not\leq H = C_P(E)$.

Let S be a genetic subgroup of H, such that $S \not\geq Z$, i.e. $S \cap Z = \mathbf{1}$. Then $N_H(S)/S$ is a group of normal p-rank 1, and SZ/S is its unique central subgroup of order p. If $E \cap S = \mathbf{1}$, then ES/S is a central subgroup of

$N_H(S)/S$ which is elementary abelian of rank 2. It follows that the group $F = S \cap E$ has order p, and $E = FZ$.

Let $x \in P$, such that $S^x \cap SZ \leq S$. Then in particular $F^x \cap FZ \leq S$. Thus $F^x \cap FZ = F^x \cap E = F^x \leq S \cap E = F$, so $x \in C_P(F) = C_P(E) = H$. This shows that $N_P(S) \leq H$, so $N_P(S)/S = N_H(S)/S$ has normal p-rank 1. In particular $Z_P(S) = Z_H(S)$. Now if $x \in P$ is such that $S^x \cap Z_P(S) \leq S$, then $S^x \cap SZ \leq S$, so $x \in H$, and $x \in N_H(S)$ since $S^x \cap Z_H(S) \leq S$ and S is a genetic subgroup of H. This shows that S is a genetic subgroup of P.

Finally, if $x, y \in P$, then S^x and S^y are genetic subgroups of H. Moreover $S^x \frown_H S^y$ if and only if $S^{xy^{-1}} \frown_H S$. Thus $S^{xy^{-1}h} \cap SZ \leq S$, for some $h \in H$, and $xy^{-1}h \in H$ by the above argument. Thus $xy^{-1} \in H$. Conversely, if $xy^{-1} \in H$, then $S^{xy^{-1}} \frown_H S$, thus $S^x \frown_H S^y$.

Conversely, let T be a genetic subgroup of P, with $Z \nleq T \leq H$. Then $T \cap E \neq \mathbf{1}$, for otherwise $ET/T \cong E$ is a normal subgroup of $N_P(T)/T$, which is elementary abelian of rank 2. And $T \cap E \neq E$, since $Z \nleq T$. So the group $F = T \cap E$ has order p, and it is normalized, hence centralized by $N_P(T)$. It follows that $N_P(T) \leq C_P(F) = C_P(E)$, since $F \neq Z$ and $E = FZ$. This shows that $N_H(T)/T = N_P(T)/T$, so it has normal p-rank 1. Similarly $Z_H(T) = Z_P(T)$, and if $y \in H$ is such that $T^y \cap Z_H(T) \leq T$, then $T^y = T$. So T is a genetic subgroup of H, and this completes the proof of Assertion 1.

For Assertion 2, let F be a subgroup of order p of E, different from Z. If T is a genetic subgroup of P with $T \ngeq Z$ and $T \nleq H$, set $S = (T \cap H)F$. If $T \cap E \neq \mathbf{1}$, then $T \cap E$ has order p, since $Z \nleq T$, and it is normalized, hence centralized, by T. Thus T centralizes $Z(T \cap E) = E$. It follows that $|T \cap E| = \mathbf{1}$. Then $|S| = |T \cap H||F| = |T|$, since moreover $|T : T \cap H| = p$.

Now $S \cap N_P(T) = (T \cap H)F \cap N_P(T) = (T \cap H)(F \cap N_P(T))$. If $F \leq N_P(T)$, then $FZ = E \leq N_P(T)$, and the group $ET/T \cong E$ is a normal subgroup of $N_P(T)/T$, which is elementary abelian of rank 2. Since $N_P(T)/T$ has normal p-rank 1, it follows that $F \nleq N_P(T)$, so $F \cap N_P(T) = \mathbf{1}$, and $S \cap N_P(T) = T \cap H \leq T$. It follows that the integer

$$a_S = |\{x \in S \backslash P / N_P(T) \mid S^x \cap N_P(T) \leq T\}|$$

is at least equal to 1, so by Lemma 9.5.3, the multiplicity of the simple $\mathbb{Q}P$-module $V(T)$ as a summand of $\mathbb{Q}P/S$ is at least equal to $\mathsf{d}_T\, a_T$.

On the other hand, if $x \in P$, then

$$Z \leq (SZ)^x \cap Z_P(T) = S^x Z \cap Z_P(T) \nleq T\ ,$$

so $V(T)$ is not a direct summand of $\mathbb{Q}P/SZ$, by Lemma 9.5.3 again. Let W denote the kernel of the projection map $\mathbb{Q}P/S \to \mathbb{Q}P/SZ$. Then the multiplicity of $V(T)$ as a summand of W is at least equal to $\mathsf{d}_T\, a_S$. But

$$\mathsf{d}_T\, a_S \dim_{\mathbb{Q}} V(T) = \mathsf{d}_T\, a_S \frac{|P:T|(p-1)}{p\,\mathsf{d}_T} = a_S|P:T|(1-\frac{1}{p}) = a_S \dim_{\mathbb{Q}} W \; .$$

It follows that $a_S = 1$, and that W is isomorphic to the direct sum of d_T copies of $V(T)$.

Hence if $x \in P$ and $S^x \cap N_P(T) \leq T$, then $x \in S \cdot N_P(T)$. In particular $N_P(S) \leq S \cdot N_P(T)$, thus

$$N_P(S) = S\big(N_P(S) \cap N_P(T)\big) \; . \tag{9.7.2}$$

Moreover $S \cap E = F$, and $N_P(S) \leq N_P(F) = C_P(F) = C_P(E) = H$. The above remarks show that $S \cap N_P(T) = T \cap H$, thus

$$N_P(S) \cap T = N_P(S) \cap H \cap T = N_P(S) \cap S \cap N_P(T) = S \cap N_P(T) \; . \tag{9.7.3}$$

Finally $N_P(T) \cap H$ has index p in $N_P(T)$, and $T \nleq N_P(T) \cap H$. Thus $N_P(T) = T \cdot \big(N_P(T) \cap H\big)$. Moreover

$$N_P(T) \cap H \leq N_P\big((T \cap H)F\big) = N_P(S) \; ,$$

so $N_P(T) \cap H = N_P(T) \cap N_P(S)$, and

$$N_P(T) = T \cdot \big(N_P(T) \cap N_P(S)\big) \; . \tag{9.7.4}$$

It follows from 9.7.2, 9.7.3, and 9.7.4 that $\big(N_P(S), S\big) — \big(N_P(T), T\big)$, so the group $N = N_P(S)/S$ is isomorphic to $N_P(T)/T$, hence it has normal p-rank 1. In particular, by Lemma 9.5.4

$$\mathbb{Q}N/\mathbf{1} - \mathbb{Q}N/\overline{Z} = \mathsf{d}\Phi_N \; , \tag{9.7.5}$$

where $\overline{Z}$ is the unique subgroup of order p in the centre of N, i.e. $\overline{Z} = ZS/S$, and d is equal to 1 if N is cyclic or generalized quaternion, or equal to 2 if N is dihedral or semi-dihedral. In other words $\mathsf{d} = \mathsf{d}_T$. Applying $\mathrm{Indinf}_{N_P(S)/S}^P$ to Equation 9.7.5 gives

$$\mathbb{Q}P/S - \mathbb{Q}P/SZ = \mathsf{d}_T \mathrm{Indinf}_{N_P(S)/S}^P \Phi_{N_P(S)/S} \; .$$

By the above remarks, this is also equal to $\mathsf{d}_T \mathrm{Indinf}_{N_P(T)/T}^P \Phi_{N_P(T)/T}$, thus

$$\mathrm{Indinf}_{N_P(S)/S}^P \Phi_{N_P(S)/S} \cong \mathrm{Indinf}_{N_P(T)/T}^P \Phi_{N_P(T)/T} = V(T) \; .$$

Hence $N_P(S)/S$ has normal p-rank 1, the module $V = \mathrm{Indinf}_{N_P(S)/S}^P \Phi_{N_P(S)/S}$ is a simple $\mathbb{Q}P$-module, and

$$\begin{aligned}\langle V, V\rangle_P &= \langle V(T), V(T)\rangle_T\\ &= \langle \Phi_{N_P(T)/T}, \Phi_{N_P(T)/T}\rangle_{N_P(T)/T}\\ &= \langle \Phi_{N_P(S)/S}, \Phi_{N_P(S)/S}\rangle_{N_P(S)/S}\ ,\end{aligned}$$

so S is a genetic subgroup of P, such that $S \frown_P T$, as was to be shown. □

9.7.6. Corollary : *In the situation of Proposition 9.7.1, there exists a genetic basis $\mathcal{G}$ of P and a decomposition*

$$\mathcal{G} = \mathcal{G}_1 \sqcup \mathcal{G}_2$$

in disjoint union such that

1. *if $S \in \mathcal{G}_2$, then $S \geq Z$, and*
2. *the set $\{{}^xS \mid S \in \mathcal{G}_1,\ x \in [P/C_P(E)]\}$ is a set of representatives of genetic subgroups T of $C_P(E)$ such that $T \not\geq Z$, for the relation $\frown_{C_P(E)}$, where $[P/C_P(E)]$ is any chosen set of representatives of $C_P(E)$-cosets in P.*

Proof: Let $\mathcal{G}_0$ be any genetic basis of P. Set

$$\mathcal{G}_2 = \{S \in \mathcal{G}_0 \mid S \geq Z\}\ .$$

Keeping the notation of the proposition, define moreover $\mathcal{G}_1 = \mathcal{G}_1' \sqcup \mathcal{G}_1''$, where

$$\begin{aligned}\mathcal{G}_1' &= \{S \in \mathcal{G}_0 \mid S \not\geq Z \text{ and } S \leq C_P(E)\}\\ \mathcal{G}_1'' &= \{\big(S \cap C_P(E)\big)F \mid S \in \mathcal{G}_0,\ S \not\geq Z,\ S \not\leq C_P(E)\}\ .\end{aligned}$$

Then Proposition 9.7.1 shows that $\mathcal{G} = \mathcal{G}_1 \sqcup \mathcal{G}_2$ is a genetic basis of P. In other words, one can suppose that the elements of a genetic basis of P which do not contain Z are contained in $C_P(E)$.

Now if T is a genetic subgroup of $C_P(E)$, which does not contain Z, then T is a genetic subgroup of P. So there exists $S \in \mathcal{G}$ and $y \in P$ such that $\big(N_P(T), T\big) — \big({}^yN_P(S), {}^yS\big)$. This implies in particular that

$$S \cap Z = S \cap Z \cap N_P(T^y) = Z \cap N_P(S) \cap T^y = Z \cap T^y = (Z \cap T)^y = \mathbf{1}\ ,$$

so $S \in \mathcal{G}_1$, and $N_P(S) \leq H = C_P(E)$. Now $y = ux$, for some $u \in C_P(E)$ and some $x \in [P/C_P(E)]$, since $C_P(E) \trianglelefteq P$. It follows that $T \frown_{C_P(E)} {}^xS$.

Conversely, if $S \in \mathcal{G}_1$, then S is a genetic subgroup of P, contained in H, and not containing Z. Proposition 9.7.1 shows that S is a genetic subgroup of H, which does not contain Z. Moreover, if $u, v \in P$ and ${}^uS \frown_H {}^vS$, then $u^{-1}v \in H$, i.e. $v \in uH$. This completes the proof. □

Chapter 10
p-Biset Functors

Definition and Notation : *Let $\mathcal{C}_p$ denote the full subcategory of $\mathcal{C}$ whose objects are finite p-groups. If R is a commutative ring,* a p-biset functor *over R is an object of $\mathcal{F}_{\mathcal{C}_p,R}$, i.e. an R-linear functor from $R\mathcal{C}_p$ to R-*Mod.

The category of p-biset functors $\mathcal{F}_{\mathcal{C}_p,\mathbb{Z}}$ will simply be denoted by $\mathcal{F}_p$.

10.1. Rational p-Biset Functors

The following result is a straightforward consequence of Corollary 6.4.5, using Lemma 9.5.2, and Theorem 9.6.1:

10.1.1. Theorem : *Let P be a finite p-group and $\mathcal{G}$ be a genetic basis of P. Then for any p-biset functor F over R, the map*

$$\mathfrak{I}_{\mathcal{G}} = \mathop{\oplus}_{S\in\mathcal{G}} \mathrm{Indinf}_{N_P(S)/S}^P : \mathop{\oplus}_{S\in\mathcal{G}} \partial F\big(N_P(S)/S\big) \to F(P)$$

is split injective. A left inverse is the map

$$\mathfrak{D}_{\mathcal{G}} = \mathop{\oplus}_{S\in\mathcal{G}} f_{\mathbf{1}}^{N_P(S)/S} \circ \mathrm{Defres}_{N_P(S)/S}^P : F(P) \to \mathop{\oplus}_{S\in\mathcal{G}} \partial F\big(N_P(S)/S\big) .$$

When $F = R_{\mathbb{Q}}$ and $R = \mathbb{Z}$, Proposition 9.6.12 shows that the map $\mathfrak{I}_{\mathcal{G}}$ is an isomorphism, for any finite p-group P and any genetic basis $\mathcal{G}$ of P. This suggests the following lemma:

10.1.2. Lemma : *Let F be a p-biset functor over R, and P be a p-group. Let $\mathcal{G}$ and $\mathcal{G}'$ be genetic bases of P. If $\mathfrak{I}_{\mathcal{G}}$ is an isomorphism, then $\mathfrak{I}_{\mathcal{G}'}$ is also an isomorphism.*

S. Bouc, *Biset Functors for Finite Groups*, Lecture Notes in Mathematics 1990, DOI 10.1007/978-3-642-11297-3_10, © Springer-Verlag Berlin Heidelberg 2010

Proof: Proving that $\mathfrak{I}_{\mathcal{G}'}$ is an isomorphism is equivalent to proving that $\mathfrak{D}_{\mathcal{G}'}$ is an isomorphism. If $\mathfrak{I}_{\mathcal{G}}$ is an isomorphism, this amounts to proving that $\mathfrak{D}_{\mathcal{G}'} \circ \mathfrak{I}_{\mathcal{G}}$ is an isomorphism. But if $T \in \mathcal{G}'$ and $S \in \mathcal{G}$, and if $T \not\frown_P S$, then $\mathcal{D}_T \circ \mathcal{I}_S = 0$, by Theorems 6.4.4 and 9.6.1. Now for each $S \in \mathcal{G}$, there exists a unique $S' \in \mathcal{G}'$ such that $S' \frown_P S$. The map $\mathfrak{D}_{\mathcal{G}'} \circ \mathfrak{I}_{\mathcal{G}}$ is equal to the direct sum of the maps

$$\mathcal{D}_{S'} \circ \mathcal{I}_S : \partial F\big(N_P(S)/S\big) \to \partial F\big(N_P(S')/S'\big) ,$$

for $S \in \mathcal{G}$. By Proposition 9.6.9, and with the same notation, the following equality holds in $RB\big(N_P(S')/S', N_P(S)/S\big)$:

$$\mathcal{D}_{S'} \circ \mathcal{I}_S = \mathrm{Iso}(\lambda_S^{S'}) \circ f_{\mathbf{1}}^{N_P(S)/S} .$$

Since $f_{\mathbf{1}}^{N_P(S)/S}$ acts trivially on $\partial F\big(N_P(S)/S\big)$, the action of $\mathcal{D}_{S'} \circ \mathcal{I}_S$ on $\partial F\big(N_P(S)/S\big)$ is equal to $\mathrm{Iso}(\lambda_S^{S'}) : \partial F\big(N_P(S)/S\big) \to \partial F\big(N_P(S')/S'\big)$. This is an isomorphism, and it follows that $\mathfrak{D}_{\mathcal{G}'} \circ \mathfrak{I}_{\mathcal{G}}$ is a direct sum of isomorphisms, hence it is an isomorphism. □

10.1.3. Definition : *A p-biset functor F over R is called* rational *if for any finite p-group P, there exists a genetic basis $\mathcal{G}$ of P such that the map $\mathfrak{I}_{\mathcal{G}}$ is an isomorphism of R-modules.*

10.1.4. Remark : By Lemma 10.1.2, if F is a rational p-biset functor, and P is a finite p-group, then $\mathfrak{I}_{\mathcal{G}}$ is an isomorphism, for *any* genetic basis $\mathcal{G}$ of P. Equivalently, a p-biset functor F is rational if and only if for any finite p-group P and any genetic basis $\mathcal{G}$ of P, the map

$$\mathfrak{I}_{\mathcal{G}} \circ \mathfrak{D}_{\mathcal{G}} = \sum_{S \in \mathcal{G}} \mathcal{I}_S \circ \mathcal{D}_S$$

is equal to the identity map of $F(P)$. Equivalently, the sum, for $S \in \mathcal{G}$, of the idempotents

$$\gamma_S^P = \mathrm{Indinf}_{N_P(S)/S}^P f_{\mathbf{1}}^{N_P(S)/S} \mathrm{Defres}_{N_P(S)/S}^P$$

of $RB(P,P)$ (see Corollary 6.4.5), acts as the identity on $F(P)$.

10.1.5. Theorem : *Let R be a commutative ring.*

1. *Let $0 \to F' \xrightarrow{a} F \xrightarrow{b} F'' \to 0$ be a short exact sequence of p-biset functors over R. Then F is rational if and only if F' and F'' are rational.*
2. *Let M be an R-module, and F be a p-biset functor over R. If F is rational, then $\mathrm{Hom}_R(F, M)$ is a rational p-biset functor over R. Conversely,*

if M is an injective cogenerator for the category of R-modules, and if $\mathrm{Hom}_R(F, M)$ *is a rational, then F is rational.*

3. *Arbitrary direct sums and direct products of rational biset functors are rational biset functors.*

Proof: Let P be a finite p-group, and $\mathcal{G}$ be a genetic basis of P. For each $S \in \mathcal{G}$, the map $a_P : F'(P) \to F(P)$ is such that $a_P \circ \mathcal{I}_S = \mathcal{I}_S \circ a_P$, since a is a morphism of biset functors from F' to F. Moreover the map $a_{N_P(S)/S} : F'\big(N_P(S)/S\big) \to F\big(N_P(S)/S\big)$ is such that

$$a_{N_P(S)/S} \circ f_{\mathbf{1}}^{N_P(S)/S} = f_{\mathbf{1}}^{N_P(S)/S} \circ a_{N_P(S)/S} \;.$$

Denoting by $\partial a_{N_P(S)/S} : \partial F'\big(N_P(S)/S\big) \to \partial F\big(N_P(S)/S\big)$ the restriction of $a_{N_P(S)/S}$ to $\partial F'\big(N_P(S)/S\big)$, there is a commutative diagram

$$\begin{array}{ccc} \partial F'\big(N_P(S)/S\big) & \xrightarrow{\mathcal{I}_S} & F'(P) \\ \downarrow{\scriptstyle \partial a_{N_P(S)/S}} & & \downarrow{\scriptstyle a_P} \\ \partial F\big(N_P(S)/S\big) & \xrightarrow{\mathcal{I}_S} & F(P) \;. \end{array}$$

A similar remark holds for the morphism $b : F \to F''$, and this yields a short exact sequence

$$0 \to \partial F'\big(N_P(S)/S\big) \xrightarrow{\partial a_{N_P(S)/S}} \partial F\big(N_P(S)/S\big) \xrightarrow{\partial b_{N_P(S)/S}} \partial F''\big(N_P(S)/S\big) \to 0 \;,$$

which is a retraction of the exact sequence

$$0 \to F'\big(N_P(S)/S\big) \xrightarrow{a_{N_P(S)/S}} F\big(N_P(S)/S\big) \xrightarrow{b_{N_P(S)/S}} F''\big(N_P(S)/S\big) \to 0 \,.$$

These exact sequences fit into a commutative diagram

$$\begin{array}{ccccccccc} & & 0 & & 0 & & 0 & & \\ & & \downarrow & & \downarrow & & \downarrow & & \\ 0 & \to & \bigoplus_{S\in\mathcal{G}} \partial F'\big(N_P(S)/S\big) & \xrightarrow{s_P} & \bigoplus_{S\in\mathcal{G}} \partial F\big(N_P(S)/S\big) & \xrightarrow{t_P} & \bigoplus_{S\in\mathcal{G}} \partial F''\big(N_P(S)/S\big) & \to & 0 \\ & & \downarrow{\scriptstyle \mathfrak{I}'_{\mathcal{G}}} & & \downarrow{\scriptstyle \mathfrak{I}_{\mathcal{G}}} & & \downarrow{\scriptstyle \mathfrak{I}''_{\mathcal{G}}} & & \\ 0 & \to & F'(P) & \xrightarrow{a_P} & F(P) & \xrightarrow{b_P} & F''(P) & \to & 0 \end{array}$$

where $s_P = \bigoplus_{S\in\mathcal{G}} \partial a_{N_P(S)/S}$ and $t_P = \bigoplus_{S\in\mathcal{G}} \partial b_{N_P(S)/S}$, and where $\mathfrak{I}'_{\mathcal{G}}$ and $\mathfrak{I}''_{\mathcal{G}}$ are the analogues of the map $\mathcal{I}_{\mathcal{G}}$ for the functors F' and F'', respectively.

The rows of this diagram are exact, as well as the columns, since the maps $\mathfrak{I}'_{\mathcal{G}}$, $\mathfrak{I}_{\mathcal{G}}$, and $\mathfrak{I}''_{\mathcal{G}}$ are split injective, by Theorem 10.1.1. By the Snake's Lemma, there is an exact sequence of R-modules

$$0 \to \operatorname{Coker} \mathfrak{I}'_{\mathcal{G}} \to \operatorname{Coker} \mathfrak{I}_{\mathcal{G}} \to \operatorname{Coker} \mathfrak{I}''_{\mathcal{G}} \to 0 \;.$$

Now the map $\mathfrak{I}_{\mathcal{G}}$ is an isomorphism if and only if $\operatorname{Coker} \mathfrak{I}_{\mathcal{G}} = \{0\}$, and this happens if and only if both $\operatorname{Coker} \mathfrak{I}'_{\mathcal{G}}$ and $\operatorname{Coker} \mathfrak{I}''_{\mathcal{G}}$ are equal to $\{0\}$. Assertion 1 follows.

For Assertion 2, observe first that $\mathcal{C}_p^{\diamond} = \mathcal{C}_p$, so the dual of a p-biset functor is again a p-biset functor. Let F^* denote the dual functor $\operatorname{Hom}_R(F, M)$, and $\mathfrak{I}^*_{\mathcal{G}}$ the analogue of the map $\mathfrak{I}_{\mathcal{G}}$ for F^*. Then clearly $\mathfrak{I}^*_{\mathcal{G}} = {}^t\mathfrak{D}_{\mathcal{G}}$. If F is rational, the map $\mathfrak{D}_{\mathcal{G}}$ is an isomorphism, so the transposed map is also an isomorphism, and F^* is rational. Conversely, if F^* is rational, then $F^{**} = \operatorname{Hom}_R(F^*, M)$ is rational by Assertion 1. If M is an injective cogenerator of R-Mod, then the natural morphism $F \to F^{**}$ is injective, so F is rational, by Assertion 1 again.

Assertion 3 follows from the fact that the functor $F \mapsto \partial F(P)$, from $\mathcal{F}_{\mathcal{C}_p,R}$ to R-Mod, commutes with arbitrary direct sums and direct products, for any finite p-group P. □

10.2. Subfunctors of Rational p-Biset Functors

Let F be a rational p-biset functor. If H is a subfunctor of F, then H is also rational, by Theorem 10.1.5. It follows that if P is a finite p-group, and $\mathcal{G}$ is a genetic basis of P, the map

$$\mathfrak{I}_{\mathcal{G}} : \mathop{\oplus}_{S \in \mathcal{G}} \operatorname{Indinf}_{N_P(S)/S}^P : \mathop{\oplus}_{S \in \mathcal{G}} \partial H\big(N_P(S)/S\big) \to H(P)$$

is an isomorphism of R-modules. Thus $H(P)$ is determined by the R-modules $\partial H\big(N_P(S)/S\big)$, for $S \in \mathcal{G}$. In particular, the value $H(P)$ is known from the data of the modules $\partial H(N)$, for each finite group $N \in \mathcal{N}$, where $\mathcal{N}$ is the class of p-groups of normal p-rank 1 introduced in Definition 9.3.1.

Conversely, suppose that for each $N \in \mathcal{N}$, an R-submodule $L(N)$ of $\partial F(N)$ is given. One can then try to define, for each finite p-group P, an R-submodule $H(P)$ of $F(P)$ by the formula

$$H(P) = \sum_{S \in \mathcal{G}} \operatorname{Indinf}_{N_P(S)/S}^P L\big(N_P(S)/S\big) \;,$$

where $\mathcal{G}$ is some chosen genetic basis of P. This raises the question of knowing when this definition yields a well defined subfunctor of F. The answer is as follows:

10.2.1. Proposition : *Let F be a rational p-biset functor.*

1. *If H is a subfunctor of F, set $L(N) = \partial H(N)$, for $N \in \mathcal{N}$. Then $L(N)$ is an R-submodule of $\partial F(N)$. Moreover $F(\alpha)\big(L(N)\big) \subseteq L(M)$, for any $M, N \in \mathcal{N}$ and any $\alpha \in f_{\mathbf{1}}^M RB(M,N) f_{\mathbf{1}}^N$.*
2. *Conversely, suppose that for each $N \in \mathcal{N}$, an R-submodule $L(N)$ of $\partial F(N)$ is given, with the property that $F(\alpha)\big(L(N)\big) \subseteq L(M)$, for each $N, M \in \mathcal{N}$ and each $\alpha \in f_{\mathbf{1}}^M RB(M,N) f_{\mathbf{1}}^N$. If P is a finite p-group, define an R-submodule $H(P)$ of $F(P)$ by*

$$H(P) = \sum_{S \in \mathcal{G}} \mathrm{Indinf}_{N_P(S)/S}^P L\big(N_P(S)/S\big) \;,$$

where $\mathcal{G}$ is some chosen genetic basis of P.
Then $H(P)$ does not depend on the choice of $\mathcal{G}$, and the correspondence $P \mapsto H(P)$ is a biset subfunctor H of F.
Moreover $\partial H(N) = L(N)$, for any $N \in \mathcal{N}$, and for each finite p-group P, the module $H(P)$ is equal to the set of elements $u \in F(P)$ such that $f_{\mathbf{1}}^{N_P(S)/S} \mathrm{Defres}_{N_P(S)/S}^P u \in L\big(N_P(S)/S\big)$, for each genetic subgroup S of P.

Proof: Assertion 1 is obvious. For Assertion 2, let P be a finite p-group, and let $\mathcal{G}$ and $\mathcal{G}'$ be two genetic bases of P. Set

$$H_{\mathcal{G}}(P) = \sum_{S \in \mathcal{G}} \mathrm{Indinf}_{N_P(S)/S}^P L\big(N_P(S)/S\big) \;.$$

Denote by $S \mapsto S'$ the unique bijection $\mathcal{G} \to \mathcal{G}'$ such that $S \mathrel{\widehat{=}_P} S'$, for $S \in \mathcal{G}$.
Any element of $H_{\mathcal{G}}(P)$ is a sum of elements of $F(P)$ of the form

$$v = \mathrm{Indinf}_{N_P(S)/S}^P u = \mathcal{I}_S(u) \;,$$

where $S \in \mathcal{G}$, and $u \in L\big(N_P(S)/S\big)$. Since F is rational and $\mathcal{G}'$ is a genetic basis of P, by Remark 10.1.4

$$v = \sum_{T \in \mathcal{G}'} \mathcal{I}_T \circ \mathcal{D}_T \circ \mathcal{I}_S(u) \;.$$

Now $\mathcal{D}_T \circ \mathcal{I}_S = 0$ if $T \neq S'$, by Proposition 6.4.4 and Lemma 9.5.2. Moreover $\mathcal{D}_{S'} \circ \mathcal{I}_S = f_{\mathbf{1}}^{N_P(S')/S'} \lambda_S^{S'} = \lambda_S^{S'} f_{\mathbf{1}}^{N_P(S)/S}$ in $B\big(N_P(S')/S', N_P(S)/S\big)$, by Proposition 9.6.9, where $\lambda_S^{S'} : N_P(S)/S \to N_P(S')/S'$ is a group isomorphism obtained by choosing $x \in P$ such that $\big(N_P(S'), S'\big) — \big(N_P({}^xS), {}^xS\big)$. The isomorphism $\lambda_S^{S'}$ is well defined up to an inner automorphism of $N_P(S)/S$, or $N_P(S')/S'$.

It follows that $\mathcal{D}_{S'} \circ \mathcal{I}_S = f_{\mathbf{1}}^{N_P(S')/S'} \lambda_S^{S'} f_{\mathbf{1}}^{N_P(S)/S}$ is an element of

$$f_{\mathbf{1}}^{N_P(S')/S'} RB\big(N_P(S')/S', N_P(S)/S\big) f_{\mathbf{1}}^{N_P(S)/S} ,$$

so the assumptions of Assertion 2 imply that $\mathcal{D}_{S'} \circ \mathcal{I}_S(u) \in L\big(N_P(S')/S'\big)$. Now

$$v = \sum_{S \in \mathcal{G}} \mathcal{I}_{S'} \circ \mathcal{D}_{S'} \circ \mathcal{I}_S(u) \in \sum_{S' \in \mathcal{G}'} \mathrm{Indinf}_{N_P(S')/S'}^P L\big(N_P(S')/S'\big) = H_{\mathcal{G}'}(P) .$$

It follows that $H_{\mathcal{G}}(P) \subseteq H_{\mathcal{G}'}(P)$, hence $H_{\mathcal{G}}(P) = H_{\mathcal{G}'}(P)$ by symmetry. This proves that $H(P)$ does not depend on the choice of the genetic basis $\mathcal{G}$.

Now let Q be a finite p-group, and $\varphi \in RB(Q, P)$. If $v \in H(P)$, then

$$v = \sum_{S \in \mathcal{G}} \mathcal{I}_S(u_S) ,$$

where $\mathcal{G}$ is some genetic basis of P, and $u_S \in L\big(N_P(S)/S\big)$, for $S \in \mathcal{G}$. Then $\varphi(v) \in F(Q)$, and choosing a genetic basis $\mathcal{H}$ of Q gives

$$\varphi(v) = \sum_{\substack{S \in \mathcal{G} \\ T \in \mathcal{H}}} \mathcal{I}_T \circ \mathcal{D}_T \circ \varphi \circ \mathcal{I}_S(u_S) .$$

Now for $S \in \mathcal{G}$ and $T \in \mathcal{H}$,

$$\mathcal{D}_T \circ \varphi \circ \mathcal{I}_S = f_{\mathbf{1}}^{N_Q(T)/T} \circ \mathrm{Defres}_{N_Q(T)/T}^Q \circ \varphi \circ \mathrm{Indinf}_{N_P(S)/S}^P \circ f_{\mathbf{1}}^{N_P(S)/S}$$

is an element of $f_{\mathbf{1}}^{N_P(T)/T} RB\big(N_P(T)/T, N_P(S)/S\big) f_{\mathbf{1}}^{N_P(S)/S}$.

Since $u_S \in L\big(N_P(S)/S\big)$, it follows that

$$\mathcal{D}_T \circ \varphi \circ \mathcal{I}_S(u_S) \in L\big(N_Q(T)/T\big) ,$$

thus $\varphi(v) \in \sum_{T \in \mathcal{H}} \mathcal{I}_T\Big(L\big(N_Q(T)/T\big)\Big) = H(Q)$.

This proves that $\varphi\big(H(P)\big) \subseteq H(Q)$, so H is a subfunctor of F.

Now obviously $\partial H(\mathbf{1}) = H(\mathbf{1}) = L(\mathbf{1})$. If $N \in \mathcal{N} - \{\mathbf{1}\}$, let Z denote the unique subgroup of order p in $Z(N)$. The trivial subgroup is the only genetic subgroup of N not containing Z. It follows that

$$\begin{aligned}\partial H(N) &= f_{\mathbf{1}}^N L(N) + f_{\mathbf{1}}^N \mathrm{Inf}_{N/Z}^N \mathrm{Def}_{N/Z}^N H(N) \\ &= f_{\mathbf{1}}^N L(N) = L(N) ,\end{aligned}$$

by Lemma 6.3.2, and since $L(N) \subseteq \partial F(N)$.

Finally if P is a finite p-group and S is a genetic subgroup of P, there is a genetic basis $\mathcal{G}$ of P containing S, and $H(P) = \sum_{T\in\mathcal{G}} \mathcal{I}_T\Big(L\big(N_P(T)/T\big)\Big)$. Thus

$$f_{\mathbf{1}}^{N_P(S)/S}\mathrm{Defres}^P_{N_P(S)/S}H(P) = \mathcal{D}_S\mathcal{I}_S\Big(L\big(N_P(S)/S\big)\Big) = L\big(N_P(S)/S\big)\ ,$$

since $\mathcal{D}_S\mathcal{I}_T = 0$ for $T \neq S$, so $f_{\mathbf{1}}^{N_P(S)/S}\mathrm{Defres}^P_{N_P(S)/S}u \in L\big(N_P(S)/S\big)$ for any genetic subgroup S of P. Conversely, if $u \in F(P)$ and if the element

$$u_S = f_{\mathbf{1}}^{N_P(S)/S}\mathrm{Defres}^P_{N_P(S)/S}u$$

lies in $L\big(N_P(S)\big)$ for any genetic subgroup S of P, then since

$$u = \sum_{S\in\mathcal{G}} \mathrm{Indinf}^P_{N_P(S)/S}u_S\ ,$$

where $\mathcal{G}$ is some chosen genetic basis of P, it follows that $u \in H(P)$. This completes the proof of Proposition 10.2.1. □

It follows from Proposition 10.2.1 that any subfunctor H of a rational functor F is determined by the data of an R-submodule $L(N)$ of $\partial F(N)$, for each $N \in \mathcal{N}$, such that for any $N, M \in \mathcal{N}$

(10.2.2)
$$f_{\mathbf{1}}^M B(M,N) f_{\mathbf{1}}^N\big(L(N)\big) \subseteq L(M)\ .$$

The structure of groups of normal p-rank 1 allows for a reduction of this condition to an explicit list of cases, particularly simple if p is odd:

10.2.3. Proposition : *Let F be a rational p-biset functor. For each $N \in \mathcal{N}$, let $L(N)$ be a submodule of $\partial F(N)$. The following conditions are equivalent:*

1. *For all $M, N \in \mathcal{N}$, the following condition $C(M,N)$ is satisfied:*
$$f_{\mathbf{1}}^M RB(M,N) f_{\mathbf{1}}^N\big(L(N)\big) \subseteq L(M)\ .$$
2. *The following three conditions are satisfied:*
   a. *If $N \in \mathcal{N}$ and $\varphi : N \to M$ is a group isomorphism, then $\mathrm{Iso}(\varphi)\big(L(N)\big) \subseteq L(M)$.*
   b. *If $M \in \mathcal{N}$ and N is a maximal subgroup of M such that $N \in \mathcal{N}$, then*
$$f_{\mathbf{1}}^N \mathrm{Res}^M_N L(M) \subseteq L(N) \qquad f_{\mathbf{1}}^M \mathrm{Ind}^M_N L(N) \subseteq L(M)\ .$$

c. If $p=2$, *if* N *is dihedral or semi-dihedral of order 16, if* $Z=Z(N)$, *and* Q *is a non-central subgroup of order 2 of* N, *then*

$$f_1^Q \mathrm{Res}_Q^N L(N) \subseteq L(Q) , \qquad f_1^N \mathrm{Ind}_Q^N L(Q) \subseteq L(N) ,$$

$$\text{and} \quad \begin{cases} f_1^{QZ/Q} \mathrm{Defres}_{QZ/Q}^N L(N) \subseteq L(QZ/Q) , \\ f_1^{QZ/Q} \mathrm{Indinf}_{QZ/Q}^N L(QZ/Q) \subseteq L(N) , \end{cases}$$

or equivalently

$$\mathrm{Defres}_{QZ/Q}^N L(N) \subseteq L(QZ/Q) , \qquad \mathrm{Indinf}_{QZ/Q}^N L(QZ/Q) \subseteq L(N) .$$

Proof: Since $L(N) \subseteq \partial F(N)$, the idempotent f_1^N acts as the identity map on $L(N)$, so Condition 1 is equivalent to

$$f_1^M RB(M,N)\big(L(N)\big) \subseteq L(M) . \tag{10.2.4}$$

In the situation of Condition (a), i.e. if $\varphi : N \to M$ is a group isomorphism, this implies that

$$f_1^M \mathrm{Iso}(\varphi)\big(L(N)\big) \subseteq L(M) .$$

Moreover $f_1^M \mathrm{Iso}(\varphi) = \mathrm{Iso}(\varphi) f_1^N$, so $f_1^M \mathrm{Iso}(\varphi)\big(L(N)\big) = \mathrm{Iso}(\varphi)\big(L(N)\big)$, and Condition 1 implies Condition (a).

Similarly, Condition 1 implies Condition (b), and the first four inclusions of Condition (c). For the last ones, observe that in $RB(QZ/Q, N)$

$$\begin{aligned} \mathrm{Defres}_{QZ/Q}^N f_1^N &= (Q\backslash N) \times_N (N - N/Z) \\ &= (QZ/Q - QZ/QZ) \times_{QZ/Q} (Q\backslash N) \\ &= f_1^{QZ/Q} \mathrm{Defres}_{QZ/Q}^N . \end{aligned}$$

Taking opposite bisets gives

$$\mathrm{Indinf}_{QZ/Q}^N f_1^{QZ/Q} = f_1^N \mathrm{Indinf}_{QZ/Q}^N . \tag{10.2.5}$$

It follows that

$$\begin{aligned} f_1^{QZ/Q} \mathrm{Defres}_{QZ/Q}^N L(N) &= \mathrm{Defres}_{QZ/Q}^N L(N) \\ f_1^N \mathrm{Indinf}_{QZ/Q}^N L(QZ/Q) &= \mathrm{Indinf}_{QZ/Q}^N L(QZ/Q) . \end{aligned}$$

This shows the equivalence claimed in Assertion (c), and that Condition 1 implies Condition (c).

Conversely, suppose that Conditions (a), (b) and (c) hold. To prove that Condition 1 holds, one has to prove that

(10.2.6) $$f_1^M \alpha f_1^N\big(L(N)\big) \subseteq L(M) ,$$

for any $N, M \in \mathcal{N}$ and any $\alpha \in RB(M,N)$. Suppose that there exists a p-group Q such that α factors through Q, i.e. that $\alpha = \theta\psi$, for some $\theta \in RB(M,Q)$ and $\psi \in RB(Q,N)$. Let $\mathcal{G}$ be a genetic basis of Q. Since F is a rational functor, the sum of the idempotents γ_X^Q, for $X \in \mathcal{G}$, acts as the identity map on $F(Q)$. It follows that

$$\begin{aligned} f_1^M \theta\psi f_1^N\big(L(N)\big) &= \sum_{X\in\mathcal{G}} f_1^M \theta\, \gamma_X^Q \psi f_1^N\big(L(N)\big) \\ &= \sum_{X\in\mathcal{G}} f_1^M \theta \,\mathrm{Indinf}_{\overline{N}_Q(X)}^Q f_1^{\overline{N}_Q(X)} \mathrm{Defres}_{\overline{N}_Q(X)}^Q \psi f_1^N\big(L(N)\big) , \end{aligned}$$

where $\overline{N}_Q(X) = N_Q(X)/X$. This shows that Condition 10.2.6 follows from Conditions $C\big(M, N_Q(X)/X\big)$ and $C\big(N_Q(X)/X, N\big)$, for $X \in \mathcal{G}$. Note moreover that $|N_Q(X)/X| \leq |Q|$, and that equality occurs if and only if $X = \mathbf{1}$, hence $Q \in \mathcal{N}$.

Proving Condition 10.2.6 for any $\alpha \in RB(M,N)$ amounts to proving it when α is a transitive (M,N)-biset, of the form $(M \times N)/P$, where P is some subgroup of $M \times N$. The factorization property 4.1.5 shows that such a biset factors through the group $Q = q(P)$. By the above remarks and an easy induction argument on $|M||N|$, it is enough to consider the case where $Q \cong M$ or $Q \cong N$. This gives two cases:

- If $Q \cong M$, then $(M \times N)/P = \mathrm{Iso}(\varphi)\mathrm{Defres}_{T/S}^N$, for a suitable section (T,S) of N, and a group isomorphism $\varphi : T/S \to Q$. If Condition (a) holds, it is enough to consider the case where $\varphi = \mathrm{Id}$, in other words to show that

(10.2.7) $$f_1^{T/S} \mathrm{Defres}_{T/S}^N L(N) \subseteq L(T/S)$$

for any section (T,S) of N such that $T/S \in \mathcal{N}$. If N is trivial, there is nothing to prove. Otherwise, let Z denote the unique subgroup of order p of $Z(N)$. If $S \geq Z$, then

$$\mathrm{Defres}_{T/S}^N = \mathrm{Defres}_{T/S}^N \mathrm{Inf}_{N/Z}^N \mathrm{Def}_{N/Z}^N ,$$

and since $\mathrm{Def}_{N/Z}^N L(N) = 0$ for $L(N) \leq \partial F(N)$, by Lemma 6.3.2, it follows that $\mathrm{Defres}_{T/S}^N L(N) = \{0\}$, and Condition 10.2.7 holds trivially in this case.

And if $S \not\geq Z$, then $S = \mathbf{1}$, or $p = 2$, the group N is dihedral or semidihedral, and S is a non central subgroup of order 2 of N, by Proposition 9.3.5.

If $S = \mathbf{1}$, then Condition 10.2.7 becomes

(10.2.8) $$f_1^T \mathrm{Res}_T^N L(N) \subseteq L(T) .$$

If $T = N$, there is nothing to prove. Otherwise T is contained in some maximal subgroup H of N. If $H \in \mathcal{N}$, then Condition 10.2.8 follows from Condition (b) by an easy induction argument. Otherwise N is dihedral or semi-dihedral of order 16, and T is a proper subgroup of N, such that $T \in \mathcal{N}$, and T is not contained in any cyclic or quaternion maximal subgroup of N. Then T is a non central subgroup of order 2 of N, and Condition 10.2.8 is the first inclusion of Condition (c) in this case.

If $S \neq \mathbf{1}$, then $p = 2$, the group N is dihedral or semi-dihedral, and S is a non central subgroup of order 2 of N. In this case the normalizer of S in N is equal to SZ, so $T \leq SZ$. Thus $T = SZ$ or $T = S$.

If $T = SZ$, then Condition 10.2.7 is the third inclusion of Condition (c). And if $T = S$, then Condition 10.2.7 becomes

$$\mathrm{Defres}^N_{S/S} L(N) \subseteq L(M) ,$$

since $f_{\mathbf{1}}^{S/S}$ is the identity element $\mathrm{Id}_{\mathbf{1}}$. But $\mathrm{Defres}^N_{SZ/S} L(N) \subseteq L(SZ/S)$ by Condition (c), and

$$\begin{aligned}
\mathrm{Defres}^N_{S/S} L(N) &= \mathrm{Res}^{SZ/S}_{S/S} \mathrm{Defres}^N_{SZ/S} L(N) \\
&\subseteq \mathrm{Res}^{SZ/S}_{S/S} L(SZ/S) \\
&\subseteq L(S/S) ,
\end{aligned}$$

where the last inclusion follows from the first inclusion of Condition (b), applied to the group $M = SZ/S \cong \mathbb{Z}/2\mathbb{Z}$ and its unique (trivial) maximal subgroup.

- Similarly, if $Q \cong N$, then $(M \times N)/P = \mathrm{Indinf}^M_{T/S} \mathrm{Iso}(\varphi)$, for a suitable section (T, S) of M and a group isomorphism $\varphi : N \to T/S$. If Condition (a) holds, then one can suppose $\varphi = \mathrm{Id}$, and it is enough to show that

$$\textbf{(10.2.9)} \qquad f_{\mathbf{1}}^M \mathrm{Indinf}^M_{T/S} L(T/S) \subseteq L(M) ,$$

for any section (T, S) of M such that $T/S \in \mathcal{N}$. Again, if M is trivial, there is nothing to prove. Otherwise let Z denote the unique subgroup of order p in $Z(M)$. If $S \geq Z$, then

$$\mathrm{Indinf}^M_{T/S} = \mathrm{Inf}^M_{T/Z} \mathrm{Def}^M_{T/Z} \mathrm{Indinf}^M_{T/S} ,$$

so $f_{\mathbf{1}}^M \mathrm{Indinf}^M_{T/S} = 0$ in this case, by Lemma 6.3.2, and Condition 10.2.9 holds trivially. Otherwise $S = \mathbf{1}$, or $p = 2$, and S is a non central subgroup of order 2 in the group M, which is dihedral or semi-dihedral, by Proposition 9.3.5.

If $S = \mathbf{1}$, then Condition 10.2.9 becomes

$$\textbf{(10.2.10)} \qquad f_{\mathbf{1}}^M \mathrm{Ind}^M_T L(T) \subseteq L(M) .$$

If $T = M$, there is nothing to prove. Otherwise T is contained in some maximal subgroup H of M. If $H \in \mathcal{N}$, then Condition 10.2.10 follows from Condition (b) by an easy induction argument. Otherwise M is dihedral or semi-dihedral of order 16, and T is a proper subgroup of M, such that $T \in \mathcal{N}$, and T is not contained in any cyclic or quaternion maximal subgroup of M. Then T is a non central subgroup of order 2 of M, and Condition 10.2.10 is the second inclusion of Condition (c) in this case.

If $S \neq \mathbf{1}$, then $p = 2$, the group M is dihedral or semi-dihedral, and S is a non central subgroup of order 2 of M. In this case the normalizer of S in M is equal to SZ, so $T \leq SZ$. Thus $T = SZ$ or $T = S$.

If $T = SZ$, then Condition 10.2.9 is the fourth inclusion of Condition (c). And if $T = S$, then Condition 10.2.9 becomes

$$f_{\mathbf{1}}^M \mathrm{Indinf}_{S/S}^M L(S/S) \subseteq L(M) ,$$

But $f_{\mathbf{1}}^M \mathrm{Indinf}_{SZ/S}^M L(SZ/S) \subseteq L(M)$ by Condition (c), and

$$\begin{aligned} f_{\mathbf{1}}^M \mathrm{Indinf}_{S/S}^M L(S/S) &= f_{\mathbf{1}}^M \mathrm{Indinf}_{SZ/S}^M \mathrm{Ind}_{S/S}^{SZ/S} L(S/S) \\ &= \mathrm{Indinf}_{SZ/S}^M f_{\mathbf{1}}^{SZ/Z} \mathrm{Ind}_{S/S}^{SZ/S} L(S/S) , \end{aligned}$$

by Equality 10.2.5. Now by Condition (b) applied to the group $N = SZ/Z \cong \mathbb{Z}/2\mathbb{Z}$ and its unique (trivial) maximal subgroup

$$f_{\mathbf{1}}^{SZ/Z} \mathrm{Ind}_{S/S}^{SZ/S} L(S/S) \subseteq L(SZ/S) = f_{\mathbf{1}}^{SZ/Z} L(SZ/S) ,$$

and by Equality 10.2.5 and the fourth inclusion of Condition (c)

$$\begin{aligned} \mathrm{Indinf}_{SZ/S}^M f_{\mathbf{1}}^{SZ/Z} L(SZ/S) &= f_{\mathbf{1}}^M \mathrm{Indinf}_{SZ/S}^M L(SZ/S) \\ &\subseteq L(M) , \end{aligned}$$

as was to be shown. This completes the proof of Proposition 10.2.3. □

10.3. The Subfunctors of $R_{\mathbb{Q}}$

In this section, the ground ring R is equal to $\mathbb{Z}$. The functor $R_{\mathbb{Q}}$ is viewed as a p-biset functor, i.e. an object of the category $\mathcal{F}_p$.

10.3.1. The functor $R_{\mathbb{Q}}$ is a rational p-biset functor, almost by definition. Thus by Propositions 10.2.1 and 10.2.3, any subfunctor H of $R_{\mathbb{Q}}$ is characterized by the data of a subgroup $L(N)$ of $\partial R_{\mathbb{Q}}(N)$, for each $N \in \mathcal{N}$, subject to Condition (a), (b), and (c) of Proposition 10.2.3.

But for $N \in \mathcal{N}$, the group $\partial R_{\mathbb{Q}}(N)$ is isomorphic to $\mathbb{Z}$, generated by Φ_N, by Proposition 9.3.5. Hence there is a well defined integer $a_N \geq 0$ such that $L(N) = <a_N\Phi_N>$.

So the subfunctor H is characterized by these integers a_N, for $N \in \mathcal{N}$. Recall that if P is a p-group, then

$$H(P) = \underset{S\in\mathcal{G}}{\oplus} <a_{N_P(S)/S}V(S)> ,$$

where $\mathcal{G}$ is a genetic basis of P. Equivalently $H(P)$ is the set of elements $u \in R_{\mathbb{Q}}(P)$ such that for any rational irreducible representation V of P, the integer $m(V, u)$ is a multiple of $a_{t(V)}$, where $t(V)$ is the type of V, i.e. $t(V) \cong N_P(S)/S$ if $S \in \mathcal{G}$ and $V \cong V(S)$.

10.3.2. Condition (a) of Proposition 10.2.3 is equivalent to $a_N = a_M$ when N and M are isomorphic, so H is determined by a function $N \mapsto a_N$ from $[\mathcal{N}]$ to $\mathbb{N}$, where $[\mathcal{N}]$ is a set of representatives of isomorphism classes of groups in $\mathcal{N}$. Moreover, the subfunctor H corresponding to the function $N \mapsto a_N$ is contained in the subfunctor H' corresponding to the function $N \mapsto a'_N$ if and only if $a'_N \mid a_N$, for any $N \in \mathcal{N}$.

10.3.3. For Condition (b) of Proposition 10.2.3, consider a group $M \in \mathcal{N}$ and a maximal subgroup N of M such that $N \in \mathcal{N}$. By Lemma 9.3.9, there exists an integer r_N^M such that

$$\mathrm{Res}_N^M \Phi_M = r_N^M \Phi_N .$$

The integer r_N^M is equal to $p-1$ if $N = \mathbf{1}$, to p if M is cyclic or generalized quaternion, or if M is dihedral or semi-dihedral and N is dihedral, and to 1 if M is dihedral or semi-dihedral and N is cyclic or generalized quaternion. Similarly:

10.3.4. Lemma : *Let $M \in \mathcal{N}$, and N be a maximal subgroup of M, such that $N \in \mathcal{N}$. Then:*

1. $\mathrm{Ind}_N^M \Phi_N = \mathbb{Q} + \Phi_M$ *if* $|M| = p$.
2. $\mathrm{Ind}_N^M \Phi_N = \Phi_M$ *if M is cyclic or generalized quaternion.*
3. $\mathrm{Ind}_N^M \Phi_N = \Phi_M$ *if M is dihedral or semi-dihedral, and N is dihedral.*
4. $\mathrm{Ind}_N^M \Phi_N = 2\Phi_M$ *if M is dihedral or semi-dihedral, and N is cyclic or generalized quaternion.*

Proof: If $N = \mathbf{1}$, then $\Phi_N = \mathbb{Q}$, so

$$\mathrm{Ind}_N^M \Phi_N = \mathbb{Q}M/\mathbf{1} = \mathbb{Q} + (\mathbb{Q}M/\mathbf{1} - \mathbb{Q}M/M) ,$$

This proves Assertion 1, since $\Phi_M = \mathbb{Q}M/\mathbf{1} - \mathbb{Q}M/M$ if M has order p.

Otherwise, let Z denote the unique central subgroup of order p of N. Then $\mathbb{Q}N/\mathbf{1} - \mathbb{Q}N/Z = \mathsf{d}_N\Phi_N$, by Lemma 9.5.4. Inducing this equality from N to M gives

$$\mathbb{Q}M/\mathbf{1} - \mathbb{Q}M/Z = \mathsf{d}_N\mathrm{Ind}_N^M\Phi_N \ ,$$

But Z is also the unique central subgroup of order p of M. Thus

$$\mathbb{Q}M/\mathbf{1} - \mathbb{Q}M/Z = \mathsf{d}_M\Phi_M \ ,$$

hence $\mathrm{Ind}_N^M\Phi_N = \frac{\mathsf{d}_M}{\mathsf{d}_N}\Phi_M$. Moreover $\mathsf{d}_M = \mathsf{d}_N$ for Assertions 2 and 3, and $\mathsf{d}_M = 2\mathsf{d}_N$ for Assertion 4. □

This lemma shows that there is an integer i_N^M such that

$$f_{\mathbf{1}}^M\mathrm{Ind}_N^M\Phi_N = i_N^M\Phi_M \ .$$

The integer i_N^M is equal to 1, except if M is dihedral or semi-dihedral, and N is cyclic or generalized quaternion, where $i_N^M = 2$.

With this notation, Condition (b) of Proposition 10.2.3 is equivalent to

$$r_N^M a_M\Phi_N \in <a_N\Phi_N> \ , \qquad\qquad i_N^M a_N\Phi_M \in <a_M\Phi_M> \ .$$

In other words

(10.3.5)

$$\forall M, N \in \mathcal{N},\ N < M,\ |M:N| = p \ \Rightarrow\ a_N \mid r_N^M a_M \text{ and } a_M \mid i_N^M a_N \ .$$

10.3.6. It remains to express Condition (c) of Proposition 10.2.3. In this case $p = 2$, and N is dihedral or semi-dihedral of order 16. Let $Z = Z(N)$, and Q be a non central subgroup of order 2 in N. Then $\mathbb{Q}N/\mathbf{1} - \mathbb{Q}N/Z = 2\Phi_N$, by Lemma 9.5.4, so

$$2\mathrm{Res}_{QZ}^N\Phi_N = \mathrm{Res}_{QZ}^N(\mathbb{Q}N/\mathbf{1} - \mathbb{Q}N/Z) = 4\mathbb{Q}QZ/\mathbf{1} - 4\mathbb{Q}QZ/Z \ .$$

It follows that

$$\mathrm{Defres}_{QZ/Q}^N\Phi_N = 2\mathbb{Q}QZ/Q - 2\mathbb{Q}QZ/QZ = 2\Phi_{QZ/Q} \ .$$

The third inclusion in Condition (c) is then equivalent to $a_{QZ/Q} \mid 2a_N$, when $N \cong D_{16}$ or $N \cong SD_{16}$. In other words

(10.3.7)
$$a_{C_2} \mid 2a_{D_{16}} \ , \qquad\qquad a_{C_2} \mid 2a_{SD_{16}} \ .$$

On the other hand $\Phi_{QZ/Q} = \mathbb{Q}QZ/Q - \mathbb{Q}QZ/QZ$, so

$$\mathrm{Indinf}_{QZ/Q}^N\Phi_{QZ/Q} = \mathbb{Q}N/Q - \mathbb{Q}N/QZ = \Phi_N \ ,$$

since Q is a basic subgroup of N and $Q \not\geq Z$, by Proposition 9.3.5. The fourth inclusion in Condition (c) is then equivalent to

$$(\mathbf{10.3.8}) \qquad a_{D_{16}} \mid a_{C_2} \;, \qquad a_{SD_{16}} \mid a_{C_2} \;.$$

Finally

$$2\mathrm{Res}_Q^N \Phi_N = \mathrm{Res}_Q^N(\mathbb{Q}N/\mathbf{1} - \mathbb{Q}N/Z) = 8\mathbb{Q}Q/\mathbf{1} - 4\mathbb{Q}Q/\mathbf{1} = 4\mathbb{Q}Q/\mathbf{1} \;,$$

thus $f_{\mathbf{1}}^Q\mathrm{Res}_Q^N\Phi_N = 2\mathbb{Q}Q/\mathbf{1} - 2\mathbb{Q}Q/Q = 2\Phi_Q$. The first inclusion in Condition (c) is then also equivalent to Condition 10.3.7. Similarly

$$\begin{aligned} f_{\mathbf{1}}^N\mathrm{Ind}_Q^N\Phi_Q &= f_{\mathbf{1}}^N(\mathbb{Q}N/\mathbf{1} - \mathbb{Q}N/Q) \\ &= (\mathbb{Q}N/\mathbf{1} - \mathbb{Q}N/Z) - (\mathbb{Q}N/Q - \mathbb{Q}N/QZ) \\ &= 2\Phi_N - \Phi_N = \Phi_N \;, \end{aligned}$$

so the second inclusion in Condition (c) is equivalent to Condition 10.3.8.

10.3.9. Conditions 10.3.5, 10.3.7 and 10.3.8 can be formulated in a more concise way, as follows: first Conditions 10.3.7 and 10.3.8 are equivalent to

$$(\mathbf{10.3.10}) \qquad a_{D_{16}} \mid a_{C_2} \mid 2a_{D_{16}} \;, \qquad a_{SD_{16}} \mid a_{C_2} \mid 2a_{SD_{16}} \;.$$

Next Condition 10.3.5 for $N = \mathbf{1}$, hence $M \cong C_p$, means that

$$a_{C_p} \mid a_{\mathbf{1}} \mid (p-1)a_{C_p} \;.$$

Finally, if N is non trivial in Condition 10.3.5, observe that either $r_N^M = 1$ and $i_N^M = p$, or $r_N^M = p$ and $i_N^M = 1$. In the first case $a_N \mid a_M \mid pa_N$, and in the second case $a_M \mid a_N \mid pa_M$. This looks very much like Condition 10.3.10, and motivates the following definition:

10.3.11. Definition : *Let $\mathfrak{N}$ denote the oriented graph whose set of vertices is $[\mathcal{N}] - \{\mathbf{1}\}$, and in which there is an arrow $N \twoheadrightarrow M$ if*

- *either N is isomorphic to a subgroup of index p of M, and $i_N^M = 1$.*
- *or M is isomorphic to a subgroup of index p of N, and $r_M^N = 1$.*
- *or $N = C_2$ and $M = D_{16}$ or $M = SD_{16}$.*

With this definition, Conditions 10.3.5, 10.3.7 and 10.3.8 can be summarized as follows:

$$(\mathbf{10.3.12}) \qquad \begin{cases} a_{C_p} \mid a_{\mathbf{1}} \mid (p-1)a_{C_p} & \\ a_M \mid a_N \mid pa_M & \text{if } N \twoheadrightarrow M \text{ in } \mathfrak{N} \;. \end{cases}$$

The graph $\mathfrak{N}$ is very simple if p is odd: in this case indeed, its vertices are labelled by the cyclic groups C_{p^n}, for $n \geq 1$. And there is an arrow $C_{p^n} \longrightarrow C_{p^m}$ if and only if $m = n+1$. Thus $\mathfrak{N}$ is the graph

In the case $p = 2$, the graph $\mathfrak{N}$ is much more complicated:

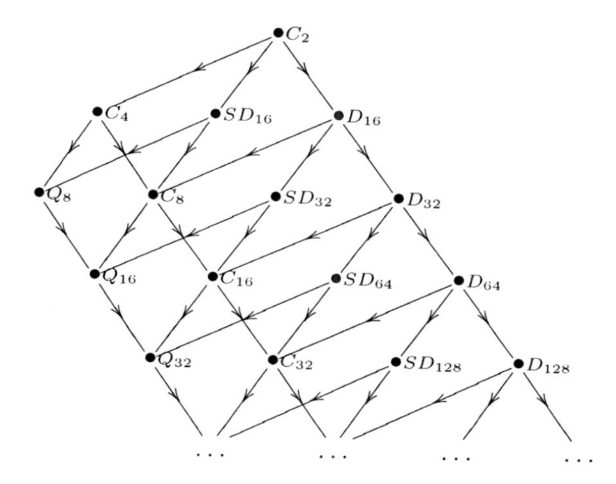

10.3.13. Notation : *Let $\mathfrak{M}$ denote the set of functions $f : \mathfrak{N} \to \mathbb{N}$ such that for any arrow $N \longrightarrow M$ in $\mathfrak{N}$*

$$f(M) \leq f(N) \leq f(M) + 1 .$$

The set $\mathfrak{M}$ is a poset for the relation

$$\forall f, f' \in \mathfrak{M}, \;\; f \leq f' \Leftrightarrow \forall N \in \mathfrak{N}, \;\; f(N) \leq f'(N) .$$

10.3.14. Theorem : *The lattice of non zero subfunctors of $R_{\mathbb{Q}}$ is isomorphic to the opposite poset of the poset $\mathfrak{T}$ of triples (d, m, f), where d and m are positive integers, with $d \mid p-1$ and $p \nmid m$, and $f \in \mathfrak{M}$, ordered by the relation*

$$(d, m, f) \leq (d', m', f') \Leftrightarrow m \mid m', \; dm \mid d'm', \; f \leq f' .$$

for $(d, m, f), (d', m', f') \in \mathfrak{T}$.

Proof: Indeed, let H be a subfunctor of $R_{\mathbb{Q}}$, and for $N \in \mathcal{N}$, let $a_N \in \mathbb{N}$ be the integer defined by $\partial H(N) = <a_N \Phi_N>$.

Suppose first that $a_N = 0$, for some $N \in \mathcal{N}$. The first condition in 10.3.12 implies that $a_{\mathbf{1}} = 0$ if and only if $a_{C_p} = 0$. The second condition in 10.3.12 implies that if $N, M \in \mathfrak{N}$, and if $N \twoheadrightarrow M$, then $a_N = 0$ if and only if $a_M = 0$. Since the graph $\mathfrak{N}$ is connected, it follows that $a_N = 0$ for any $N \in \mathcal{N}$, so $H = 0$ in this case.

Thus if $H \neq \{0\}$, then $a_N > 0$ for $N \in \mathcal{N}$. The first condition in 10.3.12 is equivalent to the existence of a positive integer d, dividing $p-1$, such that $a_{\mathbf{1}} = d a_{C_p}$.

If $l \in \mathbb{N}$, denote by l_p the largest power of p dividing l, and by $l_{p'}$ the p'-part of l, i.e. the quotient l/l_p. The second condition in 10.3.12 implies that if $N, M \in \mathfrak{N}$ with $N \twoheadrightarrow M$, then $(a_N)_{p'} = (a_M)_{p'}$. Since the graph $\mathfrak{N}$ is connected, this integer does not depend on $N \in \mathfrak{N}$. Denote it by m. It follows that there is a function $f : \mathfrak{N} \to \mathbb{N}$ such that

$$\forall N \in \mathfrak{N},\ a_N = m\, p^{f(N)}\ .$$

The second condition in 10.3.12 is now equivalent to $f \in \mathfrak{M}$. Thus, a non zero subfunctor H of $R_{\mathbb{Q}}$ yields a triple $(d, m, f) = \tau(H)$, where $d \mid (p-1)$, $p \nmid m$, and $f \in \mathfrak{M}$, defined by

$$\begin{cases} d = a_{\mathbf{1}}/a_{C_p} & \\ m = (a_N)_{p'} & \text{for } N \in \mathfrak{N} \\ p^{f(N)} = (a_N)_p & \text{for } N \in \mathfrak{N}\ . \end{cases}$$

Conversely, if such a triple $(d, m, f) \in \mathfrak{T}$ is given, then setting

$$\textbf{(10.3.15)} \qquad \begin{cases} a_N = m\, p^{f(N)} & \text{for } N \in \mathfrak{N} \\ a_{\mathbf{1}} = dm\, p^{f(C_p)} & , \end{cases}$$

yields a function $N \mapsto a_N$ on $[\mathcal{N}]$, which fulfills the conditions 10.3.12, hence yields a non zero subfunctor $H(d, m, f)$ of $R_{\mathbb{Q}}$. These two correspondences are clearly inverse to each other. Moreover, if H, H' are subfunctors of $R_{\mathbb{Q}}$, corresponding to functions $N \mapsto a_N$ and $N \mapsto a'_N$ on $\mathcal{N}$ respectively, then $H \subseteq H'$ if and only if a_N is a multiple of a'_N, for any $N \in \mathcal{N}$. It follows that for any $(d, m, f), (d', m', f') \in \mathfrak{T}$

$$H(d, m, f) \subseteq H(d', m', f') \ \Leftrightarrow\ d' \mid d,\ m' \mid m,\ f' \leq f\ ,$$

and this completes the proof of Theorem 10.3.14.

10.4. The Subfunctors of $R_{\mathbb{Q}}^*$

In this section again, the ground ring R is equal to $\mathbb{Z}$. The dual functor $R_{\mathbb{Q}}^* = \mathrm{Hom}_{\mathbb{Z}}(R_{\mathbb{Q}}, \mathbb{Z})$ is viewed as a p-biset functor, i.e. an object of the category $\mathcal{F}_p$.

10.4.1. For any finite group G, the group $R_{\mathbb{Q}}^*(G)$ is a free abelian group on the set $\{V^* \mid V \in \mathrm{Irr}_{\mathbb{Q}}(G)\}$, where V^*, for $V \in \mathrm{Irr}_{\mathbb{Q}}(G)$, is the linear form on $R_{\mathbb{Q}}(G)$ sending the class of a $\mathbb{Q}G$-module W to the multiplicity of V as a summand of W.

The functor $R_{\mathbb{Q}}^*$ is rational, by Theorem 10.1.5. Moreover if $N \in \mathcal{N}$, then $\partial R_{\mathbb{Q}}^*(N)$ is free of rank one, generated by Φ_N^*.

It follows from Propositions 10.2.1 and 10.2.3, that any subfunctor H of $R_{\mathbb{Q}}^*$ is characterized by the data of a subgroup $L(N) = <a_N\Phi_N^*>$ of $\partial R_{\mathbb{Q}}^*(N)$, where $a_N \in \mathbb{N}$, for each $N \in \mathcal{N}$, subject to Conditions (a), (b), and (c) of Proposition 10.2.3. The following lemma will be useful to expressing these conditions:

10.4.2. Lemma : *Let $M, N \in \mathcal{N}$ and $\varphi \in f_{\mathbf{1}}^N B(N,M) f_{\mathbf{1}}^M$. Then there exists a unique $r_\varphi \in \mathbb{N}$ such that $\varphi(\Phi_M) = r_\varphi \Phi_N$. Moreover $\varphi^{op} \in f_{\mathbf{1}}^M B(M,N) f_{\mathbf{1}}^N$, and $\varphi^{op}(\Phi_N^*) = r_\varphi \Phi_M^*$.*

Proof: Since $\varphi(\Phi_M) \in \partial R_{\mathbb{Q}}(N) = <\Phi_N>$, there exists a unique $r_\varphi \in \mathbb{N}$ such that $\varphi(\Phi_M) = r_\varphi \Phi_N$. Moreover if $V \in \mathrm{Irr}_{\mathbb{Q}}(M)$ and $V \not\cong \Phi_M$, then V is not faithful, and $\varphi(V) = 0$. Now there are integers c_V, for $V \in \mathrm{Irr}_{\mathbb{Q}}(M)$, such that

$$\varphi^{op}(\Phi_N^*) = \sum_{V \in \mathrm{Irr}_{\mathbb{Q}}(M)} c_V V^* \; .$$

Thus

$$\begin{aligned} c_V &= \varphi^{op}(\Phi_N^*)(V) \\ &= \Phi_N^*\big(\varphi(V)\big) \\ &= m\big(\Phi_N, \varphi(V)\big) \; . \end{aligned}$$

This is equal to 0 if $V \not\cong \Phi_M$, and to r_φ otherwise. Hence $\varphi^{op}(\Phi_N^*) = r_\phi \Phi_M^*$, as was to be shown. □

10.4.3. Condition (a) of Assertion 2 of Proposition 10.2.3 implies that if $N \in \mathcal{N}$ and $\varphi : N \to M$ is a group isomorphism, then $\mathrm{Iso}(\varphi)(a_N\Phi_N^*) \in <a_M\Phi_M^*>$. This means that $a_M \mid a_N$. Moreover $M \in \mathcal{N}$, so $a_M = a_N$ by reversing the roles of M and N. Hence a_N depends only on the isomorphism class of N. In other words the subfunctor H of $R_{\mathbb{Q}}^*$ is determined by a function $N \mapsto a_N$ from $[\mathcal{N}]$ to $\mathbb{N}$.

10.4.4. Condition (b) is the following: if $M \in \mathcal{N}$, and N is a maximal subgroup of M such that $N \in \mathcal{N}$, then

$$f_{\mathbf{1}}^N \mathrm{Res}_N^M (a_M \Phi_M^*) \in \langle a_N \Phi_N^* \rangle \;, \qquad f_{\mathbf{1}}^M \mathrm{Ind}_N^M (a_N \Phi_N^*) \in \langle a_M \Phi_M^* \rangle \;.$$

But $f_{\mathbf{1}}^N \mathrm{Res}_N^M (\Phi_M^*) = f_{\mathbf{1}}^N \mathrm{Res}_N^M f_{\mathbf{1}}^M (\Phi_M^*)$, and $(f_{\mathbf{1}}^N \mathrm{Res}_N^M f_{\mathbf{1}}^M)^{op} = f_{\mathbf{1}}^M \mathrm{Ind}_N^M f_{\mathbf{1}}^N$, it follows from Lemmas 10.3.4 and 10.4.2 that

1. $f_{\mathbf{1}}^N \mathrm{Res}_N^M \Phi_M^* = \Phi_N^*$ if $|M| = p$.
2. $f_{\mathbf{1}}^N \mathrm{Res}_N^M \Phi_M^* = \Phi_N^*$ if M is cyclic or generalized quaternion.
3. $f_{\mathbf{1}}^N \mathrm{Res}_N^M \Phi_M^* = \Phi_N^*$ if M is dihedral or semi-dihedral, and N is dihedral.
4. $f_{\mathbf{1}}^N \mathrm{Res}_N^M \Phi_M^* = 2\Phi_N^*$ if M is dihedral or semi-dihedral, and N is cyclic or generalized quaternion.

Hence $f_{\mathbf{1}}^N \mathrm{Res}_N^M (a_M \Phi_M^*) \in \langle a_N \Phi_N^* \rangle$ is equivalent to $a_N \mid a_M$ in Cases 1 to 3, and to $a_N \mid 2a_M$ in case 4.

Similarly $(f_{\mathbf{1}}^M \mathrm{Ind}_N^M f_{\mathbf{1}}^N)^{op} = f_{\mathbf{1}}^N \mathrm{Res}_N^M f_{\mathbf{1}}^M$, and it follows from Lemmas 9.3.9 and 10.4.2 that

1. $f_{\mathbf{1}}^M \mathrm{Ind}_N^M \Phi_N^* = (p-1)\Phi_M^*$ if $|M| = p$.
2. $f_{\mathbf{1}}^M \mathrm{Ind}_N^M \Phi_N^* = p\Phi_M^*$ if M is cyclic or generalized quaternion.
3. $f_{\mathbf{1}}^M \mathrm{Ind}_N^M \Phi_N^* = 2\Phi_M^*$ if M is dihedral or semi-dihedral, and N is dihedral.
4. $f_{\mathbf{1}}^M \mathrm{Ind}_N^M \Phi_N^* = \Phi_M^*$ if M is dihedral or semi-dihedral, and N is cyclic or generalized quaternion.

Hence $f_{\mathbf{1}}^M \mathrm{Ind}_N^M (a_N \Phi_N^*) \in \langle a_M \Phi_M^* \rangle$ if and only if $a_M \mid r_N^M a_N$, where $r_N^M = p-1$ in Case 1, $r_N^M = p$ in Cases 2 and 3, and $r_N^M = 1$ in Case 4. Finally, Condition (b) of Proposition 10.2.3 is equivalent to

- $a_{\mathbf{1}} \mid a_{C_p} \mid (p-1)a_{\mathbf{1}}$.
- $a_M \mid a_N \mid pa_M$ if M is dihedral or semi-dihedral, and N is cyclic or generalized quaternion.
- $a_N \mid a_M \mid pa_N$ otherwise.

10.4.5. It remains to express Condition (c): if $p = 2$, if N is dihedral or semi-dihedral of order 16, if $Z = Z(N)$, and Q is a non-central subgroup of order 2 of N, then

$$f_{\mathbf{1}}^Q \mathrm{Res}_Q^N (a_N \Phi_N^*) \in \langle a_Q \Phi_Q^* \rangle \;, \qquad f_{\mathbf{1}}^N \mathrm{Ind}_Q^N (a_Q \Phi_Q^*) \in \langle a_N \Phi_N^* \rangle \;,$$

$$\begin{cases} f_{\mathbf{1}}^{QZ/Q} \mathrm{Defres}_{QZ/Q}^N (a_N \Phi_N^*) \in \langle a_{QZ/Q} \Phi_{QZ/Q}^* \rangle \;, \\ f_{\mathbf{1}}^{QZ/Q} \mathrm{Indinf}_{QZ/Q}^N (a_{QZ/Q} \Phi_{QZ/Q}^*) \in \langle a_N \Phi_N^* \rangle \;. \end{cases}$$

It follows from Paragraph 10.3.6 that

$$f_{\mathbf{1}}^{QZ/Q} \mathrm{Defres}_{QZ/Q}^N \Phi_N = 2\Phi_{QZ/Q} \;, \qquad f_{\mathbf{1}}^N \mathrm{Indinf}_{QZ/Q}^N \Phi_{QZ/Q} = \Phi_N \;,$$

$$f_{\mathbf{1}}^Q \mathrm{Res}_Q^N \Phi_N = 2\Phi_Q \;, \qquad f_{\mathbf{1}}^N \mathrm{Ind}_Q^N \Phi_Q = \Phi_N \;.$$

Thus by Lemma 10.4.2, since $(\mathrm{Defres}_{QZ/Q}^N)^{op} = \mathrm{Indinf}_{QZ/Q}^N$,

$$f_1^N \mathrm{Indinf}_{QZ/Q}^N \Phi_{QZ/Q}^* = 2\Phi_N^* , \qquad f_1^{QZ/Q} \mathrm{Defres}_{QZ/Q}^N \Phi_N^* = \Phi_{QZ/Q}^* ,$$

$$f_1^N \mathrm{Ind}_Q^N \Phi_Q^* = 2\Phi_N^* , \qquad f_1^Q \mathrm{Res}_Q^N \Phi_N^* = \Phi_Q^* .$$

It follows that Condition (c) is equivalent to

$$a_{QZ/Q} = a_{C_2} \mid a_N \mid 2a_{C_2} , \qquad a_Q = a_{C_2} \mid a_N \mid 2a_{C_2} .$$

In other words, since N is dihedral or semi-dihedral of order 16,

$$a_{C_2} \mid a_{D_{16}} \mid 2a_{C_2} , \qquad a_{C_2} \mid a_{SD_{16}} \mid 2a_{C_2} .$$

10.4.6. Finally, the subfunctor H of $R_{\mathbb{Q}}^*$ is characterized by a function $N \mapsto a_N$ from $[\mathcal{N}]$ to $\mathbb{N}$, fulfilling the following two conditions:

$$(10.4.7) \qquad \begin{cases} a_1 \mid a_{C_p} \mid (p-1)a_1 \\ a_N \mid a_M \mid pa_N \qquad \text{if } N \rightarrowtail M \text{ in the graph } \mathfrak{N} . \end{cases}$$

10.4.8. Notation : *Let $\widehat{\mathfrak{M}}$ denote the set of functions $f : \mathfrak{N} \to \mathbb{N}$ such that for any arrow $N \rightarrowtail M$ in $\mathfrak{N}$*

$$f(N) \leq f(M) \leq f(N) + 1 .$$

The set $\widehat{\mathfrak{M}}$ is a poset for the relation

$$\forall f, f' \in \mathfrak{M}, \;\; f \leq f' \Leftrightarrow \forall N \in \mathfrak{N}, \;\; f(N) \leq f'(N) .$$

10.4.9. Theorem : *The lattice of non zero subfunctors of $R_{\mathbb{Q}}^*$ is isomorphic to the opposite poset of the poset $\widehat{\mathfrak{T}}$ of triples (d, m, f), where d and m are positive integers, with $d \mid p-1$ and $p \nmid m$, and $f \in \widehat{\mathfrak{M}}$, ordered by the relation*

$$(d, m, f) \leq (d', m', f') \Leftrightarrow m \mid m', \; dm \mid d'm', \; f \leq f' .$$

for $(d, m, f), (d', m', f') \in \widehat{\mathfrak{T}}$.

Proof: Indeed, a subfunctor H of $R_{\mathbb{Q}}^*$ is characterized by a function $N \mapsto a_N$ from $[\mathcal{N}]$ to $\mathbb{N}$, fulfilling Conditions 10.4.7. If $a_N = 0$ for some $N \in [\mathcal{N}]$, since the graph $\mathfrak{N}$ is connected, it follows that $a_N = 0$ for any $N \in [\mathcal{N}]$, and $H = \{0\}$. Hence if $H \neq \{0\}$, then $a_N > 0$ for any $N \in [\mathcal{N}]$. The first line in

Condition 10.4.7 is now equivalent to saying that the quotient $d = a_{C_p}/a_{\mathbf{1}}$ is an integer, and divides $p-1$. In particular $(a_{C_p})_p = (a_{\mathbf{1}})_p$.

Set $m = (a_{\mathbf{1}})_{p'}$. Since the graph $\mathfrak{N}$ is connected, the second line in Condition 10.4.7 is now equivalent to the existence of $f \in \widehat{\mathfrak{M}}$ such that $a_N = dmp^{f(N)}$, for any $N \in \mathfrak{N}$. So H determines a triple $(d,m,f) \in \widehat{\mathfrak{T}}$ by

$$(\mathbf{10.4.10}) \qquad \begin{cases} m = (a_{\mathbf{1}})_{p'} \\ d = a_{C_p}/a_{\mathbf{1}} \\ p^{f(N)} = a_N/dm \text{ for } N \in \mathfrak{N}\,. \end{cases}$$

Conversely, if such a triple $(d,m,f) \in \widehat{\mathfrak{T}}$ is given, define a function $N \mapsto a_N$ from $[\mathcal{N}]$ to $\mathbb{N}$ by

$$(\mathbf{10.4.11}) \qquad \begin{cases} a_{\mathbf{1}} = mp^{f(C_p)} \\ a_N = dmp^{f(N)} \text{ for } N \in \mathfrak{N}\,. \end{cases}$$

Then this function fulfills Conditions 10.4.7, and determines a subfunctor H of $R_{\mathbb{Q}}^*$. These correspondences between functions $N \mapsto a_N$ fulfilling Conditions 10.4.7 and triples (d,m,f) of $\widehat{\mathfrak{T}}$ are obviously inverse to each other, so this gives a bijection between the set of subfunctors of $R_{\mathbb{Q}}^*$ and the set $\widehat{\mathfrak{T}}$.

Now let H and H' be subfunctors of $R_{\mathbb{Q}}^*$, corresponding to functions $N \mapsto a_N$ and $N \mapsto a'_N$ respectively, and to triples (d,m,f) and (d',m',f') of $\widehat{\mathfrak{T}}$, respectively. Then $H \subseteq H'$ if and only if $a'_N \mid a_N$, for any $N \in [\mathcal{N}]$. By 10.4.10 and 10.4.11, this is equivalent to $(d',m',f') \le (d,m,f)$ in the poset $\widehat{\mathfrak{T}}$. □

10.4.12. Remark : The poset $\widehat{\mathfrak{T}}$ of Theorem 10.4.9 is very similar to the poset $\mathfrak{T}$ of Theorem 10.3.14, except that it is built from the poset $\widehat{\mathfrak{M}}$ instead of $\mathfrak{M}$, i.e. using *increasing functions* from $\mathfrak{N}$ to $\mathbb{N}$ instead of *decreasing functions*. But this is a major difference: in particular, one can show that $\mathfrak{T}$ is countable, whereas $\widehat{\mathfrak{T}}$ is not. In other words, the set of subfunctors of $R_{\mathbb{Q}}$ is countable, whereas the set of subfunctors of $R_{\mathbb{Q}}^*$ is not.

10.5. The Subfunctors of $kR_{\mathbb{Q}}$

In this section, the ring R is a field k, of characteristic $q \ge 0$. The main result is a description of the lattice of subfunctors of the p-biset functor $kR_{\mathbb{Q}}$ (see also Sect. 5.6.9).

10.5.1. Lemma : *Let k be a field of characteristic q, and let k_q denote the prime subfield of k. Then the correspondence $F \mapsto k \otimes_{k_q} F$ induces a poset isomorphism from the lattice of subfunctors of $k_qR_{\mathbb{Q}}$ and the lattice of subfunctors of $kR_{\mathbb{Q}}$.*

Proof: Let F be a subfunctor of $kR_{\mathbb{Q}}$. Then F is determined by the k-vector spaces $L(N) = \partial F(N) \subseteq \partial kR_{\mathbb{Q}}(N)$, for $N \in \mathcal{N}$. Since $\partial kR_{\mathbb{Q}}(N)$ is one dimensional, generated by the image $\overline{\Phi}_N = 1 \otimes_{\mathbb{Z}} \Phi_N$ of Φ_N in $kR_{\mathbb{Q}}(N) = k \otimes_{\mathbb{Z}} R_{\mathbb{Q}}(N)$, it follows that for each $N \in \mathcal{N}$, the vector space $L(N)$ is equal to $\{0\}$ or to $k\overline{\Phi}_N$. Let $L_q(N)$ be the subspace of $\partial k_q R_{\mathbb{Q}}(N)$ defined by $L_q(N) = \{0\}$ if $L(N) = \{0\}$, and by $L_q(N) = k_q \Phi_N$ otherwise. Then $L(N) = k \otimes L_q(N)$, for each $N \in \mathcal{N}$.

Moreover F is a subfunctor of $kR_{\mathbb{Q}}$ if and only if

$$f_{\mathbf{1}}^M kB(M,N) f_{\mathbf{1}}^N L(N) \subseteq L(M) ,$$

for any $M, N \in \mathcal{N}$. This condition is equivalent to the existence of a scalar $\nu(U) \in k$, for each finite (Q,P)-biset U, such that $f_{\mathbf{1}}^M U \overline{\Phi}_N = \nu(U) \overline{\Phi}_M$. But since $f_{\mathbf{1}}^M U \Phi_N$ is an integral multiple of Φ_M, it follows that $\nu(U) \in k_q$. Hence $f_{\mathbf{1}}^M k_q B(M,N) f_{\mathbf{1}}^N L_q(N) \subseteq L_q(M)$, so the subspaces $L_q(N)$ of $\partial k_q R_{\mathbb{Q}}(N)$, for $N \in \mathcal{N}$, determine a subfunctor F_q of $k_q R_{\mathbb{Q}}$, and F is obviously equal to $k \otimes_{k_q} F_q$. The lemma follows easily. □

10.5.2. Theorem : *Let k be a field of characteristic q.*

1. *If $q \neq p$ and $q \nmid (p-1)$, then the p-biset functor $kR_{\mathbb{Q}}$ is simple, isomorphic to $S_{\mathbf{1},k}$.*
2. *If $q \mid p-1$, then the p-biset functor $kR_{\mathbb{Q}}$ has a unique proper non zero subfunctor F, isomorphic to $S_{C_p,k}$, and the quotient $kR_{\mathbb{Q}}/F$ is isomorphic to the simple functor $S_{\mathbf{1},k}$, which is the constant functor Γ_k in this case.*
3. *If $q = p$, then the poset of subfunctors of the p-biset functor $kR_{\mathbb{Q}}$ is isomorphic to the poset of closed subsets of the graph $\mathfrak{N}$, ordered by inclusion of subsets.*

Proof: If $q = 0$, the functor $kR_{\mathbb{Q}}$ is simple, isomorphic to $S_{\mathbf{1},k}$, by Proposition 4.4.8. If q is a prime number, the lattice of subfunctors of $kR_{\mathbb{Q}}$ is isomorphic to the lattice of subfunctors of $\mathbb{F}_q R_{\mathbb{Q}}$, by Lemma 10.5.1. Moreover, the lattice of subfunctors of $\mathbb{F}_q R_{\mathbb{Q}}$ is isomorphic to the lattice of subfunctors of $R_{\mathbb{Q}}$ containing $qR_{\mathbb{Q}}$. If H is a subfunctor of $R_{\mathbb{Q}}$, corresponding to a triple $(d,m,f) \in \mathfrak{T}$, then $H \supseteq qR_{\mathbb{Q}}$ if and only if the corresponding function $N \mapsto a_N$ from $[\mathcal{N}]$ to $\mathbb{N}$ has values in $\{1,q\}$. This gives three cases:

- If $q \neq p$, and $q \nmid (p-1)$, then by 10.3.15, it follows that $f(N) = 0$ for $N \in \mathfrak{N}$, that $d = 1$, and $m \in \{1,q\}$. There are exactly two subfunctors in $\mathbb{F}_q R_{\mathbb{Q}}$, the zero subfunctor, and $\mathbb{F}_q R_{\mathbb{Q}}$ itself. In this case $\mathbb{F}_q R_{\mathbb{Q}}$ is a simple biset functor. Since $\mathbb{F}_q R_{\mathbb{Q}}(\mathbf{1}) \cong \mathbb{F}_q$, it follows that $\mathbb{F}_q R_{\mathbb{Q}}$ is isomorphic to the simple biset functor $S_{\mathbf{1},\mathbb{F}_q}$. Similarly, the functor $kR_{\mathbb{Q}}$ is simple, isomorphic to $S_{\mathbf{1},k}$.
- If $q \mid (p-1)$, then again $f = 0$, and there are three pairs (d,m) for which $d \mid (p-1)$, $p \nmid m$, and $\{dm, m\} \subseteq \{1,q\}$, namely $(1,1)$, $(1,q)$ and $(q,1)$. Moreover $(1,1,0) \leq (q,1,0) \leq (1,q,0)$ in $\mathfrak{T}$. So there is a unique proper

subfunctor H of $R_{\mathbb{Q}}$ containing $qR_{\mathbb{Q}}$ as a proper subfunctor: if P is a finite p-group, then $H(P)$ is the subgroup of $R_{\mathbb{Q}}(P)$ consisting of elements u such that $m(\mathbb{Q}, u)$ is a multiple of $a_1 = q$.

Since the dimension of every non-trivial rational irreducible representation of P is a multiple of $p-1$, by Theorem 9.2.2, this is equivalent to say that $\dim_{\mathbb{Q},P} u$ is a multiple of q, where $\dim_{\mathbb{Q},P} : R_{\mathbb{Q}}(P) \to \mathbb{Z}$ is the linear form sending a rational irreducible representation of P to its dimension over $\mathbb{Q}$.

It follows that $\mathbb{F}_q R_{\mathbb{Q}}$ has a unique proper non zero subfunctor $\overline{H}_q$. Let $\overline{H} = k \otimes_{\mathbb{F}_q} \overline{H}_q$ denote the unique proper non zero subfunctor of $kR_{\mathbb{Q}}$. For each finite p-group P, there is an exact sequence

$$0 \longrightarrow \overline{H}(P) \longrightarrow kR_{\mathbb{Q}}(P) \xrightarrow{\overline{\dim}_{\mathbb{Q},P}} k \longrightarrow 0 \; ,$$

where $\overline{\dim}_{\mathbb{Q},P} : kR_{\mathbb{Q}}(P) \to k$ is the k-reduction of the form $\dim_{\mathbb{Q},P}$. The quotient $kR_{\mathbb{Q}}/\overline{H}$ is a simple p-biset functor, and $kR_{\mathbb{Q}}/\overline{H}(\mathbf{1}) \cong k$, so $kR_{\mathbb{Q}}/\overline{H}$ is isomorphic to $S_{\mathbf{1},k}$, and in this case $S_{\mathbf{1},k}(P) \cong k$ for any finite p-group P, hence $S_{\mathbf{1},k}$ is a constant functor Γ_k.

Similarly, the functor $H/qR_{\mathbb{Q}}$ is simple, equal to $\{0\}$ when evaluated at the trivial group, and to $\mathbb{F}_q$ when evaluated at C_p. This one dimensional $\mathbb{F}_q$-vector space $V = (H/qR_{\mathbb{Q}})(C_p)$ is generated by the image of $\Phi_{C_p} \in R_{\mathbb{Q}}(C_p)$ in $\mathbb{F}_q R_{\mathbb{Q}}(C_p)$, so in particular the action of the group $\mathrm{Out}(C_p)$ on V is trivial. It follows that $\overline{H} \cong S_{C_p,k}$ in this case, and there is a non split short exact sequence of p-biset functors

$$(10.5.3) \qquad 0 \longrightarrow S_{C_p,k} \longrightarrow kR_{\mathbb{Q}} \xrightarrow{\overline{\dim}_{\mathbb{Q}}} S_{\mathbf{1},k} \longrightarrow 0 \; .$$

In other words, the linear forms $\overline{\dim}_{\mathbb{Q},P}$, when P runs through finite p-groups, fit together in a morphism of biset functors $\overline{\dim}_{\mathbb{Q}} : kR_{\mathbb{Q}} \to \Gamma_k$, and the functor $\overline{H}$ is equal to the kernel of this morphism.

- If $q = p$, then $d = m = 1$, and the function f has values in $\{0, 1\}$. Such a function f is entirely determined by the subset $f^{-1}(0)$ of the graph $\mathfrak{N}$, and $f \in \mathfrak{M}$ if and only if the set $f^{-1}(0)$ is *a closed subset* of $\mathfrak{N}$, i.e. if

$$N, M \in \mathfrak{N}, \; N \twoheadrightarrow M, \; N \in f^{-1}(0) \Rightarrow M \in f^{-1}(0) \; .$$

It follows that the poset of subfunctors of $\mathbb{F}_p R_{\mathbb{Q}}$ is isomorphic to the poset of closed subsets of the graph $\mathfrak{N}$, ordered by inclusion of subsets (see [12] Corollary 6.5). □

10.5.4. Remark : Assertions 1 and 2 of Theorem 10.5.2 were proved in [19], and Assertion 3 in [12]. The exact sequence 10.5.3 was introduced in [19] Theorem 11.2 in the case $q = 2$ and p odd. It was conjectured in this paper that $S_{C_p,\mathbb{F}_2}(P)$ is isomorphic to the torsion part $D_t(P)$ of *the Dade group of* P. This conjecture has been proved since by Carlson and Thévenaz ([26] Theorem 13.3).

10.5.5. Proposition : *Let k be a field of characteristic p.*

1. *Let $F' \subset F$ be two subfunctors of $kR_{\mathbb{Q}}$ such that F/F' is a simple functor Σ. Then there exists $N \in \mathcal{N}$, different from C_p, such that $\Sigma \cong S_{N,k}$.*
2. *Conversely, let $N \in \mathcal{N}$, different from C_p. Then the subfunctor $\overline{H}_N$ of $kR_{\mathbb{Q}}$ generated by $\bar{\Phi}_N \in kR_{\mathbb{Q}}(N)$ has a unique simple quotient, isomorphic to $S_{N,k}$.*

Proof: Let $F' \subset F \subseteq kR_{\mathbb{Q}}$, such that the quotient $\Sigma = F/F'$ is simple. Then F corresponds to a closed subset A of $\mathfrak{N}$, and F' corresponds to a closed subset $A' \subset A$. The simplicity of F/F' is equivalent to the fact that A' is a maximal proper closed subset of A.

If $N \in A - A'$, denote by $\overline{\{N\}}$ the closure of N in $\mathfrak{N}$, i.e. the intersection of closed subsets of $\mathfrak{N}$ containing N. Then $\overline{\{N\}}$ is the subset of $\mathfrak{N}$ consisting of elements M for which there exists an oriented path in $\mathfrak{N}$ from N to M.

The subset $A' \cup \overline{\{N\}}$ is a closed subset of A, and properly containing A'. Thus $A' \cup \overline{\{N\}} = A$, for any $N \in A - A'$. Thus if $N, N' \in A - A'$, then $N' \in \overline{\{N\}}$ and $N \in \overline{\{N'\}}$. So there is an oriented path in $\mathfrak{N}$ from N to N', and from N' to N. It follows that $N = N'$, since $\mathfrak{N}$ has no oriented loops. Set $A_N = \overline{\{N\}}$, and $A'_N = A_N - \{N\}$. Then $A'_N \subseteq A'$.

Let F_N (resp. F'_N) denote the subfunctor of $kR_{\mathbb{Q}}$ corresponding to the closed subset A_N (resp. A'_N). Since $A = A' \cup A_N$, it follows that $F = F' + F_N$. Moreover $A' \cap A_N = A' \cap (A_N - \{N\}) = A'_N$, thus $F' \cap F_N = F'_N$. It follows that

$$\Sigma = F/F' = (F' + F_N)/F' \cong F_N/(F' \cap F_N) = F_N/F'_N .$$

There are two cases: if $N = C_p$, then $A_N = \mathfrak{N}$, and

$$A'_N = \{M \in \mathfrak{N} \mid |M| \geq p^2\}$$

is the unique maximal closed proper subset of $\mathfrak{N}$. So $F_N = kR_{\mathbb{Q}}$, and F'_N is its unique maximal proper subfunctor. For any finite p-group P, the quotient $\Sigma(P) \cong F_N/F'_N(P)$ is isomorphic to

$$\Sigma(P) \cong kR_{\mathbb{Q}}(P)/ < \overline{V}(S) \mid S \in \mathcal{G},\ |N_P(S)/S| \geq p^2 > ,$$

where $\mathcal{G}$ is a genetic basis of P, and $\overline{V}(S)$ denotes the image of $V(S) \in R_{\mathbb{Q}}(P)$ in $kR_{\mathbb{Q}}(P)$. Thus

$$\dim_k \Sigma(P) = 1 + |\{S \in \mathcal{G} \mid N_P(S)/S \cong C_p\}| .$$

In particular $\Sigma(\mathbf{1}) \neq \{0\}$, so $\Sigma \cong S_{\mathbf{1},k}$ in this case.

And if $|N| \geq p^2$, then for any finite p-group P and any genetic basis $\mathcal{G}$ of P

$$F_N(P) = \bigoplus_{\substack{S\in\mathcal{G}\\ N_P(S)/S\in A_N}} k\overline{V}(S)\;,$$

$$F'_N(P) = \bigoplus_{\substack{S\in\mathcal{G}\\ N_P(S)/S\in A'_N}} k\overline{V}(S)\;,$$

so $\dim_k \Sigma(P) = |\{S \in \mathcal{G} \mid N_P(S)/S \cong N\}$. It follows that $\Sigma(N) \cong k$, since N has a unique rational irreducible representation of type N, namely Φ_N. Moreover, the group $\mathrm{Out}(N)$ acts trivially on this one dimensional vector space. Finally, if $|P| < |N|$, then no element S of $\mathcal{G}$ is such that $N_P(S)/S \cong N$, since $|N_P(S)/S| \leq |P| < |N|$. Thus $\Sigma(P) = \{0\}$, and it follows that $\Sigma \cong S_{N,k}$ in this case. This proves Assertion 1.

For Assertion 2, observe that the closed subset A of $\mathfrak{N}$ corresponding to the subfunctor $\overline{H}_N$ of $kR_{\mathbb{Q}}$ generated by $\overline{\Phi}_N$ is equal to $\overline{\{N\}}$ if $|N| \geq p^2$, and to $\overline{\{C_p\}} = \mathfrak{N}$ if $N = \mathbf{1}$. In any case A has a unique maximal proper closed subset, equal to $A - \{N\}$ if $|N| \geq p^2$, and to $\mathfrak{N} - \{C_p\}$ if $N = \mathbf{1}$. Thus $\overline{H}_N$ has a unique simple quotient Σ, and the previous argument shows that $\Sigma \cong S_{N,k}$. □

10.5.6. Corollary : *Let k be a field of characteristic p, and $N \in \mathcal{N}$, different from C_p. Then the simple p-biset functor $S_{N,k}$ is rational. In particular, if P is a finite p-group, and $\mathcal{G}$ is a genetic basis of P, then*

$$\dim_k S_{N,k}(P) = \begin{cases} 1 + |\{S \in \mathcal{G} \mid N_P(S)/S \cong C_p\}| & \text{if } N = \mathbf{1} \\ |\{S \in \mathcal{G} \mid N_P(S)/S \cong N\}| & \text{if } |N| \geq p^2\,. \end{cases}$$

10.6. A Characterization

The following result is a characterization of rational p-biset functors, which does not depend on the knowledge of genetic bases:

10.6.1. Theorem : [17, Theorem 3.1] *Let F be a p-biset functor. Then F is rational if and only if the following two conditions are satisfied:*

(i) If P is a p-group with non cyclic center, then $\partial F(P) = \{0\}$.
(ii) If P is any p-group, if $E \trianglelefteq P$ is a normal elementary abelian subgroup of rank 2, if Z is a central subgroup of order p of P contained in E, then the map

$$\mathrm{Res}^P_{C_P(E)} \oplus \mathrm{Def}^P_{P/Z} : F(P) \longrightarrow F\big(C_P(E)\big) \oplus F(P/Z)$$

is injective.

Proof: Suppose first that F is rational. Let P be a p-group, and $\mathcal{G}$ be a genetic basis of P. Since F is rational, the map

$$\mathcal{D}_{\mathcal{G}} : F(P) \xrightarrow{\mathop{\oplus}\limits_{S} \mathcal{D}^P_S} \mathop{\oplus}_{S\in\mathcal{G}} \partial F\big(N_P(S)/S\big)$$

is injective. Here the symbol $\mathcal{D}^P_S$ is a more precise notation for the map $\mathcal{D}_S$, since the group S may be a genetic subgroup for different groups P. Recall that for $S \in \mathcal{G} - \{P\}$, the map $\mathcal{D}^P_S$ is equal to

$$\mathcal{D}^P_S = \mathrm{Defres}^P_{N_P(S)/S} - \mathrm{Defres}^P_{N_P(S)/\hat{S}} ,$$

where $\hat{S}$ is the subgroup such that $\hat{S}/S$ is the unique subgroup of order p in the cyclic group $Z\big(N_P(S)/S\big)$. If $S = P$, then $\mathcal{D}^P_S$ is equal to $\mathrm{Def}^P_{P/P}$.

Suppose that the center $Z(P)$ of P is not cyclic. Then for any $S \in \mathcal{G}$, the group $Z = S \cap Z(P)$ is non trivial: indeed, the group $Z(P)/Z$ is isomorphic to a subgroup of the center of $N_P(S)/S$, which is cyclic since $N_P(S)/S$ has normal p-rank 1. Now for $S \neq P$, the maps $\mathrm{Defres}^P_{N_P(S)/S}$ and $\mathrm{Defres}^P_{N_P(S)/\hat{S}}$ factor as

$$\begin{aligned}
\mathrm{Defres}^P_{N_P(S)/S} &= \mathrm{Defres}^{\overline{P}}_{N_{\overline{P}}(\overline{S})/\overline{S}} \mathrm{Def}^P_{P/Z} \\
\mathrm{Defres}^P_{N_P(S)/\hat{S}} &= \mathrm{Defres}^{\overline{P}}_{N_{\overline{P}}(\overline{S})/\overline{\hat{S}}} \mathrm{Def}^P_{P/Z}
\end{aligned}$$

where $\overline{P} = P/Z$, where $\overline{S} = S/Z$, and $\overline{\hat{S}} = \hat{S}/Z$. Moreover if $S = P$, then $\mathrm{Def}^P_{P/P} = \mathrm{Def}^{\overline{P}}_{\overline{P}/\overline{P}} \mathrm{Def}^P_{P/Z}$. It follows that if $u \in \partial F(P)$, then $\mathcal{D}_{\mathcal{G}}(u) = 0$, since u is mapped to 0 by the proper deflation $\mathrm{Def}^P_{P/Z}$. Thus $u = 0$, and $\partial F(P) = \{0\}$ when the center of P is non cyclic. So if F is rational, then it satisfies condition (i) of the theorem.

For condition (ii), there is nothing to check if E is central in P, since in that case the map $\mathrm{Res}^P_{C_P(E)}$ is the identity map, hence it is injective. And if E is not central in P, then $Z = E \cap Z(P)$. In this situation, by Corollary 9.7.6, there exists a genetic basis $\mathcal{G}$ which admits a decomposition $\mathcal{G} = \mathcal{G}_1 \sqcup \mathcal{G}_2$ as a disjoint union, where

(10.6.2)
- $\mathcal{G}_2 = \{S \in \mathcal{G} \mid S \supseteq Z\}$, and
- the set $\mathcal{H}_1 = \{{}^xS \mid S \in \mathcal{G}_1,\ x \in [P/C_P(E)]\}$ is the subset of a genetic basis of $C_P(E)$ consisting of subgroups not containing Z.

Let $u \in F(P)$ be an element such that $\mathrm{Res}^P_{C_P(E)} u = 0$ and $\mathrm{Def}^P_{P/Z} u = 0$. If $S \in \mathcal{G}_2$, then $\mathcal{D}^P_S = \mathcal{D}^{\overline{P}}_{\overline{S}} \circ \mathrm{Def}^P_{P/Z}$, where $\overline{P} = P/Z$ and $\overline{S} = S/Z$. It follows that $\mathcal{D}^P_S(u) = 0$. Similarly if $S \in \mathcal{G}_1$, then $N_P(S) \subseteq C_P(E)$, and $\mathcal{D}^P_S = \mathcal{D}^{C_P(E)}_S \circ \mathrm{Res}^P_{C_P(E)}$, so $\mathcal{D}^P_S(u) = 0$ again. Hence $\mathcal{D}_\mathcal{G}(u) = 0$, and $u = 0$ since $\mathcal{D}_\mathcal{G}$ is injective when F is rational. Thus F satisfies (ii).

Conversely, suppose that F is a p-biset functor satisfying (i) and (ii). Showing that F is rational amounts to showing that the map $\mathcal{D}_\mathcal{G}$ is injective, for any p-group P and any genetic basis $\mathcal{G}$ of P. Actually, by Theorem 10.1.1 and Lemma 10.1.2, it is equivalent to show that $\mathcal{D}_\mathcal{G}$ is injective, for any p-group P and for one particular genetic basis $\mathcal{G}$ of P. The proof proceeds by induction on the order of P.

If P is the trivial group, then the map $\mathcal{D}_\mathcal{G}$ is the identity map, so there is nothing to prove in this case, and this starts induction. Suppose that P is a p-group, such that the map $\mathcal{D}_\mathcal{H}$ is injective for any p-group P' with $|P'| < |P|$ and any genetic basis $\mathcal{H}$ of P'. Let $\mathcal{G}$ be a genetic basis of P, and $u \in \mathrm{Ker}\,\mathcal{D}_\mathcal{G}$. Let N be a non-trivial normal subgroup of P. Then the set $\mathcal{H} = \{S/N \mid S \in \mathcal{G},\ S \supseteq N\}$ is a genetic basis of P/N. Moreover one checks easily that $\mathcal{D}^{\overline{P}}_{\overline{S}} \circ \mathrm{Def}^P_{P/N} = \mathcal{D}^P_S$ for $S \supseteq N$, where overlines denote quotients by N, as above. Hence $\mathcal{D}_\mathcal{H}(\mathrm{Def}^P_{P/N} u) = 0$, and by induction hypothesis, it follows that $\mathrm{Def}^P_{P/N} u = 0$, and $u \in \partial F(P)$. In other words $\mathrm{Ker}\,\mathcal{D}_\mathcal{G} \subseteq \partial F(P)$.

Hence if the center of P is not cyclic, then condition (i) implies that $\mathrm{Ker}\,\mathcal{D}_\mathcal{G} = \{0\}$, and one can suppose that the center of P is cyclic.

If P has normal p-rank 1, then the trivial subgroup is a genetic normal subgroup of P, hence is belongs to every genetic basis of P. The corresponding map $\mathcal{D}^P_{\mathbf{1}} : F(P) \to \partial F(P)$ is equal to $F(f^P_{\mathbf{1}})$. Hence it is equal to the identity map on $\partial F(P)$. Thus $u = F(f^P_{\mathbf{1}})(u) = \mathcal{D}^P_{\mathbf{1}}(u) = 0$, and the map $\mathcal{D}_\mathcal{G}$ is injective in this case.

Finally one can suppose that P admits a normal and non central subgroup E which is elementary abelian of rank 2. In this case the group $Z = E \cap Z(P)$ has order p, and by Corollary 9.7.6, there exists a genetic basis $\mathcal{G}$ of P with the two properties (10.6.2) above. With the same notation, let $\mathcal{H}$ be a genetic basis of $C_P(E)$ containing $\mathcal{H}_1$. Set $v = \mathrm{Res}^P_{C_P(E)} u$. Then for any $S \in \mathcal{G}_1$ and any $x \in [P/C_P(E)]$, one has $N_P({}^xS) \subseteq C_P(E)$ by Proposition 9.7.1, and it follows that

$$\begin{aligned}
\mathcal{D}^{C_P(E)}_{{}^xS}(v) &= \mathcal{D}^{C_P(E)}_{{}^xS} \mathrm{Res}^P_{C_P(E)} u \\
&= \mathrm{Defres}^P_{N_P({}^xS)/{}^xS} u - \mathrm{Defres}^P_{N_P({}^xS)/{}^x\hat{S}} u \\
&= {}^x\big(\mathrm{Defres}^P_{N_P(S)/S} u - \mathrm{Defres}^P_{N_P(S)/\hat{S}} u\big) \\
&= {}^x\mathcal{D}^P_S u = 0 \ ,
\end{aligned}$$

since ${}^xu = u$ for $x \in P$ and $u \in F(P)$. Thus $\mathcal{D}^{C_P(E)}_R(v) = 0$ for any $R \in \mathcal{H}_1$. And if $R \in \mathcal{H} - \mathcal{H}_1$, then $R \supseteq Z$, and

$$\mathcal{D}_R^{C_P(E)}(v) = \mathcal{D}_{R/Z}^{C_P(E)/Z}\mathrm{Res}_{C_P(E)/Z}^{P/Z}\mathrm{Def}_{P/Z}^{P}u = 0 \ .$$

It follows that $\mathcal{D}_{\mathcal{H}}(v) = 0$, hence $v = 0$ by induction hypothesis. Now $\mathrm{Res}_{C_P(E)}^{P}u = 0$ and $\mathrm{Def}_{P/Z}^{P}u = 0$. Condition (ii) implies that $u = 0$, so the map $\mathcal{D}_{\mathcal{G}}$ is injective. □

10.6.3. Corollary : *Let F be a p-biset functor. Then F is rational if and only if the following two conditions are satisfied:*

(i) If P is a p-group with non cyclic center, then $\partial F(P) = \{0\}$.
(ii) If P is any p-group, if $E \trianglelefteq P$ is a normal elementary abelian subgroup of rank 2, if Z is a central subgroup of order p of P contained in E, then

$$F(P) = \mathrm{Ind}_{C_P(E)}^{P}F\big(C_P(E)\big) + \mathrm{Inf}_{P/Z}^{P}F(P/Z) \ .$$

Proof: Let I be an injective cogenerator of the category of $\mathbb{Z}$-modules (for example $I = \mathbb{Q}/\mathbb{Z}$). Then set $F^* = \mathrm{Hom}_{\mathbb{Z}}(F, I)$. The functor F^* fulfills the two conditions of Theorem 10.6.1, so F^* is rational by Theorem 10.1.5. Thus F is rational, by Theorem 10.1.5 again. □

10.7. Yoneda–Dress Construction

This section is devoted to showing that the class of rational p-biset functors is closed under Yoneda–Dress construction relative to a p-group (see Definition 8.2.3).

10.7.1. Lemma : *Let P, Q and H be p-groups. If U is a finite (P,P)-biset and V is a finite $(P \times H, Q)$-biset, then there is an isomorphism of $(P \times H, Q)$-bisets*

$$(U \times H) \times_{(P\times H)} V \cong U \times_P V \ ,$$

where the double action on $U \times_P V$ is induced by

$$(x, h)(u, v)y = \big(xu, (1, h)vy\big) \ ,$$

for $x \in P$, $h \in H$, $y \in Q$, $u \in U$, and $v \in V$.

In particular, if for any $v \in V$, the intersection of the stabilizer of v in $P \times H$ with $Z(P) \times \{1\}$ is non trivial, then

$$(f_1^P \times H) \times_{(P\times H)} V = 0$$

in $B(P \times H, Q)$.

Proof: For Assertion 1, one checks easily that the maps

$$\big((u,h),_{P\times H} v\big) \mapsto \big(u,_P (1,h)v\big) \qquad \text{and} \qquad (u,_P v) \mapsto \big((u,1),_{P\times H} v\big)$$

are well defined inverse isomorphisms of $(P \times H, Q)$-bisets.

The hypothesis of Assertion 2 means that V is a disjoint union of bisets V_i, for $i = 1 \ldots n$, such that for each i, there exists a non trivial central subgroup T_i of P with

$$V_i = \mathrm{Inf}_{P/T_i\times H}^{P\times H} \circ \mathrm{Def}_{P/T_i\times H}^{P\times H} \circ V_i$$

in $B(P \times H, Q)$. Now

$$(f_{\mathbf{1}}^P \times H) \circ \mathrm{Inf}_{P/T_i\times H}^{P\times H} \circ \mathrm{Def}_{P/T_i\times H}^{P\times H} = (f_{\mathbf{1}}^P \circ \mathrm{Inf}_{P/T_i}^P \circ \mathrm{Def}_{P/T_i}^P) \times H = 0$$

by Lemma 6.3.2. □

10.7.2. Proposition : *Let H be a finite p-group. If F is a rational p-biset functor, then F_H is a rational p-biset functor.*

Proof: The proof consist in checking that F_H fulfills the conditions (i) and (ii) of Theorem 10.6.1.

Let P be a p-group with non cyclic center. Let S be a genetic subgroup of $P \times H$. Since $Z(P) \times \{1\} \subseteq Z(P \times H)$, the group $Z(P) \times \{1\}$ is contained in the center of $N_{P\times H}(S)$, hence it maps to the center of $N_{P\times H}(S)/S$, which is cyclic. Thus if the center of P is not cyclic, then $Z(P) \times \{1\}$ intersects S non trivially. In other words $Z(P) \cap k_1(S) \neq \mathbf{1}$.

Now, setting $T = P \times H$, there are two cases: either $S = T$. In this case the idempotent γ_S^T introduced in Corollary 6.4.5 (see also Remark 10.1.4) is just a (T,T)-biset $\bullet$ of cardinality 1. Thus

$$(f_{\mathbf{1}}^P \times H) \times_T \gamma_S^T = f_{\mathbf{1}}^P \times_P \bullet = 0 ,$$

since P is non trivial. And if $S \neq T$, then

$$\gamma_S^T = (T \times T)/\Delta_{N_T(S),S} - (T \times T)/\Delta_{N_T(S),\hat{S}} ,$$

where $\hat{S}$ is defined by the condition that $\hat{S}/S$ is the only subgroup of order p in the center of $N_T(S)/S$, and where

$$\Delta_{N_T(S),S} = \{(r,s) \in T \times T \mid r \in N_T(S),\ rs^{-1} \in S\}$$

$$\Delta_{N_T(S),\hat{S}} = \{(r,s) \in T \times T \mid r \in N_T(S),\ rs^{-1} \in \hat{S}\} .$$

The left stabilizer in T of the point $\Delta_{N_T(S),S}$ of $(T \times T)/\Delta_{N_T(S),S}$ is equal to S, which intersects $Z(P) \times \{1\}$ non trivially by the previous remarks. Similarly, the left stabilizer in T of the point $\Delta_{N_T(S),\hat{S}}$ of $(T \times T)/\Delta_{N_T(S),\hat{S}}$

is equal to $\hat{S}$, which also intersects $Z(P)\times\{1\}$ non trivially. By Lemma 10.7.1, it follows that

$$(f_1^P \times H) \times_{(P\times H)} \gamma_S^T = 0 ,$$

for any genetic subgroup S of T. But since F is rational, the sum of idempotents γ_S^T, for S in a genetic basis of $P \times H$, is the identity map of $F(P \times H)$. Thus $f_1^P \times H$ acts by 0 on $F(P \times H)$. In other words, the idempotent f_1^P acts by 0 on $F_H(P)$. Hence $\partial F_H(P) = \{0\}$, and F_H fulfills condition (i) of Theorem 10.6.1.

For condition (ii), suppose that E is a normal subgroup of P, and that E is elementary abelian of rank 2. Let Z be a central subgroup of order p in P, contained in E. It is easy to see that the restriction map from $F_H(P)$ to $F_H\big(C_P(E)\big)$ is equal to the restriction map $\mathrm{Res}^{P\times H}_{C_P(E)\times H}$, after the identifications $F_H(P) = F(P \times H)$ and $F_H\big(C_P(E)\big) = F\big(C_P(E) \times H\big)$. Similarly, the deflation map from $F_H(P)$ to $F_H(P/Z)$ is equal to the map $\mathrm{Def}^{P\times H}_{(P\times H)/(Z\times\{1\})}$ from $F(P \times H)$ to $F\big((P/Z) \times H\big)$, after the identification $(P \times H)/(Z \times \{1\}) = (P/Z) \times H$. Now $E' = E \times \{1\}$ is a normal subgroup of $P \times H$, which is elementary abelian of rank 2. Its centralizer in $P \times H$ is $C_P(E) \times H$. Moreover E' contains the subgroup $Z' = Z \times \{1\}$, which is central of order p in $P \times H$. Condition (ii) for the functor F implies that

$$\mathrm{Ker}\,\mathrm{Res}^{P\times H}_{C_P(E)\times H} \cap \mathrm{Ker}\,\mathrm{Def}^{P\times H}_{(P/Z)\times H} = \{0\} .$$

Thus F_H fulfills condition (ii) of Theorem 10.6.1, and F_H is rational. □

10.7.3. Proposition : *Let M and N be objects of $\mathcal{F}_{\mathcal{C}_p,R}$.*

1. *If M is rational, then $M \otimes N$ is rational.*
2. *If N is rational, then $\mathcal{H}(M, N)$ is rational.*
3. *If M is rational, then $\mathcal{H}(M, N)$ is rational.*

Proof: Let Y be a finite p-group. Propositions 8.2.7 and 8.3.4 show that $\mathrm{Hom}_{\mathcal{F}_{\mathcal{C}_p,R}}(RB_Y, M) \cong M(Y)$. It follows that choosing a set $\mathcal{S}$ of representatives of isomorphism classes of finite p-groups, and for each $Y \in \mathcal{S}$, a set I_Y of generators of $N(Y)$ as an R-module, yields a surjective morphism of biset functors

$$\mathop{\oplus}_{Y\in\mathcal{S}} (RB_Y)^{(I_Y)} \to N .$$

Tensoring with M, since $M \otimes -$ is right exact by Corollary 8.4.4, and commutes with direct sums, this yields a surjective morphism

$$\mathop{\oplus}_{Y\in\mathcal{S}} (M \otimes RB_Y)^{(I_Y)} \to M \otimes N .$$

Now $M \otimes RB_Y \cong (M \otimes RB)_Y \cong (RB \otimes M)_Y \cong M_Y$ by Corollary 8.4.12 and Proposition 8.4.6. Moreover M_Y is rational, if M is rational, by

Proposition 10.7.2. Since a direct sum of rational functors is rational, and since a quotient of a rational functor is rational, it follows that $M \otimes N$ is rational when M is rational. Assertion 1 follows.

The proof of Assertion 2 is similar: since for any finite p-group Y

$$\mathcal{H}(M,N)(Y) = \mathrm{Hom}_{\mathcal{F}_{C_p,R}}(M, N_Y) ,$$

and since the Yoneda–Dress construction is an exact functor, it follows that the functor $\mathcal{H}(-,N)$ is left exact (and contravariant), and changes direct sums into direct products. As above, there is a surjective morphism

$$\bigoplus_{Y\in\mathcal{S}} (RB_Y)^{(J_Y)} \to M ,$$

for suitable sets J_Y. This gives an injective morphism

$$\mathcal{H}(M,N) \to \mathcal{H}\big((RB_Y)^{(J_Y)}, N\big) \cong \prod_{Y\in\mathcal{S}} \mathcal{H}(RB_Y, N)^{J_Y} .$$

Now $\mathcal{H}(RB_Y, N) \cong N_Y$ is rational if N is rational. Since any product of rational biset functors and any subfunctor of a rational functor is rational, it follows that $\mathcal{H}(M,N)$ is rational.

For Assertion 3, let I denote an injective cogenerator of the category of R-modules (e.g. $I = \mathrm{Hom}_{\mathbb{Z}}(R, \mathbb{Q}/\mathbb{Z})$). As above, there are sets L_Y, for $Y \in \mathcal{S}$, and a surjective morphism of biset functors

$$\bigoplus_{Y\in\mathcal{S}} (RB_Y)^{(L_Y)} \to \mathrm{Hom}_R(N, I) .$$

Taking duals over R yields an injective morphism of biset functors

$$\mathrm{Hom}_R\big(\mathrm{Hom}_R(N,I), I\big) \to \mathrm{Hom}_R\big(\bigoplus_{Y\in\mathcal{S}} (RB_Y)^{(L_Y)}, I\big) .$$

Moreover the canonical morphism $N \to \mathrm{Hom}_R\big(\mathrm{Hom}_R(N,I), I\big)$ is injective, since I is an injective cogenerator of the category of R-modules. Also

$$\mathrm{Hom}_R\big(\bigoplus_{Y\in\mathcal{S}} (RB_Y)^{(L_Y)}, I\big) \cong \prod_{Y\in\mathcal{S}} \mathrm{Hom}_R(RB_Y, I)^{L_Y} .$$

So there is an injective morphism of biset functors

$$N \to \prod_{Y\in\mathcal{S}} \mathrm{Hom}_R(RB_Y, I)^{L_Y} .$$

Applying the functor $\mathcal{H}(M,-)$, which is left exact by Corollary 8.4.4, and commutes with direct products, this gives an injective morphism

$$\begin{aligned}\mathcal{H}(M,N) &\to \mathcal{H}(M, \prod_{Y\in\mathcal{S}} \mathrm{Hom}_R(RB_Y, I)^{L_Y}) \\ &\cong \prod_{Y\in\mathcal{S}} \mathcal{H}\big(M, \mathrm{Hom}_R(RB_Y, I)\big)^{L_Y} .\end{aligned}$$

Finally $\mathcal{H}\big(M, \mathrm{Hom}_R(RB_Y, I)\big) \cong \mathrm{Hom}_R(M \otimes RB_Y, I) \cong \mathrm{Hom}_R(M_Y, I)$, by Corollary 8.4.13. If M is rational, then M_Y is rational, and $\mathrm{Hom}_R(M_Y, I)$ is rational, for any $Y \in \mathcal{S}$. Thus $\mathcal{H}(M,N)$ is a subfunctor of a product of rational functors, hence it is rational. □

Chapter 11
Applications

This chapter exposes two applications of rational p-biset functors to the description of some groups attached to a finite p-group P: the first one concerns the kernel $K(P)$ of the linearization morphism $\chi_P = \chi_{\mathbb{Q},P} : B(P) \to R_{\mathbb{Q}}(P)$ (see Remark 1.2.3), and the second one the group $B^{\times}(P)$ of units of the Burnside ring of P.

11.1. The Kernel of the Linearization Morphism

This section addresses the old standing following problem: when G is a finite group, describe all pairs (X, Y) of finite G-sets such that the corresponding permutation $\mathbb{Q}G$-modules $\mathbb{Q}X$ and $\mathbb{Q}Y$ are isomorphic? For an arbitrary finite group, this question is essentially open, but for a p-group, a rather complete answer can be given (see Corollary 11.1.16). For this reason, unless otherwise specified, all finite groups considered in this section will be assumed to be p-groups, for a fixed prime number p.

The above isomorphism $\mathbb{Q}X \cong \mathbb{Q}Y$ is equivalent to requiring that $X - Y$ is in the kernel of the linearization morphism $\chi_G : B(G) \to R_{\mathbb{Q}}(G)$. This linearization morphism is a morphism of biset functors, and this motivates the following notation:

11.1.1. Notation : *Let K denote the kernel of the linearization morphism $\chi : B \to R_{\mathbb{Q}}$, considered as a p-biset functor.*

By the Ritter–Segal theorem 9.2.1, there is a short exact sequence of p-biset functors

$$0 \to K \to B \xrightarrow{\chi} R_{\mathbb{Q}} \to 0 \,. \tag{11.1.2}$$

S. Bouc, *Biset Functors for Finite Groups*, Lecture Notes in Mathematics 1990,
DOI 10.1007/978-3-642-11297-3_11, © Springer-Verlag Berlin Heidelberg 2010

11.1.3. Remark : Let P be a finite p-group. Since $R_{\mathbb{Q}}(P)$ is a free abelian group, it follows that the exact sequence of abelian groups

$$0 \to K(P) \to B(P) \stackrel{\chi_P}{\to} R_{\mathbb{Q}}(P) \to 0$$

is split, i.e. $B(P) \cong K(P) \oplus R_{\mathbb{Q}}(P)$. But this splitting is not compatible with the biset functor structure. In other words, the exact sequence 11.1.2 is a non split exact sequence of p-biset functors.

11.1.4. Notation : *If G is a finite group, let $nc(G)$ denote the number of conjugacy classes of non-cyclic subgroups of G.*

11.1.5. Lemma : *Let P be a finite p-group. Then the group $K(P)$ is a free abelian group of rank $nc(P)$. It is a pure submodule of $B(P)$.*

Proof: The group $R_{\mathbb{Q}}(P)$ is a free group of rank equal to the number of conjugacy classes of cyclic subgroups of P, and $B(P)$ is a free group of rank equal to the number of conjugacy classes of subgroups of P. Since the map χ_P is surjective by the Ritter–Segal theorem, it admits a section, and the group $K(P)$ is a direct summand of $B(P)$. So it is a pure submodule of $B(P)$, of rank equal to the difference $nc(P)$ of the rank of $B(P)$ and the rank of $R_{\mathbb{Q}}(P)$. □

11.1.6. Remark : Lemma 11.1.5 is more general: by Artin's theorem, for any finite group G, the cokernel of the linearization map $\chi_G : B(G) \to R_{\mathbb{Q}}(G)$ is always finite, so the image of χ_G is always free, of rank equal to the number of conjugacy classes of cyclic subgroups of G. Hence the kernel $K(G)$ of χ_G is a pure submodule of $B(G)$, of rank equal to $nc(G)$.

11.1.7. Lemma : *Let P be a finite p-group.*

1. *If P is cyclic, then $K(P) = \{0\}$.*
2. *If $P \cong C_p \times C_p$, then $K(P)$ is free of rank 1, generated by the element*

$$\varepsilon_P = f_{\mathbf{1}}^P P/\mathbf{1} = P/\mathbf{1} - \Big(\sum_{\mathbf{1}<Q<P} P/Q\Big) + pP/P \,. \tag{11.1.8}$$

Proof: Assertion 1 is obvious. For Assertion 2, since $K(Q) = \{0\}$ for any proper quotient Q of P, it follows that $K(P) = \partial K(P) \subseteq \partial B(P)$. Now if S is a non trivial subgroup of P, then $f_{\mathbf{1}}^P P/S = 0$, by Lemma 6.3.2, since $P/S \in \mathrm{Inf}_{P/S}^P B(P/S)$. Thus $\partial B(P)$ is generated by the element ε_P, and $K(P)$ is generated by $m\varepsilon_P$, for some positive integer m. Since $K(P)$ is a pure $\mathbb{Z}$-submodule of $B(P)$, it follows that $m = 1$, i.e. $\varepsilon_P \in K(P)$, and $K(P) = \mathbb{Z}\varepsilon_P$ in this case. □

11.1.9. Lemma : *Let P be an extraspecial p-group of order p^3 and exponent p, when p is odd, or a dihedral 2-group of order at least 8, when $p = 2$. Let Z denote the center of P. Let I and J be two subgroups of order p of P, different from Z, and not conjugate in P. Define an element $\delta_{P,I,J}$ of $B(P)$ by*

$$\delta_{P,I,J} = (P/I - P/IZ) - (P/J - P/JZ) .$$

Then $\delta_{P,I,J} \in K(P)$.

Proof: If $p > 2$, then the group P is extraspecial of order p^3 and exponent p. The normalizer of I in P is equal to IZ, so $N_P(I)/I$ is cyclic of order p, hence it has normal p-rank 1. Moreover $Z_P(I) = IZ$ is a normal subgroup of P. So $I^x \leq Z_P(I)$, for any $x \in P$. Thus $I^x \cap Z_P(I) = I^x$ is contained in I if and only if $I^x = I$, and I is a genetic subgroup of P. Similarly J is a genetic subgroup of P.

Moreover $I \cap Z_P(J) = I \cap JZ = \mathbf{1} \leq J$, and $J \cap Z_P(I) = \mathbf{1} \leq J$, so $I \,\widehat{=}_P\, J$. It follows that the simple $\mathbb{Q}P$-modules $V(I)$ and $V(J)$ are isomorphic.

Let $H = N_P(I)/I \cong C_p$. The unique irreducible faithful $\mathbb{Q}H$-module is the kernel of the augmentation map $\mathbb{Q}H/\mathbf{1} \to \mathbb{Q}H/H$. Applying $\mathrm{Indinf}_{N_P(I)/I}^P$, it follows that there is an exact sequence of $\mathbb{Q}P$-modules

$$0 \to V(I) \to \mathbb{Q}P/I \to \mathbb{Q}P/IZ \to 0 .$$

In other words $V(I) = \mathbb{Q}P/I - \mathbb{Q}P/IZ$ in $R_\mathbb{Q}(P)$. So the image of $\delta_{P,I,J}$ by the linearization morphism χ_P is equal to $V(I) - V(J) = 0$, proving that $\delta_{P,I,J} \in K(P)$.

Now if $p = 2$, the group P is a dihedral 2-group, and I and J are basic subgroups of P, by Proposition 9.3.5. The corresponding simple $\mathbb{Q}P$-modules V_I and V_J introduced in Notation 9.2.5 are faithful, since $I \cap Z$ and $J \cap Z$ are both trivial (see Example 9.2.8). Since P has a unique faithful irreducible rational representation Φ_P, by Proposition 9.3.5, it follows that $V_I \cong V_J$. Moreover $V_I = \mathbb{Q}P/I - \mathbb{Q}P/IZ$ in $R_\mathbb{Q}(P)$. This shows that $\chi_P(\delta_{P,I,J}) = V_I - V_J = 0$, so $\delta_{P,I,J} \in K(P)$. □

11.1.10. Notation :

1. *Let E denote a fixed elementary abelian p-group of rank 2. Denote by ε the element ε_E of $K(E)$ defined in 11.1.8.*
2. *Let X_n, for $n \geq 3$, be a fixed dihedral 2-group of order 2^n. Choose some subgroups I_n and J_n as in Lemma 11.1.9, and denote by δ_n the corresponding element δ_{X_n,I_n,J_n} of $K(X_n)$.*
   If $n = 3$, simply set $X = X_3$ and $\delta = \delta_3$.
3. *When $p > 2$, let X denote a fixed extraspecial p-group of order p^3 and exponent p. Choose some subgroups I and J as in Lemma 11.1.9, and denote by δ the corresponding element $\delta_{X,I,J}$ of $K(X)$.*

So the group X is extraspecial of order p^3 and exponent p, when p is odd, or dihedral of order 8, when $p = 2$, and there is a distinguished element $\delta \in K(X)$ in each case.

11.1.11. Notation : *Let B_δ (resp. B_ε) denote the subfunctor of B generated (see 3.2.9) by the element $\delta \in K(X) \subseteq B(X)$ (resp. by the element $\varepsilon \in K(E) \subseteq B(E)$).*

11.1.12. Lemma :

1. *The functors B_δ and B_ε are subfunctors of K. So B_δ is equal to the subfunctor K_δ of K generated by δ, and similarly $B_\varepsilon = K_\varepsilon$.*
2. *Moreover $pB_\delta \subseteq B_\varepsilon \subseteq B_\delta$.*

Proof: Assertion 1 is obvious, since $\delta \in K(X)$ and $\varepsilon \in K(E)$. For Assertion 2, denote by Y the subgroup JZ of X. Then Y is elementary abelian of rank 2, so there exists a group isomorphism $\varphi : Y \to E$. Moreover, by the Mackey formula, since $IY = X$ and $Y \cap I = \mathbf{1}$

$$\begin{aligned}\mathrm{Res}_Y^X \delta &= (Y/\mathbf{1} - Y/Z) - \sum_{x \in X/Y} (Y/{}^xJ - Y/Y)\\ &= Y/\mathbf{1} - (\sum_{\mathbf{1}<F<Y} Y/F) + pY/Y = \varepsilon_Y \ .\end{aligned}$$

It follows that $\varepsilon = \mathrm{Iso}(\varphi)(\mathrm{Res}_Y^X \delta) \in B_\delta(E)$, so $B_\varepsilon \subseteq B_\delta$.
Similarly $\varepsilon_Y = \mathrm{Iso}(\varphi^{-1}) \in B_\varepsilon(Y)$, and

$$\begin{aligned}\mathrm{Ind}_Y^X \varepsilon_Y &= X/\mathbf{1} - (\sum_{\mathbf{1}<F<Y} X/F) + pX/Y\\ &= X/\mathbf{1} - X/Z - pX/J + pX/JZ \ .\end{aligned}$$

It follows that

$$\begin{aligned}\mathrm{Ind}_{JZ}^X \varepsilon_{JZ} - \mathrm{Ind}_{IZ}^X \varepsilon_{IZ} &= (pX/I - pX/IZ) - (pX/J - pX/JZ)\\ &= p\delta \ ,\end{aligned}$$

so $p\delta \in B_\varepsilon(X)$, and $pB_\delta \subseteq B_\varepsilon$. □

11.1.13. Lemma : *Let P be a finite p-group.*

1. *If Q is a subgroup of P such that $Q \cap Z(P) \neq \mathbf{1}$, then $f_{\mathbf{1}}^P P/Q = 0$.*

2. *The group $\partial B(P)$ has a basis consisting of the elements $f_{\mathbf{1}}^P P/Q$, where Q runs through a set of representatives of conjugacy classes of subgroups of P such that $Q \cap Z(P) = \mathbf{1}$.*
3. *If the center of P is not cyclic, then $\partial B(P) \subseteq B_\varepsilon(P)$.*

Proof: If Q is a subgroup of P and $Z = Q \cap Z(P) \neq \mathbf{1}$, then $P/Q = \mathrm{Inf}_{P/Z}^P\big((P/Z)/(Q/Z)\big)$, thus $f_{\mathbf{1}}^P P/Q = 0$ by Lemma 6.3.2. This proves Assertion 1.

It follows that $\partial B(P)$ is generated by the set $\mathcal{S}$ of elements $f_{\mathbf{1}}^P P/Q$, where Q runs through the subgroups of P such that $Q \cap Z(P) = \mathbf{1}$, up to conjugation. Moreover

$$f_{\mathbf{1}}^P P/Q = \sum_{\mathbf{1} \leq Z \leq \Omega_1 Z(P)} \mu(\mathbf{1}, Z) P/QZ \ ,$$

and all the terms in this summation except $\mu(\mathbf{1},\mathbf{1})P/Q = P/Q$ are multiple of elements P/H such that $H \cap Z(P) \neq \mathbf{1}$. It follows easily that the elements of $\mathcal{S}$ are linearly independent, and this proves Assertion 2.

Now proving Assertion 3 amounts to showing that $f_{\mathbf{1}}^P P/Q \in B_\varepsilon(P)$, for any subgroup Q of P. Consider first the case $Q = \mathbf{1}$. Since the center $Z(P)$ of P is not cyclic, there exists a subgroup $E_0 \leq Z(P)$ which is elementary abelian of rank 2. In particular the element

$$\varepsilon_{E_0} = E_0/\mathbf{1} - \sum_{\substack{F \in E_0 \\ |F|=p}} E_0/F + pE_0/E_0$$

of $B(E_0)$ belongs to $B_\varepsilon(E_0)$. Inducing up to P gives the element

$$e = P/\mathbf{1} - \sum_{\substack{F \leq E_0 \\ |F|=p}} P/F + pP/E_0$$

of $B_\varepsilon(P)$. But $f_{\mathbf{1}}^P e = f_{\mathbf{1}}^P P/\mathbf{1}$ by Assertion 1, since $E_0 \leq Z(P)$. Hence $f_{\mathbf{1}}^P P/\mathbf{1} \in B_\varepsilon(P)$.

Now the proof that $f_{\mathbf{1}}^P P/Q \in B_\varepsilon(P)$ for any subgroup Q of P can be completed by induction on the index $|P:Q|$. If $Q = P$, then $f_{\mathbf{1}}^P P/P = 0$, since the center of P is non trivial, so $f_{\mathbf{1}}^P P/Q \in B_\varepsilon(P)$ in this case. Now let Q be any subgroup of P, and suppose that $f_{\mathbf{1}}^P P/R \in B_\varepsilon(P)$ for any subgroup R of P with $|R| > |Q|$. If $Q \cap Z(P) \neq \mathbf{1}$, then $f_{\mathbf{1}}^P P/Q = 0 \in B_\varepsilon(P)$. If $Q \cap Z(P) = \mathbf{1}$, then $Z(P)$ embeds in the center of $\overline{N} = N_P(Q)/Q$, so this center is not cyclic. The special case above and Lemma 6.2.10 show that the element

$$f_{\mathbf{1}}^{\overline{N}} \overline{N}/\mathbf{1} = \sum_{\overline{Z} \leq \Omega_1 Z(\overline{N})} \mu(\mathbf{1}, \overline{Z}) \overline{N}/\overline{Z}$$

belongs to $B_\varepsilon(\overline{N})$. Taking inflation from $\overline{N}$ to N, and then induction from N to P gives the element

$$w = \mathrm{Indinf}_{N_P(Q)/Q}^{P} f_{\mathbf{1}}^{\overline{N}} \overline{N}/\mathbf{1} = \sum_{Q'/Q \leq \Omega_1 Z(\overline{N})} \mu(\mathbf{1}, Q'/Q) P/Q'$$

of $B_\varepsilon(P)$. It follows that

$$f_{\mathbf{1}}^{P} w = f_{\mathbf{1}}^{P} P/Q + \sum_{\substack{Q'/Q \leq \Omega_1 Z(\overline{N}) \\ Q'/Q \neq \mathbf{1}}} \mu(\mathbf{1}, Q'/Q) f_{\mathbf{1}}^{P} P/Q' \in B_\varepsilon(P) \ .$$

By induction hypothesis, all terms in the summation are in $B_\varepsilon(P)$, and by difference $f_{\mathbf{1}}^{P} P/Q \in B_\varepsilon(P)$, as was to be shown. □

11.1.14. Proposition : *The functor B/B_δ is a rational p-biset functor.*

Proof: By Theorem 10.1.5, showing that B/B_δ is rational is equivalent to showing that the functor $F = \mathrm{Hom}_{\mathbb{Z}}(B/B_\delta, I)$ is rational, where I is some injective cogenerator of $\mathbb{Z}$-Mod (e.g. $I = \mathbb{Q}/\mathbb{Z}$). Or equivalently, to showing that F fulfills the conditions of Theorem 10.6.1.

Since $f_{\mathbf{1}}^{P}$ is an idempotent, it follows that

$$\partial F(P) = \mathrm{Hom}_{\mathbb{Z}}\big(\partial(B/B_\delta), I\big) \ .$$

But by Lemmas 11.1.13 and 11.1.12, if the center of P is not cyclic, then

$$\partial B(P) \subseteq B_\varepsilon(P) \subseteq B_\delta(P) \ ,$$

so $\partial(B/B_\delta)(P) = \{0\}$ in this case, and $\partial F(P) = \{0\}$. Hence F fulfills Condition (i) of Theorem 10.6.1.

Now let P be any finite p-group, and let E be a normal subgroup of P, which is elementary abelian of rank 2. Let moreover Z be a central subgroup of order p of P, contained in E. Let $f \in F(P)$, such that $\mathrm{Res}_{C_P(E)}^{P}(f) = 0$ and $\mathrm{Def}_{P/Z}^{P} f = 0$. Checking condition (ii) of Theorem 10.6.1 amounts to showing that $f = 0$.

If E is central in P, there is nothing to do, since $C_P(E) = P$ and $\mathrm{Res}_{C_P(E)}^{P}(f) = f$. Otherwise $C_P(E)$ has index p in P. Let $\tilde{f} : B(P) \to I$ be the homomorphism obtained by composing f with the projection $B(P) \to (B/B_\delta)(P)$. The hypothesis on f means that $\tilde{f}(P/Q) = 0$ if $Q \subseteq C_P(E)$ or if $Q \supseteq Z$. Let Q be any subgroup of P with $Q \not\subseteq C_P(E)$ and $Q \not\supseteq Z$. Then $Q \not\supseteq E$. Moreover $QC_P(E) = P$. If the intersection $Q \cap E$ has order p, it is normalized, hence centralized by Q, and also by $C_P(E)$, hence $Q \cap E \subseteq Z(P)$. Since $Q \cap E \neq Z$, it follows that $E = Z{\cdot}(Q \cap E)$ is central in P, a contradiction. Hence $Q \cap E = \mathbf{1}$.

Now the group $C_Q(E) = Q \cap C_P(E)$ is a normal subgroup of QE. The factor group $R = QE/C_Q(E)$ has order p^3, and is the semidirect product of the group $\overline{Q} = Q/C_Q(E)$, of order p, by the normal elementary abelian subgroup group $\overline{E} = EC_Q(E)/C_Q(E) \cong E$. Moreover R is not abelian, since otherwise $[Q,E] \subseteq C_Q(E) \cap E = \mathbf{1}$, so $Q \subseteq C_P(E)$. It follows that R is extraspecial of order p^3 and exponent p if p is odd, or dihedral of order 8 if $p = 2$.

The center of R is the group $\overline{Z} = ZC_Q(E)/C_Q(E)$. The group $\overline{Q}$ is a non central subgroup of order p of R. If S is any subgroup of order p of E, different from Z, then $\overline{S} = SC_Q(E)/C_Q(E)$ is another non central subgroup of order p of R, not conjugate to $\overline{Q}$ in R. So in the group R, the element

$$d = (R/\overline{Q} - R/\overline{Q}\overline{Z}) - (R/\overline{S} - R/\overline{E})$$

is in $B_\delta(R)$, since it is the image of δ by a suitable group isomorphism. Taking induction-inflation from the section $(QE, C_Q(E))$ of P gives the element

$$\tilde{d} = \mathrm{Indinf}_{QE/C_Q(E)}^P d = (P/Q - P/QZ) - \big(P/SC_Q(E) - P/EC_Q(E)\big) ,$$

of $B_\delta(P)$. Now $\tilde{f}$ vanishes on $B_\delta(P)$, thus $\tilde{f}(\tilde{d}) = 0$, and

$$\tilde{f}(P/Q) = \tilde{f}(P/QZ) + \tilde{f}\big(P/SC_Q(E)\big) - \tilde{f}\big(P/EC_Q(E)\big) = 0 ,$$

since $QZ \supseteq Z$ and since $SC_Q(E) \subseteq EC_Q(E) \subseteq C_P(E)$.

This shows that $\tilde{f} = 0$, hence $f = 0$, and the functor F satisfies both conditions of Theorem 10.6.1. Hence F is rational, as was to be shown. □

11.1.15. Theorem : [15, Theorem 6.12]

1. *If $p \neq 2$, then the functor K is generated by δ: in other words for any p-group P*
$$K(P) = \mathrm{Hom}_{\mathcal{C}_p}(X_{p^3}, P) \times_{X_{p^3}} \delta .$$
2. *If $p = 2$, then the functor K is generated by the elements δ_n, for $n \geq 3$: in other words for any 2-group P*
$$K(P) = \sum_{n \geq 3} \mathrm{Hom}_{\mathcal{C}_2}(D_{2^n}, P) \times_{D_{2^n}} \delta_n .$$

Proof: Let K' denote the subfunctor of B generated by δ, when $p > 2$, or by all the elements δ_n, for $n \geq 3$, when $p = 2$. Since $\delta_n \in K(X_n)$ for any n, it follows that K' is contained in K. Moreover K' contains B_δ.

Hence the functor K/K' is a quotient of K/B_δ, which is a subfunctor of B/B_δ. As B/B_δ is rational by Proposition 11.1.14, it follows from Theorem 10.1.5 that K/B_δ and K/K' are also rational. Thus, proving that

$K/K' = \{0\}$ is equivalent to proving that $\partial K(P) \subseteq K'(P)$, whenever P is a p-group of normal p-rank 1.

If p is odd, there is not much to say, for P is cyclic in this case, so $K(P) = \{0\}$ by Lemma 11.1.7, and $\partial K(P) = \{0\}$.

If $p = 2$, then P is cyclic, generalized quaternion, dihedral or semi-dihedral. If P is cyclic, the argument is the same as for p odd, and $K(P) = \{0\} = \partial K(P)$.

If P is generalized quaternion, then $\partial B(P)$ is free of rank 1, generated by the element $a = f_{\mathbf{1}}^P P/\mathbf{1} = P/\mathbf{1} - P/Z$, by Lemma 11.1.13. Since $\chi_P(a) = \Phi_P$, the only multiple of a which belongs to $K(P)$ is equal to 0, and $\partial K(P) = \{0\}$ again in this case.

If P is semi-dihedral, then $\partial B(P)$ is freely generated by $a = f_{\mathbf{1}}^P P/\mathbf{1}$ and $b = f_{\mathbf{1}}^P P/Q$, where Q is a non central subgroup of order 2 in P. Again here $\chi_P(a) = \Phi_P$, and $\chi_P(b)$ is a linear combination of faithful rational irreducible representations of P, i.e. a multiple of Φ_P. Thus $\chi_P(b) = 2\Phi_P$, for dimension reasons. It follows that $\partial K(P)$ is generated by the element

$$b - 2a = P/\mathbf{1} - P/Z - 2P/Q + 2P/QZ = \mathrm{Ind}_{QZ}^P \varepsilon_{QZ} \ ,$$

where Z is the centre of P. Thus $b - 2a \in B_\varepsilon(P) \subseteq B_\delta(P) \subseteq K'(P)$ also in this case.

Finally, if P is dihedral, of order 2^n,then $\partial B(P)$ is freely generated by the elements $a = f_{\mathbf{1}}^P P/Q$, $b = f_{\mathbf{1}}^P P/R$ and $c = f_{\mathbf{1}}^P P/\mathbf{1}$, where Q and R are two non conjugate non central subgroups of order 2 of P. In this case also $\chi_P(a) = \chi_P(b) = \Phi_P$, and $\chi_P(c)$ is a multiple of Φ_P, by the same argument as before. Thus $\chi_P(c) = 2\Phi_P$ for dimension reasons. It follows that $\partial K(P)$ is generated by $a - b$ and $c - 2a$. But

$$a - b = (P/Q - P/QZ) - (P/R - P/RZ)$$

is equal to the image of δ_n under a suitable group isomorphism $D_{2^n} \to P$. Thus $a - b \in K'(P)$. Finally

$$b - 2a = P/\mathbf{1} - P/Z - 2P/Q + 2P/QZ = \mathrm{Ind}_{QZ}^P \varepsilon_{QZ}$$

belongs to $B_\varepsilon(P) \subseteq B_\delta(P) \subseteq K'(P)$. Thus $\partial K(P) \subseteq K'(P)$ in this case also, as was to be shown. □

11.1.16. Corollary : *Let P be a finite p-group. Then $K(P)$ is equal to the set of linear combinations of elements of the form* $\mathrm{Indinf}_{T/S}^P \theta(\kappa)$, *where (T, S) is a section of P, where*

- *(if $p > 2$) θ is a group isomorphism from E or X to T/S, and $\kappa = \varepsilon$ or $\kappa = \delta$, respectively.*
- *(if $p = 2$) θ is a group isomorphism from E or some X_n, for $n \geq 3$, to T/S, and $\kappa = \varepsilon$ or $\kappa = \delta_n$, respectively.*

Proof: If U is a transitive (P,X)-biset, then the factorization property 4.1.5 shows that there is a section (T,S) of P such that T/S is isomorphic to a subquotient of X, and $U(\delta)$ is equal to $\mathrm{Indinf}_{T/S}^{P} v$, for some element $v \in K(T/S)$. If $U(\delta) \neq 0$, this implies that $K(T/S) \neq \{0\}$, hence that T/S is not cyclic.

Now when p is odd, any non cyclic subquotient of an extraspecial group X of order p^3 and exponent p is either isomorphic to X, or elementary abelian of rank 2. Similarly, any non cyclic subquotient of a dihedral 2-group is either dihedral or elementary abelian of rank 2. Corollary 11.1.16 now follows from an easy induction argument. □

11.2. Units of Burnside Rings

This section exposes an application of biset functors to the structure of the group of multiplicative units of the Burnside ring $B(G)$ of a finite group G. When G is a p-group, this will lead to a complete description of this group of units, in terms of genetic subgroups of G.

The first definition and results of this section deal with an arbitrary finite group.

11.2.1. Notation : *When G is a finite group, let $B(G)^\times$ denote the group of invertible elements of the Burnside ring $B(G)$ of G. Let $1_B = G/G$ denote the identity element of $B(G)$.*

11.2.2. Lemma : *Let G be a finite group, and $u \in B(G)$. The following conditions are equivalent:*

1. $u \in B(G)^\times$.
2. $|u^H| \in \{\pm 1\}$, *for any subgroup H of G.*
3. $u^2 = 1_B$.

In particular $B(G)^\times$ is a finite elementary abelian 2-group.

Proof: Let H be a subgroup of G. Since the map $\phi_H : u \mapsto |u^H|$ is a ring homomorphism from $B(G)$ to $\mathbb{Z}$, it follows that if $u \in B(G)$, then $\phi_H(u)$ is invertible in $\mathbb{Z}$, so $\phi_H(u) \in \{\pm 1\}$. Thus *1* implies *2*.

Now if *2* holds, then setting $v = u^2$, it follows that $\phi_H(v) = 1$ for any $H \leq G$. Thus $v = 1$, by Burnside's Theorem 2.4.5, and *2* implies *3*.

Finally, if $u^2 = 1_B$, then u is its own inverse in $B(G)$, so *3* implies *1*.

Burnside's Theorem shows that $B(G)^\times$ is isomorphic to a subgroup of a direct product of copies of $\mathbb{Z}^\times = \{\pm 1\}$, so it is a finite elementary abelian 2-group □

11.2.3. Remark : The determination of $B(G)^\times$ for an arbitrary finite group is extremely difficult, as can be seen from the following argument, due to tom Dieck ([52] Proposition 1.5.1): the first observation is that the mappings

$u \mapsto (1_B - u)/2$ and $e \mapsto 1_B - 2e$ are mutual inverse bijections between $B(G)^\times$ and the set of idempotents e of $\mathbb{Q}B(G)$ such that $2e \in B(G)$.

Suppose now that G has odd order. If $u \in B(G)^\times$, the idempotent $e = (1_B - u)/2$ of $\mathbb{Q}B(G)$ is such that $2e \in B(G)$. But $|G|e$ is also in $B(G)$, by the formulae of Theorem 2.5.2 for primitive idempotents in $\mathbb{Q}B(G)$. Since $|G|$ is odd, it follows that $e \in B(G)$.

Conversely, if e is an idempotent in $B(G)$, then $1_B - 2e \in B(G)^\times$. In other words, when G is odd, there is a one to one correspondence between $B(G)^\times$ and the set of idempotents of $B(G)$. Now by a theorem of Dress [33], the spectrum of $B(G)$ is connected if and only if G is solvable. In other words the group G is solvable if and only if the only idempotents of $B(G)$ are 0 and 1. So when $|G|$ is odd, proving that G is solvable is equivalent to proving that $B(G)^\times = \{\pm 1_B\}$. Thus:

11.2.4. Theorem : [tom Dieck] *Feit–Thompson's theorem is equivalent to the statement that if G has odd order, then $B(G)^\times = \{\pm 1_B\}$.*

So even the question of knowing when $B(G)^\times = \{\pm 1_B\}$ is highly non-trivial. Note however that for an odd order *p-group*, this becomes rather obvious:

11.2.5. Lemma : *Let P be a finite p-group.*

1. *If $a \in B(P)$, then $|a^P| \equiv |a|$ (mod.p).*
2. *If p is odd, then $B(P)^\times = \{\pm 1_B\}$.*

Proof: Assertion 1 follows from the case where a is the class of a finite P-set X: in this case, all the non trivial P-orbits on X have length multiple of p, so $|X|$ is congruent to $|X^P|$ modulo p.

Now if p is odd and $u \in B(P)^\times$, then $|u^Q|$ is congruent to $|u|$ modulo p, by Assertion 1, for any subgroup Q of P. Since $|u|$ and $|u^Q|$ are both equal to ± 1, it follows that $|u^Q| = |u|$ for any $Q \leq P$. Hence $u = \pm 1_B$. □

11.2.6. Lefschetz Invariants of G-Posets. One of the problems in handling elements of the Burnside ring $B(G)$ of a finite group G, is that they are not represented by combinatorial objects, but by *differences* $X - Y$ of two such objects, namely G-sets. Replacing G-sets by G-posets yields a nice tool to get around this problem, via the notion of *Lefschetz invariant*. The idea of considering these invariants goes back to tom Dieck [52], but their formal definition is due to Thévenaz [48]. Details on Lefschetz invariants can be found in Sect. 4 of [8].

11.2.7. Definition : *A partially order set, or* poset, *is a set X equipped with an order relation $\leq$. If X is a poset, and if $n \in N$, let $Sd_n(X)$ denote the set of chains $x_0 < \ldots < x_n$ of elements of X of cardinality $n+1$. If X is finite, then the* Euler–Poincaré *characteristic $\chi(X)$ of X is the integer defined by*

$$\chi(X) = \sum_{n\in\mathbb{N}} (-1)^n |Sd_n(X)| .$$

The reduced Euler–Poincaré characteristic *is the integer defined by*

$$\tilde{\chi}(X) = \chi(X) - 1 .$$

*Let G be a finite group. A G-*poset *X is a G-set equipped with an order relation $\leq$ compatible to the G-action: if $x \leq x'$ are elements of X and if $g \in G$, then $gx \leq gx'$. In particular, if X is a G-poset, then for any $n \in \mathbb{N}$, the set $Sd_n(X)$ is a G-set.*

If X and Y are G-posets, a morphism of G-posets *$f : X \to Y$ is a map from X to Y such that $f(gx) = gf(x)$ if $g \in G$ and $x \in X$, and such that $f(x) \leq f(x')$ in Y, whenever $x \leq x'$ in X. The category of G-posets is denoted by G-*Poset*, while G-*poset *denote the full subcategory of finite G-posets.*

If X is a finite G-poset, the Lefschetz invariant *Λ_X of X is the element of $B(G)$ defined by*

$$\Lambda_X = \sum_{n\geq 0} (-1)^n Sd_n(X) .$$

The reduced Lefschetz invariant *$\tilde{\Lambda}_X$ is the element of $B(G)$ defined by*

$$\tilde{\Lambda}_X = \Lambda_X - G/G .$$

It follows from those definitions that $|\Lambda_X|$ is equal to the Euler–Poincaré characteristic of X. One can say more:

11.2.8. Lemma : *Let G be a finite group.*

1. *If X is a finite G-poset, then for any subgroup H of G*

   $$(\Lambda_X)^H = \Lambda_{X^H}$$

   in $B\big(N_G(H)/H\big)$. In particular $|(\Lambda_X)^H| = \chi(X^H)$.
2. *If X and Y are finite G-posets, then $\Lambda_X = \Lambda_Y$ in $B(G)$ if and only if $\chi(X^H) = \chi(Y^H)$ for any subgroup H of G.*

Proof: The first assertion is obvious, since $Sd_n(X)^H = Sd_n(X^H)$ for all $n \in \mathbb{N}$. The second one follows from Burnside's Theorem 2.4.5. □

11.2.9. Lemma : *Let G be a finite group. If X and Y are finite G-posets, denote by $X \times Y$ the cartesian product of X and Y, endowed with diagonal G-action, and ordered by the product order*

$$\forall x, x' \in X,\ \forall y, y' \in Y,\ \ (x, y) \leq (x', y') \Leftrightarrow x \leq x' \text{ and } y \leq y' .$$

Then $\chi(X \times Y) = \chi(X)\chi(Y)$.

Proof: If $s \in Sd_n(X \times Y)$, let s_X denote the projection of s on X, and s_Y denote its projection on Y. Then $s_X \in Sd_i(X)$, for some integer $i \leq n$, and $s_Y \in Sd_j(Y)$, for some integer $j \leq n$.

With this notation

$$\begin{aligned}\chi(X \times Y) &= \sum_{n \in \mathbb{N}} \sum_{s \in Sd_n(X \times Y)} (-1)^n \\ &= \sum_{i,j \in \mathbb{N}} \sum_{\substack{a \in Sd_i(X) \\ b \in Sd_j(Y)}} \sum_{n \in \mathbb{N}} \sum_{\substack{s \in Sd_n(X \times Y) \\ s_X = a \\ s_Y = b}} (-1)^n .\end{aligned}$$

Now when $a \in Sd_i(X)$ and $b \in Sd_j(Y)$ are given, the set of chains s in $Sd_n(X \times Y)$ for which $s_X = a$ and $s_Y = b$ is in one to one correspondence with the set $C_{i,j,n}$ of strictly increasing maps $\{0, \ldots, n\} \to \{0, \ldots, i\} \times \{0, \ldots, j\}$, for which the maps $\{0, \ldots, n\} \to \{0, \ldots, i\}$ and $\{0, \ldots, n\} \to \{0, \ldots, j\}$, obtained by composition with the projection maps, are surjective. In particular, the cardinality $c_{i,j,n}$ of this set depends only on i, j, and n. Setting $c_{i,j} = \sum_{n \in \mathbb{N}} (-1)^n c_{i,j,n}$, it follows that

$$\chi(X \times Y) = \sum_{i,j \in \mathbb{N}} |Sd_i(X)||Sd_j(Y)| c_{i,j} .$$

Now if $f \in C_{i,j,n}$, then $f(n) = (i,j)$: indeed, the first component of $f(n)$ is greater than i, hence equal to i, because the map $p_1 \circ f$ is increasing and surjective, where p_1 is the projection on $\{0, \ldots, i\}$. Similarly, the second component of $f(n)$ is equal to j.

Now the conditions on f imply that $f(n-1)$ is one of $(i-1, j)$, $(i-1, j-1)$, or $(i, j-1)$, and the restriction of f to $\{0, \ldots, n-1\}$ is an element of $C_{i-1,j,n-1}$, $C_{i-1,j-1,n-1}$, or $C_{i,j-1}$ respectively. Conversely, any element f' in one of these sets can be extended to an element f of $C_{i,j,n}$ by setting $f(n) = (i,j)$.

It follows that

(11.2.10) $$c_{i,j} = -c_{i-1,j} - c_{i-1,j-1} - c_{i,j-1} .$$

It is clear moreover that $c_{i,0} = (-1)^i = c_{0,i}$, for any $i \in \mathbb{N}$. An easy induction argument on $i+j$, using Equation 11.2.10, shows that $c_{i,j} = (-1)^{i+j}$, for any $i, j \in \mathbb{N}$.

It follows that

$$\begin{aligned}\chi(X \times Y) &= \sum_{i,j \in \mathbb{N}} |Sd_i(X)||Sd_j(Y)|(-1)^{i+j} \\ &= \Big(\sum_{i \in \mathbb{N}} (-1)^i |Sd_i(X)|\Big)\Big(\sum_{j \in \mathbb{N}} (-1)^j |Sd_j(Y)|\Big) \\ &= \chi(X)\chi(Y) ,\end{aligned}$$

as was to be shown. □

11.2.11. Lemma : *Let G be a finite group.*

1. *Let X be a finite G-set, ordered by the equality relation (i.e. $x \leq y$ in X $\Leftrightarrow x = y$). Then $\Lambda_X = X$ in $B(G)$.*
2. *Let X and Y be finite G-posets, and $X \sqcup Y$ be their disjoint as G-sets, ordered by*

$$\forall z, t \in X \sqcup Y,\quad z \leq t \Leftrightarrow \left[\begin{array}{c} z, t \in X \text{ and } z \leq t \text{ in } X \\ \text{or} \\ z, t \in Y \text{ and } z \leq t \text{ in } Y\ . \end{array}\right.$$

Then $\Lambda_{X \sqcup Y} = \Lambda_X + \Lambda_Y$ in $B(G)$.

3. *Let X and Y be finite G-posets. Then $\Lambda_{X \times Y} = \Lambda_X \Lambda_Y$ in $B(G)$.*
4. *Let X be a finite G-poset. Denote by $X * \{0,1\}$ the disjoint union $X \sqcup \{0,1\}$, where $\{0,1\}$ is a set with two elements on which G acts trivially, ordered by*

$$\forall z, t \in X * \{0,1\},\quad z \leq t \Leftrightarrow \left[\begin{array}{c} z, t \in X \text{ and } z \leq t \text{ in } X \\ \text{or} \\ z \in X \text{ and } t \in \{0,1\} \\ \text{or} \\ z = t \in \{0,1\}\ . \end{array}\right.$$

Then $\Lambda_{X * \{0,1\}} = 2{\cdot}1_B - \Lambda_X$ in $B(G)$.

Proof: Assertions 1 and 2 are straightforward.

Assertion 3 follows from Lemma 11.2.9, and from the fact that for any subgroup H of G, the posets $(X \times Y)^H$ and $X^H \times Y^H$ are isomorphic. The equality $\Lambda_{X \times Y} = \Lambda_X \Lambda_Y$ now follows from Lemma 11.2.8.

For Assertion 4, let $s = \{z_0 < z_1 < \cdots < z_n\}$ be a chain in $X * \{0,1\}$. Then either $z_n \in X$, and then s is a chain in X, or $s = \{0\}$, or $s = \{1\}$, or $s = s' \sqcup \{m\}$, where s' is a chain in X and $m \in \{0,1\}$. In the latter case, the stabilizer G_s of s in G is equal to the stabilizer of s'. Thus

$$\begin{aligned} \Lambda_{X * \{0,1\}} &= \sum_{n \in \mathbb{N}} (-1)^n Sd_n(X) + 2G/G + 2 \sum_{n \in \mathbb{N}} (-1)^{n+1} Sd_n(X) \\ &= 2{\cdot}1_B - \sum_{n \in \mathbb{N}} (-1)^n Sd_n(X) \\ &= 2{\cdot}1_B - \Lambda_X\ , \end{aligned}$$

as was to be shown. □

11.2.12. Corollary : *Let G be a finite group. If $a \in B(G)$, then there exists a finite G-poset X such that $\Lambda_X = a$.*

Proof: Any element $a \in B(G)$ is equal to $X - Y$, for suitable finite G-sets X and Y. Let E be a set with 3 elements, with trivial G-action, ordered by the equality relation. It follows that $\Lambda_E = 3G/G = 3{\cdot}1_B$. Thus if $F = E * \{0,1\}$, then $\Lambda_F = -1_B$, and $\Lambda_{F \sqcup F} = -2{\cdot}1_B$. Hence, setting $Z = (Y * \{0,1\}) \sqcup (F \sqcup F)$

$$\begin{aligned}\Lambda_{X \sqcup Z} &= \Lambda_X + \Lambda_{(Y*\{0,1\})\sqcup(F\sqcup F)}\\ &= \Lambda_X + \Lambda_{Y*\{0,1\}} - 2{\cdot}1_B\\ &= \Lambda_X - \Lambda_Y\\ &= X - Y = a\ ,\end{aligned}$$

as was to be shown. □

11.2.13. Bisets and Generalized Tensor Induction.

11.2.14. Notation : *Let G and H be finite groups, and U be an (H,G)-biset. If X is a G-poset, denote by*

$$t_U(X) = \mathrm{Hom}_G(U^{op}, X)$$

the set of maps $f : U \to X$ such that $f(ug) = g^{-1}f(u)$, for any $g \in G$ and $u \in U$. The group H acts on $t_U(X)$ by $(hf)(u) = f(h^{-1}u)$, for any $h \in H$, any $f \in t_U(X)$, and any $u \in U$. The relation

$$\forall f, f' \in t_U(X),\ \ f \le f' \Leftrightarrow \forall u \in U,\ f(u) \le f'(u) \text{ in } X\ ,$$

endows $t_U(X)$ with an H-poset structure.

This construction $X \mapsto t_U(X)$ is called the *generalized tensor induction* of posets, associated to U (see [9] for other aspects of this construction).

11.2.15. Lemma : *Let G and H be finite groups, and U be a finite (H,G)-biset.*

1. *If X is a finite G-poset, then $t_U(X)$ is a finite H-poset.*
2. *There exists a unique map $T_U : B(G) \to B(H)$ such that*

$$T_U(\Lambda_X) = \Lambda_{t_U(X)}$$

for any finite G-poset X.

Assertion 1 is obvious. Proving Assertion 2 amounts to showing that if X and Y are finite G-posets such that $\Lambda_X = \Lambda_Y$ in $B(G)$, then $\Lambda_{t_U(X)} = \Lambda_{t_U(Y)}$ in $B(H)$. By Lemma 11.2.8, this is equivalent to $\chi\big(t_U(X)^K\big) = \chi\big(t_U(Y)^K\big)$, for any subgroup K of H. Now

$$t_U(X)^K = \{f : U \to X \mid f(kug) = g^{-1}f(u),\ \forall k \in K, \forall u \in U, \forall g \in G\} .$$

Thus, an element f of $t_U(X)^K$ is determined by its values $f(u)$, when u runs through a set $[K\backslash U/G]$ of representatives of $K\backslash U/G$. Conversely, if these values are given, one can define the value $f(v)$ for any $v \in U$ by $f(v) = g^{-1}f(u)$, if v can be written $v = kug$, for $k \in K$, $g \in G$, and $u \in [K\backslash U/G]$. This definition is coherent if $g^{-1}f(u) = f(u)$, whenever $kug = u$. In other words $f(u)$ must be invariant by the subgroup

$$K^u = \{g \in G \mid \exists k \in K,\ ku = ug\}$$

of G. This shows that

$$t_U(X)^K \cong \prod_{u\in[K\backslash U/G]} X^{K^u} ,$$

and this bijection is actually an isomorphism of posets. In particular, by Lemma 11.2.9

(11.2.16)
$$\chi\big(t_U(X)^K\big) = \prod_{u\in[K\backslash U/G]} \chi(X^{K^u}) .$$

This shows that if $\Lambda_X = \Lambda_Y$, then $\Lambda_{t_U(X)} = \Lambda_{t_U(Y)}$. So one can define a map $T_U : B(G) \to B(H)$ by setting $T_U(a) = \Lambda_{t_U(X)}$, where X is a finite G-poset such that $\Lambda_X = a$, as in Corollary 11.2.12. □

11.2.17. Corollary : *Let G and H be finite groups, and U be a finite (H,G)-biset. If $a \in B(G)$, then for any subgroup K of H*

$$|T_U(a)^K| = \prod_{u\in[K\backslash U/G]} |a^{K^u}| ,$$

where $[K\backslash U/G]$ is a set of representatives of $K\backslash U/G$.

Proof: This follows from 11.2.16. □

11.2.18. Example (Inflation) : Let G be a finite group, and N be a normal subgroup of G. Let U denote the $(G,G/N)$-biset G/N, for the obvious biset structure. If $u \in B(G/N)$ and $H \le G$, then

$$|T_U(u)^H| = |u^{HN/N}| .$$

It follows that T_U coincides with $\mathrm{Inf}_{G/N}^G$ in this case. Note in particular that $\mathrm{Inf}_{G/G}^G(-1_B) = -1_B$.

In the same situation, if now $V = U^{op}$ is the $(G/N,G)$-biset G/N, with obvious biset structure, if $v \in B(G)$, and K/N is a subgroup of G/N, then

$$|T_V(v)^{K/N}| = |v^K| = |(v^N)^{K/N}| \,.$$

It follows that T_V is the map $v \mapsto v^N$ from $B(G)$ to $B(G/N)$ in this case.

11.2.19. Proposition : *Let G and H be finite groups, and U be a finite (H,G)-biset. Then $T_U(G/G) = H/H$, and $T_U(ab) = T_U(a)T_U(b)$, for any $a,b \in B(G)$. In particular, the restriction of T_U to $B(G)^\times$ is a group homomorphism $B(G)^\times \to B(H)^\times$.*

Proof: Let $\bullet$ be a G-set of cardinality 1. Then $\Lambda_\bullet = G/G$. So $T_U(G/G) = \Lambda_{t_U(\bullet)}$. But there is a unique map from U to $\bullet$, so $t_U(\bullet)$ is an H-set of cardinality 1. It follows that $T_U(G/G) = H/H$.

Now if $a,b \in B(G)$, let X and Y be finite G-sets such that $\Lambda_X = a$ and $\Lambda_Y = b$. Then

$$T_U(ab) = T_U(\Lambda_X\Lambda_Y) = T_U(\Lambda_{X\times Y}) = \Lambda_{t_U(X\times Y)} \,.$$

Moreover there are isomorphisms of posets

$$\begin{aligned} t_U(X\times Y) &= \mathrm{Hom}_G(U^{op}, X\times Y)\\ &\cong \mathrm{Hom}_G(U^{op},X)\times \mathrm{Hom}_G(U^{op},Y)\\ &= t_U(X)\times t_U(Y)\,, \end{aligned}$$

thus $\Lambda_{t_U(X\times Y)} = \Lambda_{t_U(X)\times t_U(Y)} = \Lambda_{t_U(X)}\Lambda_{t_U(Y)}$, thus $T_U(ab) = T_U(a)T_U(b)$. In particular, if $a \in B(G)^\times$, then there exists $b \in B(G)$ such that $ab = G/G$. Hence $T_U(ab) = T_U(a)T_U(b) = T_U(G/G) = H/H$, so $T_U(a) \in B(H)^\times$, and the restriction of T_U is a group homomorphism $B(G)^\times \to B(H)^\times$. □

11.2.20. Proposition : *Let G, H, and K be finite groups.*

1. *If U is a finite (H,G)-biset and V is a finite (K,H)-biset, then*

$$T_V \circ T_U = T_{V\times_H U} \,.$$

2. *If U and U' are finite (H,G)-bisets, then $T_{U\sqcup U'}(a) = T_U(a)T_{U'}(a)$, for any $a \in B(G)$.*
3. *Let Id_G denote the identity (G,G)-biset. Then T_{Id_G} is the identity map of $B(G)$.*

Proof: Let X be a finite G-poset. For Assertion 1, observe that there are isomorphisms of K-posets

$$\begin{aligned} t_V \circ t_U(X) &= \mathrm{Hom}_H\big(V^{op}, \mathrm{Hom}_G(U^{op}, X)\big) \\ &\cong \mathrm{Hom}_G(U^{op} \times_H V^{op}, X) \\ &\cong \mathrm{Hom}_G\big((V \times_H U)^{op}, X\big) \\ &= t_{V\times_H U}(X) , \end{aligned}$$

by Proposition 2.3.14. Thus

$$\begin{aligned} T_V \circ T_U(\Lambda_X) &= T_V(\Lambda_{t_U(X)}) \\ &= \Lambda_{t_V \circ t_U(X)} \\ &= \Lambda_{t_{V\times_H U}(X)} \\ &= T_{V\times_H U}(\Lambda_X) , \end{aligned}$$

as was to be shown.

Similarly, for Assertion 2, there are isomorphisms of H-posets

$$\begin{aligned} t_{U\sqcup U'}(X) &= \mathrm{Hom}_G(U^{op} \sqcup U'^{op}, X) \\ &\cong \mathrm{Hom}_G(U^{op}, X) \times \mathrm{Hom}_G(U'^{op}, X) \\ &= t_U(X) \times t_{U'}(X) . \end{aligned}$$

Thus $T_{U\sqcup U'}(\Lambda_X) = \Lambda_{t_U(X)\times t_{U'}(X)} = \Lambda_{t_U(X)}\Lambda_{t_{U'}(X)} = T_U(X)T_{U'}(X)$.

Finally, for Assertion 3, the following isomorphism of G-posets

$$t_{\mathrm{Id}_G}(X) = \mathrm{Hom}_G(G^{op}, X) \cong X$$

implies that $T_{\mathrm{Id}_G}(\Lambda_X) = \Lambda_X$. □

11.2.21. Corollary : *There exists a unique biset functor $B^\times$ such that $B^\times(G) = B(G)^\times$, for any finite group G, and*

$$B^\times(U) = T_U : B^\times(G) \to B^\times(H)$$

for any finite groups G and H, and any finite (H,G)-biset U.

Proof: This follows from Proposition 11.2.20, using the universal property of Burnside groups (see Remark 2.4.3). □

11.2.22. Remark : It follows easily from Corollary 11.2.17 that the operations of restriction to a subgroup, and inflation from a factor group, for the functor $B^\times$, coincide with the similar operations for the Burnside functor. On the other hand, the operation of deflation $\mathrm{Def}^G_{G/N}$ is the *fixed points operation* $v \mapsto v^N$, instead of the *orbit operation* $v \mapsto N\backslash v$ for B.

11.2.23. Remark (faithful elements in $B^\times(G)$) : In particular, if G is a finite group, then an element $u \in B^\times(G)$ belongs to $\partial B^\times(G)$ if and only if $|u^H| = 1$, whenever H is a subgroup of G containing a non trivial normal subgroup of G.

11.2.24. Notation : *When H is a subgroup of G, the induction operation $B^\times(H) \to B^\times(G)$ is called* multiplicative induction *or* tensor induction, *and denoted by* Ten^G_H. *It is different from ordinary* additive induction *for B.*

Similarly, when (T,S) is a section of G, the map

$$B^\times(G/S) : B^\times(T/S) \to B^\times(G) ,$$

associated to the $(G, T/S)$-biset G/S, will be denoted by $\mathrm{Teninf}^G_{T/S}$. *It is equal to the composition of* $\mathrm{Inf}^T_{T/S}$, *followed by* Ten^G_T.

11.2.25. A Morphism of Biset Functors. The main result of this subsection states that $B^\times$ is a biset subfunctor of the dual of the Burnside functor with values in $\mathbb{Z}/2\mathbb{Z}$.

11.2.26. Lemma : *Let G and H be groups, and U be an (H,G)-biset. Then for any subgroup K of G, there is an isomorphism of H-sets*

$$U \times_G (G/K) \cong \bigsqcup_{u \in [H\backslash U/K]} H/{}^uK ,$$

where $[H\backslash U/K]$ is a set of representatives of $H\backslash U/L$, and ${}^uK$ is the subgroup of H defined in Notation 2.3.16 by ${}^uK = \{h \in H \mid \exists k \in K,\ \ hu = uk\}$.

Proof: Obviously $U \times_G (G/K) \cong U/K$, and for $u \in U$, the stabilizer of uK in H is equal to ${}^uK$. □

11.2.27. Notation : *Let $z \mapsto z_+$ denote the unique group isomorphism from $\mathbb{Z}^\times = \{\pm 1\}$ to $\mathbb{Z}/2\mathbb{Z}$.*

11.2.28. Proposition : *Let G be a finite group, and $\epsilon_G : B^\times(G) \to \mathrm{Hom}_{\mathbb{Z}}(B(G), \mathbb{Z}/2\mathbb{Z})$ denote the map defined by*

$$\forall v \in B^\times(G),\ \forall K \leq G,\ \ \epsilon_G(u)(G/K) = |u^K|_+ .$$

Then the maps ϵ_G form an injective morphism of biset functors

$$\epsilon : B^\times \rightarrow \mathrm{Hom}_{\mathbb{Z}}(B, \mathbb{Z}/2\mathbb{Z}) .$$

Proof: Let G and H be finite groups, and U be a finite (H,G)-biset. Denote by F the biset functor $\mathrm{Hom}_{\mathbb{Z}}(B, \mathbb{Z}/2\mathbb{Z})$. Then, for $v \in B^\times(G)$, and for a subgroup L of H

$$\begin{aligned}
\epsilon_H\big(T_U(v)\big)(H/L) &= |T_U(v)^L|_+ \\
&= \Big(\prod_{u\in[L\backslash U/G]} |v^{L^u}|\Big)_+ = \sum_{u\in[L\backslash U/G]} |v^{L^u}|_+ \\
&= \sum_{u\in[L\backslash U/G]} \epsilon_G(v)(G/L^u) = \epsilon_G(v)\Big(\sum_{u\in[L\backslash U/G]} G/L^u\Big) \\
&= \epsilon_G(v)\big(U^{op}\times_H (H/L)\big) \\
&= F(U)\big(\epsilon_G(v)\big)(H/L) ,
\end{aligned}$$

where the last equality follows from the definition of the map $F(U)$, and the next to last equality follows from Lemma 11.2.26, applied with G instead of H, H instead of G, U^{op} instead of U, and L instead of K.

So $\epsilon : B^\times \rightarrow \mathrm{Hom}_{\mathbb{Z}}(B, \mathbb{Z}/2\mathbb{Z})$ is a morphism of biset functors. It is clearly injective, since if $v \in B^\times(G)$ and if $|v^K|_+ = 0$ for any $K \leq G$, then $|v^K| = 1$ for any $K \leq G$, hence $v = 1_B$. □

11.2.29. The Case of p-Groups. Let p be a prime number. This subsection describes the structure of the restriction of $B^\times$ to the subcategory $\mathcal{C}_p$ of the biset category, i.e. the restriction of $B^\times$ to finite p-groups. This restriction is still denoted by $B^\times$.

By the Ritter–Segal Theorem 9.2.1, there is an exact sequence of p-biset functors

(11.2.30)
$$0 \rightarrow K \rightarrow B \xrightarrow{\chi} R_{\mathbb{Q}} \rightarrow 0 ,$$

where χ is the linearization morphism. Moreover, for each p-group P, since $R_{\mathbb{Q}}(P)$ is a free abelian group, the sequence

$$0 \rightarrow K(P) \rightarrow B(P) \rightarrow R_{\mathbb{Q}}(P) \rightarrow 0$$

is a split exact sequence of abelian groups. Hence applying $\mathrm{Hom}_{\mathbb{Z}}(-, \mathbb{Z}/2\mathbb{Z})$ to the exact sequence 11.2.30 yields a short exact sequence of p-biset functors

(11.2.31)
$$0 \rightarrow \mathrm{Hom}_{\mathbb{Z}}(R_{\mathbb{Q}}, \mathbb{Z}/2\mathbb{Z}) \xrightarrow{\iota} \mathrm{Hom}_{\mathbb{Z}}(B, \mathbb{Z}/2\mathbb{Z}) \xrightarrow{j} \mathrm{Hom}_{\mathbb{Z}}(K, \mathbb{Z}/2\mathbb{Z}) \rightarrow 0 ,$$

where $\iota = \mathrm{Hom}_{\mathbb{Z}}(\chi, \mathbb{Z}/2\mathbb{Z})$, and, for a finite p-group P, the map j_P is given by restriction of $\mathbb{Z}/2\mathbb{Z}$-valued linear forms from $B(P)$ to $K(P)$.

It turns out that the morphism $\epsilon : B^\times \to \mathrm{Hom}_{\mathbb{Z}}(B, \mathbb{Z}/2\mathbb{Z})$ factors through the map ι:

11.2.32. Proposition : *There exists an injective homomorphism of p-biset functors $\bar{\epsilon} : B^\times \to \mathrm{Hom}_{\mathbb{Z}}(R_{\mathbb{Q}}, \mathbb{Z}/2\mathbb{Z})$ such that $\epsilon = \iota \circ \bar{\epsilon}$. In particular $B^\times$ is a rational p-biset functor.*

Proof: Since the sequence 11.2.31 is an exact sequence of p-biset functors, the existence of $\bar{\epsilon}$ is equivalent to $j \circ \epsilon = 0$. In other words

$$\forall P,\ \forall v \in B^\times(P),\ \forall k \in K(P),\ \ \epsilon(v)(k) = 0\ , \tag{11.2.33}$$

where P is a finite p-group. By Theorem 11.1.15, if $p > 2$, the p-biset functor K is generated by one element $\delta \in K(X)$, where X is an extraspecial p-group of order p^3 and exponent p, and if $p = 2$, it is generated by the series of elements $\delta_n \in K(X_n)$, for $n \geq 3$, where X_n is a dihedral group of order 2^n. Hence proving 11.2.33 is equivalent to proving that

$$\epsilon_P(v)\big(B(U)(\delta_n)\big) = 0\ ,$$

for $n = 3$ if $p > 2$ (with the above convention $X_3 = X$ and $\delta_3 = \delta$ in this case), or for $n \geq 3$ if $p = 2$, and any finite (P, X_n)-biset U. But again, setting $F = \mathrm{Hom}_{\mathbb{Z}}(B, \mathbb{Z}/2\mathbb{Z})$,

$$\begin{aligned}\epsilon(v)\big(B(U)(\delta_n)\big) &= F(U^{op})\big(\epsilon_P(v)\big)(\delta_n)\\ &= \epsilon_{X_n}\big(B^\times(U^{op})(v)\big)(\delta_n)\ .\end{aligned}$$

Since $B^\times(U^{op})(v) \in B^\times(X_n)$, it is enough to show that $\epsilon_{X_n}(w)(\delta_n) = 0$, for any $w \in B^\times(X_n)$.

Now each of the groups X_n has a unique minimal (non trivial) normal subgroup, equal to its center Z, of order p. This shows that

$$B^\times(X_n) = \partial B^\times(X_n) \oplus \mathrm{Inf}_{X_n/Z}^{X_n} B^\times(X_n/Z)\ .$$

Moreover

$$\epsilon_{X_n}(\mathrm{Inf}_{X_n/Z}^{X_n} t)(\delta_n) = \epsilon_{X_n/Z}(t)(\mathrm{Def}_{X_n/Z}^{X_n} \delta_n)\ ,$$

and it is clear from the definition of δ_n that $\mathrm{Def}_{X_n/Z}^{X_n} \delta_n = 0$. So it is enough to show that $\epsilon_{X_n}(w)(\delta_n) = 0$, for any $w \in \partial B^\times(X_n)$.

If $p > 2$, there is nothing more to do, since $\partial B^\times(X_n)$ is the trivial group, by Lemma 11.2.5 and Example 11.2.18.

If $p = 2$, then X_n is a dihedral group of order 2^n, with $n \geq 3$. Then $w \in \partial B^\times(X_n)$ if and only if $w \in B^\times(X_n)$ and w is in the kernel of $\mathrm{Def}_{X_n/Z}^{X_n}$.

By Remark 11.2.23, this is equivalent to $|w^Q| = 1$, for any subgroup Q of X_n containing Z. And if $Q \leq X_n$ with $Z \not\leq Q$, then $Q = \mathbf{1}$, or $Q = I_n$, or $Q = J_n$, up to conjugation in X_n. Since

$$w = \sum_{Q \in S} |w^Q| e_Q^{X_n} ,$$

where S is a set of representatives of conjugacy classes of subgroups of X_n, it follows that

$$\begin{aligned} w &= \Big(\sum_{\substack{Q \in S \\ Q \geq Z}} e_Q^{X_n} \Big) \pm e_{I_n}^{X_n} \pm e_{J_n}^{X_n} \pm e_{\mathbf{1}}^{X_n} \\ &= 1_B - 2a e_{I_n}^{X_n} - 2b e_{J_n}^{X_n} - 2c e_{\mathbf{1}}^{X_n} , \end{aligned}$$

where a, b, c are equal to 0 or 1. Since

$$e_{I_n}^{X_n} = \frac{1}{4}(-X_n/\mathbf{1} + 2X_n/I_n) , \qquad e_{J_n}^{X_n} = \frac{1}{4}(-X_n/\mathbf{1} + 2X_n/J_n) ,$$

by Theorem 2.5.2, and since $e_{\mathbf{1}}^{X_n} = \frac{1}{2^n} X_n/\mathbf{1}$, it follows that

$$w = 1_B - \frac{a}{2}(-X_n/\mathbf{1} + 2X_n/I_n) - \frac{b}{2}(-X_n/\mathbf{1} + 2X_n/J_n) - \frac{c}{2^{n-1}} X_n/\mathbf{1} .$$

The coefficient of $X_n/\mathbf{1}$ in w is equal to $\frac{a+b}{2} - \frac{c}{2^{n-1}}$. This is an integer, so in particular $c \equiv 2^{n-2}(a+b)$ modulo 2^{n-1}. Since $n \geq 3$, it follows that c is even, hence $c = 0$ since $c \in \{0, 1\}$. Now $\frac{a+b}{2}$ is an integer, so $a = b$, since a and b are equal to 0 or 1. So either $a = b = 0$, and then $w = 1_B$, or $a = b = -1$, and w is equal to

$$\upsilon_n = (X_n/X_n + X_n/\mathbf{1}) - (X_n/I_n + X_n/J_n) .$$

It follows that $\partial B^\times(X_n)$ has order 2 in this case, generated by υ_n. The linear form $\epsilon_{X_n}(\upsilon_n)$ takes the following values

$$\epsilon_{X_n}(\upsilon_n)(X_n/Q) = \begin{cases} 1_{\mathbb{Z}/2\mathbb{Z}} \text{ if } Q =_{X_n} I_n \text{ or } Q =_{X_n} J_n \\ 0_{\mathbb{Z}/2\mathbb{Z}} \text{ otherwise.} \end{cases}$$

Thus $\epsilon_{X_n}(\upsilon_n)(\delta_n) = 1_{\mathbb{Z}/2\mathbb{Z}} - 1_{\mathbb{Z}/2\mathbb{Z}} = 0_{\mathbb{Z}/2\mathbb{Z}}$, as was to be shown.

It follows that the functor $B^\times$ is isomorphic to a subfunctor of the functor $\mathrm{Hom}_{\mathbb{Z}}(R_{\mathbb{Q}}, \mathbb{Z}/2\mathbb{Z})$, which is a rational p-biset functor, by Theorem 10.1.5. Hence $B^\times$ is a rational p-biset functor. □

The main consequence of this fact is that $B^\times$ is determined by the groups $\partial B^\times(N)$, for $N \in \mathcal{N}$. These groups have been determined by Yalçın (with a different terminology, see [58] Lemmas 4.6 and 5.2):

11.2.34. Lemma : *Let N be a p-group of normal p-rank 1. Then $\partial B^\times(N)$ is trivial, except if N is*

- *the trivial group, and $\partial B^\times(N)$ is the group of order 2 generated by $\upsilon_N = -N/N$.*
- *cyclic of order 2, and $\partial B^\times(N)$ is the group of order 2 generated by*

$$\upsilon_N = N/N - N/\mathbf{1} .$$

- *dihedral of order at least 16, and then $\partial B^\times(N)$ is the group of order 2 generated by the element*

$$\upsilon_N = N/N + N/1 - N/I - N/J ,$$

where I and J are non-central subgroups of order 2 of N, not conjugate in N.

Proof: Recall from Remark 11.2.23 that $u \in \partial B^\times(N)$ if and only if u is an element of $B^\times(N)$ such that $|u^H| = 1$ whenever H is a subgroup of N such that $H \cap Z(N) \neq \mathbf{1}$.

If N is trivial, then obviously $\partial B^\times(N) = B^\times(N) = \{\pm 1_B\}$. It is generated by $\upsilon_N = -1_B = -N/N$. So one can assume that N is non trivial.

If N is cyclic or generalized quaternion, of order $p^n > 1$, then the only subgroup H of N such that $H \cap Z(N) = \mathbf{1}$ is the trivial group. It follows that the only possibly non trivial element in $\partial B^\times(N)$ is equal to $1_B - 2e_\mathbf{1}^N = 1_N - \frac{2}{p^n}N/\mathbf{1}$. The coefficient of $N/\mathbf{1}$ is this expression is equal to $\frac{2}{p^n}$. If $p > 2$, this is not an integer (since $n \geq 1$). If $p = 2$, this is an integer only if $n = 1$, i.e. if N is cyclic of order 2. In this case $\partial B^\times(N)$ has order 2. It is generated by the element

$$\upsilon_N = 1 - 2e_\mathbf{1}^N = N/N - N/\mathbf{1} .$$

If $p = 2$ and N is semi-dihedral of order 2^n, with $n \geq 4$, and if H is a subgroup of N such that $H \cap Z(N) = \mathbf{1}$, then H is a non central subgroup of N of order 2, and there is a single conjugacy class of such subgroups in N. Let I denote one of these subgroups. The elements of $\partial B^\times(N)$ are equal to

$$1_B - 2ae_\mathbf{1}^N - 2be_I^N , \qquad \textbf{(11.2.35)}$$

where a and b are equal to 0 or 1. Since

$$e_I^N = \frac{1}{4}(-N/\mathbf{1} + 2N/I) , \qquad e_\mathbf{1}^N = \frac{1}{2^n}N/\mathbf{1} ,$$

the coefficient of $N/\mathbf{1}$ in 11.2.35 is equal to $\frac{a}{2} - \frac{b}{2^{n-1}}$. This is an integer, so b is even, hence $b = 0$ since $b \in \{0, 1\}$. Now $\frac{a}{2}$ is an integer, so $a = 0$ for the same reason. It follows that $\partial B^\times(N)$ is trivial in this case.

Finally, the case of dihedral groups follows from the proof of Proposition 11.2.32.

□

11.2.36. Theorem : [16, Theorem 8.5] *Let P be a finite p-group. Then $B^\times(P)$ is an elementary abelian 2-group of rank equal to the number of isomorphism classes of rational irreducible representations of P whose type is trivial, cyclic of order 2, or dihedral. More precisely:*

1. *If $p \neq 2$, then $B^\times(P) = \{\pm 1\}$.*
2. *If $p = 2$, then let $\mathcal{G}$ be a genetic basis of P, and let $\mathcal{H}$ be the subset of $\mathcal{G}$ consisting of elements S such that $N_P(S)/S$ is trivial, cyclic of order 2, or dihedral. If $S \in \mathcal{H}$, then $\partial B^\times\big(N_P(S)/S\big)$ has order 2, generated by $\upsilon_{N_P(S)/S}$. Then the set*

$$\{\mathrm{Teninf}_{N_P(S)/S}^P \upsilon_{N_P(S)/S} \mid S \in \mathcal{H}\}$$

is an $\mathbb{F}_2$-basis of $B^\times(P)$.

Proof: This follows from Proposition 11.2.32 and Lemma 11.2.34, by definition of a rational p-biset functor.

□

11.2.37. Theorem :

1. *If $p \neq 2$, then the p-biset functor $B^\times$ is simple, isomorphic to $S_{\mathbf{1},\mathbb{F}_2}$, i.e. the constant functor $\Gamma_{\mathbb{F}_2}$.*
2. *If $p = 2$, then the p-biset functor $B^\times$ is uniserial: the set of its subfunctors $\{0\} = F_0 \subset F_1 \subset F_2 \subset \ldots \subset F_n \subset \ldots \subset F_\infty = B^\times$ is linearly ordered. The subfunctor F_1 is generated by $\upsilon_\mathbf{1} = -1 \in B^\times(\mathbf{1})$, and for $n \geq 2$, the subfunctor F_n is generated by $\upsilon_{n+2} \in B^\times(D_{2^{n+2}})$. Moreover, for $n \in \mathbb{N}$,*

$$F_{n+1}/F_n \cong \begin{cases} S_{\mathbf{1},\mathbb{F}_2} & \text{if } n = 0 \\ S_{D_{2^{n+3}},\mathbb{F}_2} & \text{if } n > 0\,. \end{cases}$$

Proof: Since $R_\mathbb{Q}(P)$ is a free abelian group, for any finite p-group P, applying the functor $\mathrm{Hom}_\mathbb{Z}(R_\mathbb{Q}, -)$ to the exact sequence $0 \to \mathbb{Z} \overset{\times 2}{\to} \mathbb{Z} \to \mathbb{Z}/2\mathbb{Z} \to 0$ of abelian groups yields a short exact sequence of functors

$$0 \to R_\mathbb{Q}^* \overset{\times 2}{\to} R_\mathbb{Q}^* \to \mathrm{Hom}_\mathbb{Z}(R_\mathbb{Q}, \mathbb{Z}/2\mathbb{Z}) \to 0\,.$$

This shows that $\mathrm{Hom}_\mathbb{Z}(R_\mathbb{Q}, \mathbb{Z}/2\mathbb{Z}) \cong R_\mathbb{Q}^*/2R_\mathbb{Q}^*$. Since $B^\times$ is isomorphic to a subfunctor of $\mathrm{Hom}_\mathbb{Z}(R_\mathbb{Q}, \mathbb{Z}/2\mathbb{Z})$, by Proposition 11.2.32, it follows that there exists a subfunctor L of $R_\mathbb{Q}^*$, containing $2R_\mathbb{Q}^*$, such that $B^\times \cong L/2R_\mathbb{Q}^*$.

By Theorem 10.4.9, the lattice of non zero subfunctors of $R_{\mathbb{Q}}^*$ is isomorphic to the opposite lattice of the poset $\widehat{\mathfrak{T}}$ of triples (d,m,f), where d and m are integers with $d \mid p-1$ and $p \nmid m$, and $f : \mathfrak{N} \to \mathbb{N}$ is a function such that $f(N) \leq f(M) \leq f(N)+1$ whenever $N \twoheadrightarrow M$ in $\mathfrak{N}$. The ordering on $\widehat{\mathfrak{T}}$ was defined by $(d,m,f) \leq (d',m',f')$ if and only if $m \mid m'$, $dm \mid d'm'$, and $f \leq f'$.

Recall that when L is a non zero subfunctor of $R_{\mathbb{Q}}^*$, the corresponding triple (d,m,f) is obtained by first defining integers $a_N \geq 0$, for $N \in \mathcal{N}$, by $\partial R_{\mathbb{Q}}^*(N) = \langle a_N \Phi_N^* \rangle$, and then using relations 10.4.10

$$\begin{cases} m = (a_{\mathbf{1}})_{p'} \\ d = a_{C_p}/a_{\mathbf{1}} \\ p^{f(N)} = a_N/dm \text{ for } N \in \mathfrak{N} \,. \end{cases}$$

The integers a_N associated to the subfunctor $2R_{\mathbb{Q}}^*$ of $R_{\mathbb{Q}}^*$ are all equal to 2. Hence there are two cases:

• If $p \neq 2$, then the triple corresponding to $2R_{\mathbb{Q}}^*$ is equal to $(1,2,0)$, where 0 denotes the zero function. If L is a subfunctor of $R_{\mathbb{Q}}^*$ containing $2R_{\mathbb{Q}}^*$, then L corresponds to a triple (d,m,f) such that $(d,m,f) \leq (1,2,0)$ in $\widehat{\mathfrak{T}}$. Thus $f=0$, $m \mid 2$, and $dm \mid 2$.

If $m=2$, then $d=1$, and $L = 2R_{\mathbb{Q}}^*$. Otherwise $m=1$, and $d \mid 2$. If $d=1$, then $L = R_{\mathbb{Q}}^*$. It follows that there is a unique proper subfunctor L strictly containing $2R_{\mathbb{Q}}^*$, corresponding to the triple $(2,1,0)$. The integers a_N for this subfunctor are given by the relations 10.4.11

$$\begin{cases} a_{\mathbf{1}} = mp^{f(C_p)} \\ a_N = dmp^{f(N)} \text{ for } N \in \mathfrak{N} \,. \end{cases}$$

Thus $a_{\mathbf{1}} = 1$, and $a_N = 2$, for $N \in \mathfrak{N}$. Now the functor $L/2R_{\mathbb{Q}}^*$ is a simple p-biset functor, whose evaluation at the trivial group is isomorphic to $(a_{\mathbf{1}}\mathbb{Z})/2\mathbb{Z} = \mathbb{Z}/2\mathbb{Z}$. Thus $L/2R_{\mathbb{Q}}^* \cong S_{\mathbf{1},\mathbb{F}_2}$. Moreover since $B^\times(P)$ has order 2 for any finite p-group P, the functor $B^\times$ is non zero, and different from $R_{\mathbb{Q}}^*/2R_{\mathbb{Q}}^*$. Thus $B^\times \cong S_{\mathbf{1},\mathbb{F}_2}$, which is also the constant functor $\Gamma_{\mathbb{F}_2}$.

• If $p=2$, then the relations 10.4.10 show that the triple associated to the subfunctor $2R_{\mathbb{Q}}^*$ is equal to $(1,1,1)$, where the third 1 denotes the constant function equal to 1. If L is the subfunctor such that $L/R_{\mathbb{Q}}^* = \bar{\epsilon}(B^\times)$, then L corresponds to a triple (d,m,f) of $\widehat{\mathfrak{T}}$ such that $(d,m,f) \leq (1,1,1)$. Thus $d=m=1$, and f has values equal to 0 or 1.

Since moreover $\partial B^\times(N) = \{0\}$ by Lemma 11.2.34, for $N \in \mathcal{N}$, except if N is trivial, cyclic of order 2, or dihedral, where $\partial B^\times(N) \cong \mathbb{Z}/2\mathbb{Z}$, it follows that the integer a_N for the subfunctor L is equal to 1 if N is trivial, cyclic of order 2, or dihedral, and equal to 2 otherwise. In other words, the values of the function f on the graph $\mathfrak{N}$ are given by the following graph

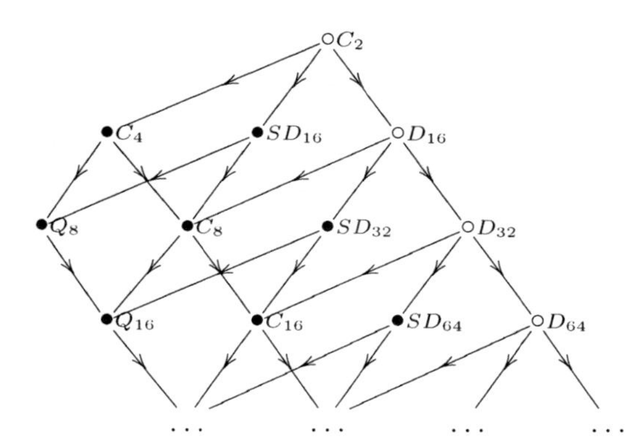

where f is equal to 1 on a $\bullet$ vertex, and to 0 on a $\circ$ vertex.

Now the lattice of subfunctors of $B^\times$ is isomorphic to the lattice of triples (d', m', f') of $\widehat{\mathfrak{T}}$ such that $(d, m, f) \leq (d', m', f') \leq (1, 1, 1)$, i.e. to the poset of functions $f' : \mathfrak{N} \to \mathbb{N}$ such that $f'(N) \leq f'(M) \leq f'(N) + 1$ whenever $N \rightarrowtail M$ in $\mathfrak{N}$, and $f \leq f' \leq 1$. Such a function f' is entirely determined by the set $f'^{-1}(1) - f^{-1}(1)$, which is a closed subset of the graph

$$\underset{C_2}{\circ} \longrightarrow \underset{D_{16}}{\circ} \longrightarrow \cdots \longrightarrow \underset{D_{2^n}}{\circ} \longrightarrow \underset{D_{2^{n+1}}}{\circ} \longrightarrow \underset{D_{2^{n+2}}}{\circ} \longrightarrow \cdots$$

It follows that $B^\times$ has a unique minimal (non zero) subfunctor F_1, corresponding to the set of $\bullet$ vertices in the following graph

$$\underset{C_2}{\circ} \longrightarrow \underset{D_{16}}{\bullet} \longrightarrow \cdots \longrightarrow \underset{D_{2^n}}{\bullet} \longrightarrow \underset{D_{2^{n+1}}}{\bullet} \longrightarrow \underset{D_{2^{n+2}}}{\bullet} \longrightarrow \cdots$$

The integers a_N corresponding to this subfunctor are equal to 1 if N is trivial or cyclic of order 2, and to 2 otherwise. In particular $F_1(\mathbf{1}) \cong (a_\mathbf{1}\mathbb{Z})/2\mathbb{Z} = \mathbb{Z}/2\mathbb{Z}$. Since F_1 is a simple functor, it follows that $F_1 \cong S_{\mathbf{1},\mathbb{F}_2}$.

More generally, the poset of subfunctors of $B^\times$ is linearly ordered

$$\{0\} = F_0 \subset F_1 \subset F_2 \ldots \subset F_n \subset \ldots ,$$

where F_n, for $n > 0$, corresponds to the subset

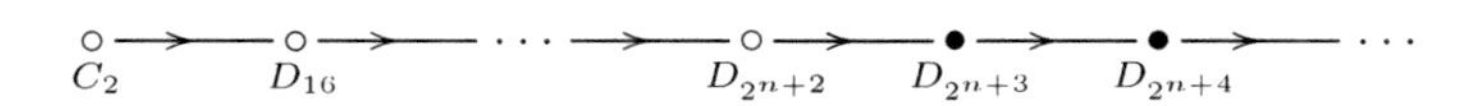

The functor F_{n+1}/F_n is simple, isomorphic to $S_{Q,V}$ for some seed (Q, V). Moreover it is a quotient of a subfunctor of a rational p-biset functor, so it is a rational p-biset functor. Since Q is a minimal group for $S_{Q,V}$, and since $S_{Q,V}(Q) \cong V$ is isomorphic to the direct sum of the groups $\partial S_{Q,V}\big(N_Q(S)/S\big)$, where S runs through a genetic basis $\mathcal{G}$ of Q, it follows that $S = \mathbf{1}$ belongs to $\mathcal{G}$, in other words $Q \in \mathcal{N}$. Moreover $V \cong \partial S_{Q,V}(Q)$.

Now clearly for $N \in \mathcal{N}$, the group $\partial(F_{n+1}/F_n)(N)$ is equal to $\{0\}$, except if $N \cong D_{2^{n+3}}$, and $\partial(F_{n+1}/F_n)(N) \cong \mathbb{Z}/2\mathbb{Z}$ in this case. Since $(F_{n+1}/F_n)(N)$ is isomorphic to the direct sum of the groups $\partial(F_{n+1}/F_n)(N/M)$, for $M \trianglelefteq N$, by Proposition 6.3.3, it follows that $D_{2^{n+3}}$ is a minimal group for F_{n+1}/F_n, and that $(F_{n+1}/F_n)(D_{2^{n+3}}) \cong \mathbb{Z}/2\mathbb{Z}$. Hence $F_{n+1}/F_n \cong S_{D_{2^{n+3}},\mathbb{F}_2}$, for $n > 0$, as was to be shown.

Since $F_1 \cong S_{\mathbf{1},\mathbb{F}_2}$ is clearly generated by $-1 \in F_1(\mathbf{1})$, it remains to prove that for $n \geq 2$, the functor F_n is generated by the element $v_{n+2} \in B^\times(D_{2^{n+2}})$.

Recall from Lemma 11.2.34 that if $n \geq 4$, then $\partial B^\times(D_{2^n}) \cong \mathbb{Z}/2\mathbb{Z}$, generated by v_n. Hence, there exists an integer k_n such that $\partial F_m(D_{2^n}) = \{0\}$ if $m < k_n$, and $\partial F_m(D_{2^n}) = \partial B^\times(D_{2^n}) = <v_n>$ otherwise. Since $F_m/F_{m-1} \cong S_{D_{2^{m+2}},\mathbb{F}_2}$, for $m \geq 2$, it follows that $\partial F_m(D_{2^{m+2}}) \neq \partial F_{m-1}(D_{2^{m+2}})$, thus $k_n = n - 2$.

Now let $m \in \mathbb{N} \sqcup \{\infty\}$, and set $F_\infty = B^\times$. The subfunctor of $B^\times$ generated by v_n is equal to F_m if and only if v_n belongs to $F_m(D_{2^n})$. Equivalently $\partial B^\times(D_{2^n}) \subseteq \partial F_m(D_{2^n})$, which is equivalent to $m \geq n - 2$. This shows that the subfunctor of $B^\times$ generated by v_n is equal to F_{n-2}, for $n \geq 4$. So for $n \geq 2$, the subfunctor F_n is generated by v_{n+2}. □

Chapter 12
The Dade Group

This chapter exposes how the formalism of biset functors, and more precisely of p-biset functors, can be used to describe *the Dade group* $D(P)$ of a finite p-group P. This group was introduced by Dade [30, 31] in order to classify the *endo-permutation modules* for a p-group. This definition will be quickly recalled (see [49] Chap. 5 Sects. 28 and 29 for details), and then an equivalent construction will be given, which is better suited to define biset operations between various Dade groups.

One of the major problems with these operations is that, even if for any finite p-groups P and Q, and any finite (Q, P)-biset U, there is a well defined group homomorphism $D(U) : D(P) \to D(Q)$, this does not endow the correspondence $P \mapsto D(P)$ with a biset functor structure. The reason for this is that in general (with obvious notation), the composition $D(V) \circ D(U)$ is *not equal* to $D(V \times_Q U)$.

However, for each finite p-group P, there is an important subgroup $D^{\Omega}(P)$ of $D(P)$, called *the group of relative syzygies*, for which these problems disappear, so one can define a genuine p-biset functor $P \mapsto D^{\Omega}(P)$. This functor is closely related to the Burnside functor, and to the functor of rational representations.

This functor D^{Ω} is a major tool in the description of the structure of $D(P)$, in particular because the quotient $D(P)/D^{\Omega}(P)$ is always finite (actually $D(P) = D^{\Omega}(P)$ when $p > 2$, but this is far from obvious, and can only be proved at a late stage). The other major tool is the classification of *endo-trivial modules*, which has been completed by J. Carlson and J. Thévenaz in a series of three fundamental papers [25–27]. For a very complete survey of endo-permutation modules, see J. Thévenaz article [50].

The present chapter is intended to be an overview of the biset functor approach to the Dade group. This means in particular that some straightforward proofs will only be sketched, and that the most technical ones will be omitted. The exposition does not always respect the chronology, and some proofs have been reorganized and (hopefully) simplified.

S. Bouc, *Biset Functors for Finite Groups*, Lecture Notes in Mathematics 1990,
DOI 10.1007/978-3-642-11297-3_12,

12.1. Permutation Modules and Algebras

This section briefly recalls basic properties of permutation modules and permutation algebras over p-groups.

12.1.1. Permutation Modules. Let p be a prime number, let P be a finite p-group, and k be a field of characteristic p. Recall that *a permutation kP-module M* is a kP-module admitting a P-invariant k-basis X. Equivalently M is isomorphic to kX, where X is some P-set. In this chapter, all permutation kP-modules will be supposed finitely generated, or equivalently, finite dimensional over k.

A permutation kP-module is isomorphic to a direct sum of modules of the form kP/Q, for subgroups Q of P. Such kP-modules are indecomposable, because they have simple socle, and kP/Q has vertex Q. It follows that any direct summand of a permutation kP-module is a permutation kP-module. Also, the Krull–Schmidt theorem implies that if V is a permutation kP-module, and if X and Y are P-invariant k-bases of V, then the P-sets X and Y are isomorphic.

If V and W are permutation kP-modules, with respective P-invariant k-bases X and Y, then $V \oplus W \cong k(X \sqcup Y)$ is a permutation kP-module. Similarly $V \otimes W \cong k(X \times Y)$ is a permutation kP-module (where $\otimes$ denotes $\otimes_k$).

Let P and Q be finite p-groups, and U be a finite (Q,P)-biset. If V is a permutation kP-module, with P-invariant k-basis X, then $kU \otimes_{kP} V$ is isomorphic to $k(U \times_P X)$, so it is a permutation kQ-module. Thus, the class of permutation modules is closed under the usual biset operations on modules, namely induction, restriction, inflation and deflation. It is also closed under the following construction:

12.1.2. Notation : *Let k be a field, and P be a group. If V is a kP-module, and $Q \leq P$, the* Brauer quotient *$V[Q]$ of V at Q is the $kN_P(Q)/Q$-module defined by*

$$V[Q] = V^Q / \sum_{S<Q} \mathrm{tr}_S^Q V^S,$$

where $\mathrm{tr}_S^Q : V^S \to V^Q$ is the trace map $v \mapsto \sum_{x \in Q/S} xv$.

The correspondence $V \mapsto V[Q]$ is a functor from kP-mod to $kN_P(Q)/Q$-mod, denoted by Br_Q^P (or Br_Q if P is clear from context).

(the standard notation for this Brauer quotient is $V(Q)$. The brackets notation $V[Q]$ is preferred here, to avoid confusion with evaluations of biset functors).

It is easy to show that when $V = kX$ is a permutation kP-module, where X is some P-set, then the image of the set X^Q in $V[Q]$ is a k-basis of $V[Q]$. In this case $V[Q]$ is a permutation $kN_P(Q)/Q$-module.

12.1.3. Permutation Algebras. *A P-algebra over k* is a (finite dimensional unital) k-algebra endowed with an action of P by algebra automorphisms. *A permutation P-algebra over k* is a P-algebra over k, which is a permutation kP-module, i.e. which admits a P-invariant k-basis.

A P-algebra A is called *primitive* if the identity element 1_A is a primitive idempotent of the algebra A^P.

It is straightforward to check that if A is a P-algebra over k, and if $Q \leq P$, then $\sum_{S<Q} \mathrm{tr}_S^Q A^S$ is a two sided ideal of A^Q, so $A[Q]$ is a $N_P(Q)/Q$-algebra over k. Moreover, if A is a permutation P-algebra, then $A[Q]$ is a permutation $N_P(Q)/Q$-algebra.

In particular, if A is a permutation P-algebra over k, with P-invariant basis X, and if $Q \trianglelefteq P$, then the natural bijection $(X^Q)^{P/Q} \cong X^P$ induces an algebra isomorphism $A[Q][P/Q] \to A[P]$ (see, e.g. [23] for details).

12.2. Endo-Permutation Modules

The notion of *endo-permutation module* was introduced by Dade [30,31] as a generalization of the notion of *endo-trivial module*. The first examples of such modules were sources of simple modules, and this is the reason to consider first the case of p-groups, as in the following definition:

12.2.1. Definition : *Let k be a field of characteristic $p > 0$, and P be a finite p-group. A finitely generated kP-module M is called an* endo-permutation *module if the kP-module* $\mathrm{End}_k(M)$ *is a permutation module, i.e. if it admits a P-invariant k-basis.*

12.2.2. Remark : In this definition, the action of P on $\mathrm{End}_k(M)$ is given by $(xf)(m) = xf(x^{-1}m)$, for $x \in P$, $f \in \mathrm{End}_k(M)$, and $m \in M$.

12.2.3. Examples :

• Permutation kP-modules are trivial examples of endo-permutation kP-modules.

• Let X be a non empty finite P-set. Let $\epsilon_X : kX \to k$ be the augmentation map, defined by $\epsilon_X(x) = 1$, for $x \in X$. The kernel $\Omega_X(k)$ of ϵ_X is called *the relative syzygy* of the trivial module with respect to X. So there is a short exact sequence

$$0 \longrightarrow \Omega_X(k) \longrightarrow kX \xrightarrow{\epsilon_X} k \longrightarrow 0 \; .$$

It has been shown by J. L. Alperin ([2], see also Lemma 2.3.3 of [10]) that $\Omega_X(k)$ is an endo-permutation kP-module. Details on relative syzygies will be given in Sect. 12.6.

• If M is an endo-permutation kP-module, then its k-dual $M^* = \mathrm{Hom}_k(M,k)$ is also an endo-permutation kP-module: indeed $\mathrm{End}_k(M) \cong M \otimes M^* \cong \mathrm{End}_k(M^*)$, as M is finite dimensional over k.

• Any direct summand M' of an endo-permutation kP-module M is an endo-permutation kP-module: indeed $M' \otimes M'^*$ is a direct summand of $M \otimes M^*$.

• Let M be an endo-permutation kP-module, and P_M be a projective (hence free, but this does not matter here) cover of M. Then there is a short exact sequence E_M of kP-modules

$$0 \longrightarrow \Omega(M) \longrightarrow P_M \longrightarrow M \longrightarrow 0 ,$$

where $\Omega(M)$ is *the first Heller translate* of M.

On the other hand, tensoring with M the sequence E_k yields the exact sequence

$$0 \longrightarrow \Omega(k) \otimes M \longrightarrow P_k \otimes M \longrightarrow M \longrightarrow 0 ,$$

and $P_k \otimes M$ is projective. It follows that $\Omega(M)$ is a direct summand of $\Omega(k) \otimes M$. Since $P_k = kP$, the module $\Omega(k)$ is isomorphic to Ω_P, and in particular it is an endo-permutation kP-module. Thus $\Omega(k) \otimes M$ is an endo-permutation kP-module, and $\Omega(M)$ is an endo-permutation kP-module. Hence the class of endo-permutation kP-modules is closed under taking Heller translates.

• If M and N are endo-permutation kP-modules, then so is $M \otimes N$: indeed $(M \otimes N) \otimes (M \otimes N)^* \cong (M \otimes M^*) \otimes (N \otimes N^*)$ is a permutation kP-module.

• If M and N are endo-permutation kP-module, then

$$(M \oplus N) \otimes (M \oplus N)^* \cong (M \otimes M^*) \oplus (M \otimes N^*) \oplus (N \otimes M^*) \oplus (N \otimes N^*) .$$

In particular, if $M \oplus N$ is an endo-permutation kP-module, then $M \otimes N^*$ is a permutation kP-module. Conversely, if $M \otimes N^*$ is a permutation kP-module, then $N \otimes M^* \cong (M \otimes N^*)^*$ is a permutation module, and $M \oplus N$ is an endo-permutation kP-module.

This leads to the following definition:

12.2.4. Definition : *Let k and P be as in Definition 12.2.1. Two endo-permutation kP-modules M and N are said to be* compatible *(which is denoted by $M \sim N$) if $M \oplus N$ is an endo-permutation kP-module, or, equivalently, if $M \otimes N^*$ is a permutation kP-module.*

12.2.5. Definition : *Let k be a field of characteristic p, and P be a finite p-group. An endo-permutation kP-module M is said to be* capped *if it admits an indecomposable summand with vertex P.*

12.2.6. Lemma : *Let k, P, and M as in Definition 12.2.5. The following conditions are equivalent:*

1. *The module M is capped.*
2. *The Brauer quotient* $\mathrm{End}_k(M)[P]$ *is non zero.*
3. *The trivial module k appears as a direct summand of the kP-module* $\mathrm{End}_k(M) \cong M \otimes M^*$.

Proof: An arbitrary kP-module M admits an indecomposable summand with vertex P if and only if $\mathrm{End}_k(M)[P] \neq \{0\}$, so Assertions 1 and 2 are equivalent. Obviously, Assertion 3 implies Assertion 2, since $k[P] \cong k \neq \{0\}$. And Assertion 2 implies Assertion 3 because $\mathrm{End}_k(M)$ is a permutation kP-module, and because $(kP/Q)[P] = \{0\}$ for any proper subgroup Q of P. □

12.2.7. Lemma : *Let k and P be as in Definition 12.2.5. If M is a capped endo-permutation module, so is M^*. If M and N are capped endo-permutation kP-modules, so is $M \otimes N$.*

Proof: The first assertion is obvious. For the second assertion, observe that $(M \otimes N) \otimes (M \otimes N)^* \cong (M \otimes M^*) \otimes (N \otimes N^*)$ has a direct summand isomorphic to k, since both $M \otimes M^*$ and $N \otimes N^*$ have. □

12.2.8. Theorem and Definition : [Dade] *Let k be a field of characteristic $p > 0$, and P be a finite p-group.*

1. *The relation $\sim$ is an equivalence relation on the class of capped endo-permutation kP-modules. Let $D_k(P)$ denote the set of equivalence classes for this relation.*
2. *Let M and N be capped endo-permutation kP-modules. Then $M \sim N$ if and only if $M^* \sim N^*$.*
3. *If M, N, M' and N' are capped endo-permutation kP-modules such that $M \sim N$ and $M' \sim N'$, then $M \otimes M' \sim N \otimes N'$.*
4. *The tensor product of modules induces an addition on $D_k(P)$, defined by*

$$[M] + [N] = [M \otimes N] ,$$

where $[M]$ denotes the equivalence class of the capped endo-permutation kP-module M.
Then $D_k(P)$ is an abelian group for this addition law, called the Dade group of P over k. *The zero element of $D_k(P)$ is the class $[k]$ of the trivial module, and the opposite of the class of M is the class of M^*.*

Proof: For Assertion 1, the relation $\sim$ is reflexive, for $M \otimes M^*$ is a permutation kP-module, for any endo-permutation kP-module M. It is symmetric, since $M \oplus N \cong N \oplus M$, for any endo-permutation kP-modules M and N. It is transitive: indeed, if L, M, and N are capped endo-permutation kP-modules, if $L \sim M$ and $M \sim N$, then $L \otimes M^*$ and $M \otimes N^*$ are permutation kP-modules. It follows that the module

$$T = (L \otimes M^*) \otimes (M \otimes N^*) \cong L \otimes (M^* \otimes M) \otimes N^*$$

is a permutation kP-module. But since k is a direct summand of $M^* \otimes M$, the module $L \otimes N^*$ is a summand of T, hence it is a permutation kP-module, and $L \sim N$.

Assertion 2 follows from the fact that $(M \oplus N)^* \cong M^* \oplus N^*$. Moreover the dual of an endo-permutation kP-module is an endo-permutation kP-module.

For Assertion 3, observe that

$$(M \otimes N) \otimes (M' \otimes N')^* \cong (M \otimes M'^*) \otimes (N \otimes N'^*) ,$$

and that the tensor product of two permutation kP-modules is a permutation kP-module.

The addition on $D_k(P)$ is well defined by Assertion 3. It is obviously commutative and associative, and $[k]$ is a zero element, since $M \otimes k \cong M$ for any kP-module M. The modules in the class of k are exactly the permutation kP-modules admitting k as a direct summand. In particular, for any capped endo-permutation module M, the module $M \otimes M^*$ is in the class $[k]$, so $[M] + [M^*] = [k]$. This completes the proof of Theorem 12.2.8. □

12.2.9. Lemma : [Dade] *Let k be a field of characteristic $p > 0$, let P be a finite p-group, and let M be a capped endo-permutation kP-module.*

1. *If V is a capped indecomposable endo-permutation kP-module, then $V \sim M$ if and only if V is isomorphic to a direct summand of M.*
2. *In particular, if V and W are indecomposable summands of M with vertex P, then $V \cong W$.*

Proof: If V is a direct summand of M, then $V \otimes M^*$ is a direct summand of $M \otimes M^*$, hence it is a permutation module, and $V \sim M$. Conversely, if $V \sim M$, then $M \otimes V^*$ is a permutation module, and it is a capped endo-permutation module by Lemma 12.2.7. Hence $M \otimes V^*$ has a direct summand isomorphic to k. Then V is isomorphic to a direct summand of $M \otimes V \otimes V^*$.

But $V \otimes V^*$ is a permutation kP-module, admitting k as a direct summand. So $V \otimes V^* \cong k^n \oplus W$, for some positive integer n (actually $n = 1$, but this doesn't matter here), and some module W which is a direct sum of kP-modules induced from proper subgroups of P. It follows that $M \otimes V \otimes V^* \cong M^n \oplus (M \otimes W)$, and $M \otimes W$ is a direct sum of kP-modules induced from proper subgroups of P.

Since V has vertex P, it follows that V is not a direct summand of $M \otimes W$. So V is a direct summand of M^n, hence of M. This proves Assertion 1.

Assertion 2 follows, because if V and W are capped indecomposable summands of M, then $V \sim M$ and $W \sim M$ by Assertion 1, thus $V \sim W$ since the relation $\sim$ is transitive on the class of capped endo-permutation kP-modules, so V is a direct summand of W by Assertion 1 again, and $V \cong W$ since W is indecomposable. □

12.2.10. Definition : *Let k and P as above. If M is a capped endo-permutation kP-module,* a cap *of M is an indecomposable summand of M with vertex P.*

12.2.11. Remark : By Lemma 12.2.9, the cap of a capped endo-permutation kP-module is unique, up to isomorphism, and two capped endo-permutation kP-modules are compatible if and only if they have isomorphic caps. It means that $D_k(P)$ is the set of isomorphism classes of capped indecomposable endo-permutation kP-modules.

12.2.12. Example : the group $\mathbb{Z}/p\mathbb{Z}$. Let P be a cyclic group of order p. Then the group algebra kP is isomorphic to $k[T]/{<}(T-1)^p{>}$. Via this isomorphism, the indecomposable kP-modules are the modules $E_d = k[T]/{<}(T-1)^d{>}$, for $1 \leq d \leq p$. The module E_d is self dual. It has vertex P if $1 \leq d < p$, and trivial vertex if $d = p$.

The indecomposable permutation kP-modules are $E_1 \cong k$, and $E_p \cong kP$. The module E_d is an endo-permutation kP-module if and only if there exists non negative integers n and m such that

$$\text{(12.2.13)} \qquad \mathrm{End}_k(E_d) \cong E_d \otimes E_d^* \cong k^n \oplus (kP)^m \; .$$

It follows in particular that $d^2 = n + mp$. Now there is a projective resolution of E_d

$$\cdots \xrightarrow{b_d} kP \xrightarrow{a_d} kP \xrightarrow{b_d} \cdots kP \xrightarrow{a_d} E_d \longrightarrow 0 \; ,$$

where a_d is the multiplication by $(T-1)^d$ and b_d is the multiplication by $(T-1)^{p-d}$. Applying the functor $\mathrm{Hom}_{kP}(-, E_d)$ to this complex (truncated at E_d) gives the complex of k-vector spaces

$$0 \longrightarrow \mathrm{Hom}_{kP}(kP, E_d) \xrightarrow{\alpha_d} \mathrm{Hom}_{kP}(kP, E_d) \xrightarrow{\beta_d} \mathrm{Hom}_{kP}(kP, E_d) \xrightarrow{\alpha_d} \cdots \; ,$$

whose cohomology is $\mathrm{Ext}^*_{kP}(E_d, E_d)$. Here $\alpha_d = \mathrm{Hom}_{kP}(a_d, E_d)$ and similarly $\beta_d = \mathrm{Hom}_{kP}(b_d, E_d)$. Since $\mathrm{Hom}_{kP}(kP, E_d) \cong E_d$, this complex is isomorphic to

$$0 \longrightarrow E_d \xrightarrow{\alpha_d} E_d \xrightarrow{\beta_d} E_d \xrightarrow{\alpha_d} \cdots \; ,$$

where α_d is the multiplication by $(T-1)^d$ and β_d is the multiplication by $(T-1)^{p-d}$. It follows in particular that $\alpha_d = 0$.

If $p - d \geq d$, i.e. if $d \leq p/2$, then β_d is also equal to zero. In this case, all the groups $\mathrm{Ext}^j_{kP}\,(E_d, E_d)$ have dimension d. Since

$\mathrm{Ext}^j_{kP}(E_d, E_d) \cong H^j\big(P, \mathrm{End}_k(E_d)\big)$, Equation 12.2.13 implies that $d = \dim_k H^j\big(P, k^n \oplus (kP)^m\big)$. This is equal to $n+m$ if $j=0$, and to n if $j>0$. It follows that $m=0$ and $n=d$. Thus $d^2 = n+pm = n = d$, and $d=1$. This shows that if $d \leq p/2$, the module E_d is an endo-permutation kP-module if and only if $d=1$ (i.e. $E_d \cong k$).

And if $p-d<d$, then $\mathrm{Ext}^j_{kP}(E_d, E_d)$ has dimension d if $j=0$, and $p-d$ if $j>0$. Equation 12.2.13 implies that $d=n+m$ and $p-d=n$. It follows that $m=2d-p$, and that $d^2 = p-d+p(2d-p)$. This gives $d=p$ or $d=p-1$ (note that $p-1>p/2$ if and only if $p>2$). The module $E_p \cong kP$ is a permutation module, and the module E_{p-1} is the relative syzygy of the trivial module with respect to the P-set P, hence it is an endo-permutation kP-module.

In other words, if P has order $p>2$, the indecomposable endo-permutation kP-module are $E_0 = k$, $E_{p-1} = \Omega_P$, and $E_p = kP$. If $p=2$, then all kP-modules are permutation modules. In any case, if W is an indecomposable non projective endo-permutation kP-module, then $\mathrm{End}_k(W) \cong k \oplus (kP)^m$, for some integer $m \geq 0$.

Thus $D_k(\mathbb{Z}/2\mathbb{Z})$ is trivial. And if $p>2$, then $D_k(\mathbb{Z}/p\mathbb{Z})$ is generated by the class of the module $E_{p-1} = \Omega_P$. Since E_{p-1} is self dual, the class $[E_{p-1}]$ has order 2 in the Dade group. It follows that $D_k(\mathbb{Z}/p\mathbb{Z})$ is cyclic of order 2 if $p>2$.

12.3. Dade P-Algebras

There is another equivalent definition of the Dade group $D_k(P)$, which is due to L. Puig [40]. The main idea consists in replacing an endo-permutation kP-module M by its endomorphism algebra $A = \mathrm{End}_k(M)$. This algebra has the following properties:

- There is an action of P on A by algebra automorphisms. In other words A is a P-algebra over k.
- The algebra A is a permutation P-algebra over k: in other words A admits a P-invariant basis over k.
- The algebra A is a split simple central k-algebra: in other words A is isomorphic to a full matrix algebra $M_n(k)$, for some integer n.
- The module M is capped if and only if the Brauer quotient $A[P]$ is non zero.

This leads to the following definition:

12.3.1. Definition : [Puig] *Let k be a field of characteristic $p>0$, and P be a finite p-group.* A Dade P-algebra over k *is a split simple central k-algebra, which is a permutation P-algebra over k, and such that $A[P] \neq \{0\}$.*

12.3.2. Remark : The condition $A[P] \neq \{0\}$ is equivalent to $X^P \neq \emptyset$, where X is any P-invariant k-basis of A.

12.3.3. Theorem : *Let k be a field of characteristic $p > 0$, let P be a finite p-group, and let A be a Dade P-algebra over k. If $N \trianglelefteq P$, then $A[N]$ is a Dade P/N-algebra over k.*

Proof: The proof goes by induction on the order of N: if $N = \mathbf{1}$, there is nothing to prove, since $A[N] \cong A$ in this case. Otherwise let Q be a subgroup of order p contained in $N \cap Z(P)$.

Using a P-invariant k-basis of A, it is easy to see that there is an isomorphism of P/N-algebras over k (i.e. an isomorphism of algebra which is compatible with the actions of P/N)

$$A[N] \cong A[Q][N/Q] \; .$$

If $Q < N$, then $A[Q]$ is a Dade P/Q-algebra over k, by induction hypothesis. Similarly, since $|N/Q| < |N|$, the algebra $A[Q][N/Q] \cong A[N]$ is a Dade P/N-algebra over k. So it suffices to consider the case $Q = N$, i.e. the case where N has order p.

Let X be a P-stable basis of A. Then the image of X^N in $A[N]$ is a P/N-stable k-basis of $A[N]$, thus $A[N]$ is a permutation P/N-algebra. Moreover $A[N][P/N] \cong A[P]$ is non zero. So the only thing left to prove is that $A[N]$ is a full matrix algebra over k.

The group N is cyclic of order p, and acts on the algebra $A \cong M_n(k)$, for some integer n. Since all the automorphisms of $M_n(k)$ are inner, this means that a generator of N acts on $M_n(k)$ by conjugation by some invertible matrix M. This automorphism has order 1 or p. This means that M^p is a scalar matrix, i.e. $M^p = \lambda \mathrm{Id}_n$, for some $\lambda \in k^\times$, where Id_n is the identity matrix.

Now the vector space $V = k^n$ is a $k[T]$-module, where T acts via the matrix M. So it splits as a direct sum of indecomposable $k[T]$-modules of the form $k[T]/<R^m>$, where $R \in k[T]$ is some irreducible factor of $T^p - \lambda$. If $\lambda \notin (k^\times)^p$ (which can happen only when k is not a perfect field), then $T^p - \lambda$ is irreducible, and V is isomorphic to a sum of copies of $k[T]/<T^p - \lambda>$. In other words, up to replacing M by some conjugate matrix, one can suppose that M is equal to the block diagonal matrix

$$\begin{pmatrix} C & 0 & \cdots & 0 \\ 0 & C & \ddots & 0 \\ \vdots & \ddots & \ddots & \vdots \\ 0 & \cdots & \cdots & C \end{pmatrix} ,$$

where C is the matrix

$$C = \begin{pmatrix} 0 & 0 & \cdots & \cdots & \lambda \\ 1 & 0 & \ddots & \cdots & 0 \\ 0 & 1 & 0 & \cdots & 0 \\ \vdots & \ddots & \ddots & \ddots & \vdots \\ 0 & \cdots & 0 & 1 & 0 \end{pmatrix} . \quad \textbf{(12.3.4)}$$

Let E denote the $p \times p$-matrix with coefficient 1 on the top left corner, and 0 elsewhere. It is easy to check that

$$\sum_{l=0}^{p-1} C^{-l} E C^{l} = \mathrm{Id}_p \ .$$

It follows that if F is the $n \times n$-matrix, with block diagonal consisting of matrices equal to E, and 0 matrices elsewhere, then

$$\sum_{l=0}^{p-1} M^{-l} F M^{l} = \mathrm{Id}_n \ .$$

Equivalently $1_A = \mathrm{tr}_1^N(F)$, so $A[N] = \{0\}$, which is impossible since $A[P] \cong A[N][P/N] \neq \{0\}$. This contradiction shows that $\lambda \in (k^\times)^p$, say $\lambda = \mu^p$, for some $\mu \in k^\times$. Up to replacing M by $\frac{1}{\mu}M$, which does not change the action of N on A, one can suppose that $M^p = \mathrm{Id}_n$.

In this case, the vector space $V = k^n$ becomes a kN-module, and $A = \mathrm{End}_k(V)$ is a permutation module. In other words V is an endo-permutation kN-module, and the action of N on A comes from this structure. It follows that there are non negative integers m and n such that

$$V \cong W^m \oplus (kN)^n \ ,$$

where W is an indecomposable endo-permutation kN-module with vertex N. Thus

$$\mathrm{End}_k(V) = A \cong V \otimes V^* \cong M_m\big(\mathrm{End}_k(W)\big) \oplus L \ ,$$

where L is some free kN-module. It follows that $A[N] \cong M_m\big(\mathrm{End}_k(W)[N]\big)$. Now it was shown in Sect. 12.2.12 that W is isomorphic to k or to Ω_N, and that in each case $\mathrm{End}_k(W)[N] \cong k$. It follows that $A[N] \cong M_m(k)$, as was to be shown. □

12.3.5. Remark : Theorem 12.3.3 and its proof are essentially due to Dade: starting with a capped endo-permutation kP-module V, Dade shows that up to isomorphism, there is a unique capped endo-permutation kP/N-module W, called the *slashed module*, such that $(\mathrm{End}_k V)[N] \cong \mathrm{End}_k W$ as P/N-algebras. The slight difference is that Dade rather considers modules than algebras (this point of view is due to Puig). In other words, Dade's theorem deals with the case where A is an *interior* P-algebra, which means that there is a group homomorphism $r : P \to A^\times$, and that the action of $g \in P$ on A is given by conjugation by $r(g)$. In fact, it turns out that any Dade P-algebra over k is actually interior:

12.3.6. Theorem : *Let k be a field of characteristic p, let P be a finite p-group, and let A be a Dade P-algebra over k. Then A is an interior P-algebra. More precisely, there exists a unique capped endo-permutation kP-module M, up to isomorphism, such that A is isomorphic to* $\mathrm{End}_k(M)$ *as P-algebra.*

Proof: Suppose that A is a Dade P-algebra over k. Then P acts on A by algebra automorphisms. Equivalently, there is a group homomorphism α from P to the group of algebra automorphisms $\mathrm{Aut}_{k\text{-alg}}(A)$ of A. Since A is isomorphic to a full matrix k-algebra $M_n(k)$, for some $n > 0$, all its automorphisms are inner, and $\mathrm{Aut}_{k\text{-alg}}(A)$ is isomorphic to the quotient of the group of invertible elements of A, which is isomorphic to the general linear group $GL_n(k)$, by its centre, which consists of non zero scalar multiples of the identity matrix. In other words $\mathrm{Aut}_{k\text{-alg}}(A)$ is isomorphic to the projective linear group $PGL_n(k)$, and there is a commutative diagram

$$\begin{array}{ccccccccc} & & & & & & P & & \\ & & & & & \swarrow^{\tilde{\alpha}} & \downarrow{\scriptstyle\alpha} & & \\ \mathbf{1} & \longrightarrow & k^\times & \longrightarrow & GL_n(k) & \xrightarrow{\pi} & PGL_n(k) & \longrightarrow & \mathbf{1} \end{array}$$

where $\pi : GL_n(k) \to PGL_n(k)$ is the projection map, whose kernel $k^\times$ consists of scalar invertible matrices.

Saying that A is an interior P-algebra is equivalent to saying that there exists a lifting of α to a group homomorphism $\tilde{\alpha} : P \to GL_n(k)$. This in turn depends on an element of $H^2(P, k^\times)$, obtained by choosing a section $s : PGL_n(k) \to GL_n(k)$, and considering the corresponding 2-cocycle γ on P, with values in $k^\times \cong \mathrm{Ker}\,\pi$ defined by

$$\textbf{(12.3.7)} \qquad \forall x, y \in P, \ \ \gamma(x,y)\mathrm{Id}_n = s(xy)s(y)^{-1}s(x)^{-1} \ .$$

Then α can be lifted to $\tilde{\alpha}$ if and only if this cocycle is a coboundary.

Now on the one hand, the group $H^2(P, k^\times)$ has exponent dividing $|P|$, by an elementary cohomological argument. So $|P|\gamma$ (in additive notation) is a coboundary. On the other hand, taking determinant of both sides of Equation 12.3.7 gives

$$\gamma(x,y)^n = u(xy)u(y)^{-1}u(x)^{-1} \ ,$$

where $u(x)$ is the determinant of $s(x)$, for $x \in P$. This shows that $n\gamma$ is a coboundary. So if n is prime to $|P|$, i.e. if $p \nmid n$, then γ is a coboundary.

Suppose that A is not primitive. Then $1_A = i + j$, where i and j are non zero idempotents of A^P. In this case, the algebras iAi and jAj are P-algebras. Moreover they are both full matrix algebras over k (the idempotent i of $A = M_k(k)$ is a projector on some subspace I of k^n, and $iAi \cong \mathrm{End}_k(I)$). Both are also permutation P-algebra, because iAi is a direct summand of A as a kP-module. Finally, at least one of the algebras iAi or jAj has non zero Brauer quotient at P (i.e. it is a Dade P-algebra over k), since otherwise both i and j belong to the ideal $\sum_{Q<P} \mathrm{tr}_Q^P(A^Q)$, and so does $i+j = 1_A$, and then $A[P] = \{0\}$.

Each of the P-algebras iAi and jAj corresponds to some 2-cocycle on P: up to conjugation in A, one can suppose that for any $x \in P$, the matrix $s(x)$ has the form

$$s(x) = \begin{pmatrix} s_1(x) & 0 \\ 0 & s_2(x) \end{pmatrix} ,$$

where the action of x on iAi is conjugation by the matrix $s_1(x)$, and the action of x on jAj is conjugation by $s_2(x)$.

Since for $x, y \in P$, the matrix $s(xy)s(y)^{-1}s(x)^{-1}$ is the scalar matrix $\gamma(x,y)\mathrm{Id}_n$, it follows that $s_1(xy)s_1(y)^{-1}s_1(x)^{-1} = \gamma(x,y)\mathrm{Id}_{n_1}$, where n_1 is the rank of i, and $s_2(xy)s_2(y)^{-1}s_2(x)^{-1} = \gamma(x,y)\mathrm{Id}_{n_2}$, where n_2 is the rank of j, i.e. $n_2 = n - n_1$. In other words, the P-algebras A, iAi and jAj define the same 2-cocycle on P. Thus, to prove that this cocycle is a coboundary, it is enough to consider the case where A is primitive.

In this case A^P is local, so $A[P]$ is local. But $A[P]$ is a full matrix algebra over k, by Theorem 12.3.3, thus $A[P] \cong k$. Now if X is a P-stable k-basis of A, the image of X^P in $A[P]$ is a k-basis of $A[P]$. Thus $|X^P| = 1$, and since $|X| - |X^P|$ is a multiple of p, for $X - X^P$ is a union of non trivial P-orbits, it follows that $\dim_k A \equiv 1$ modulo p. In particular $\dim_k A$ is prime to p, so the corresponding 2-cocycle is a coboundary, as was to be shown.

The uniqueness assertion of the theorem follows from the fact that there are no non trivial group homomorphisms from P to $k^\times$ (i.e. $H^1(P, k^\times) = \{0\}$), since P is a p-group and k has characteristic p. □

12.3.8. Corollary : *Let k be a field of characteristic p, and P be a finite p-group.*

1. *Let A be a primitive Dade P-algebra over k. Then $A[P] \cong k$, and $\dim_k A \equiv 1$ modulo p.*
2. *Let M be an indecomposable capped endo-permutation kP-module. Then $\dim_k M \equiv \pm 1$ modulo p.*

Proof: Assertion 1 was proved at the end of the proof of Theorem 12.3.6. Assertion 2 follows, since $\dim_k \mathrm{End}_k M = (\dim_k M)^2$. □

12.3.9. Remark : Theorem 12.3.6 is false without the hypothesis that $A[P] \neq 0$, when the ground field is not perfect: for example, if $\lambda \in k^\times$, then the cyclic group P of order p acts on $A = M_p(k)$ via the matrix C defined in 12.3.4, and A is a free kP-module (by inspection, or by [44] IX Théorème 5, since the zero-th Tate cohomology group $\widehat{H}^0(P, A) = A[P]$ vanishes), so A is a permutation P-algebra, but A is not an interior P-algebra if $\lambda \notin (k^\times)^p$.

12.3.10. Alternative Definition of $D_k(P)$. Theorem 12.3.6 allows for a translation of the definition of the Dade group of a finite p-group P in terms of Dade P-algebras. First observe that:

- If M is a kP-module, then the P-algebra $\mathrm{End}_k(M^*)$ is isomorphic (as a P-algebra) to the opposite algebra $\mathrm{End}_k(M)^{op}$.
- If M and M' are kP-modules, then $\mathrm{End}_k(M \otimes M')$ is isomorphic to $\mathrm{End}_k(M) \otimes \mathrm{End}_k(M')$ as P-algebra over k.
- If A and A' are Dade P-algebras over k, then $A \otimes A'$ is a Dade P-algebra over k: clearly $A \otimes A'$ is a permutation P-algebra, and it is isomorphic to a full matrix algebra over k. Moreover, the natural map $A^P \otimes A'^P \to (A \otimes A')^P$ induces a map $A[P] \otimes A'[P] \to (A \otimes A')[P]$, which is an isomorphism of algebras when A and A' are permutation P-algebras.

This leads to the following equivalent definition of $D_k(P)$:

12.3.11. Definition : *Let k be a field of characteristic p, and P be a finite p-group. Two Dade P-algebras A and A' over k are said to be* equivalent *(notation $A \sim A'$) if there exists a permutation kP-module M and an isomorphism $A' \otimes A^{op} \cong \mathrm{End}_k(M)$ of P-algebras over k. This is an equivalence relation on the class of Dade P-algebras over k.*

The Dade group $D_k(P)$ is the set of equivalence classes of Dade P-algebras over k for this relation, endowed with the abelian group law defined by $[A] + [A'] = [A \otimes A']$, where A and A' are Dade P-algebras over k, and $[A]$ denotes the equivalence class of $[A]$.

12.4. Bisets and Permutation Modules

12.4.1. Permutation Modules Revisited. Let $\mathsf{Perm}_k(P)$ denote the full subcategory of the category of kP-modules consisting of finitely generated kP-modules. The natural framework to define biset operations for the Dade group is the following equivalent category $\underline{\mathsf{Perm}}_k(P)$, introduced in Sect. 2 of [19]:

12.4.2. Definition : *Let k be a field of characteristic p, and P be a p-group. Let $\underline{\mathsf{Perm}}_k(P)$ denote the following category:*

- *The objects of $\underline{\mathsf{Perm}}_k(P)$ are the finite P-sets.*
- *If X and Y are finite P-sets, then $\mathrm{Hom}_{\underline{\mathsf{Perm}}_k(P)}(X, Y)$ is the k-vector space of P-invariant maps $m : Y \times X \to k$, i.e. the k-vector space of*

matrices m indexed by $Y\times X$, with coefficients in k, such that $m(gy,gx) = m(y,x)$ for all $g\in P$, and all $(y,x)\in Y\times X$.

- *The composition of morphism is given by the usual product of matrices: if X, Y, and Z are finite P-sets, if $m\in \mathrm{Hom}_{\underline{\mathsf{Perm}}_k(P)}(X,Y)$ and $n\in \mathrm{Hom}_{\underline{\mathsf{Perm}}_k(P)}(Y,Z)$, then $n\circ m$ is the matrix defined by*

$$\forall z\in Z,\ \forall x\in X,\quad (n\circ m)(z,x) = \sum_{y\in Y} n(z,y)m(y,x)\ .$$

- *The identity morphism of the finite P-set X is the identity matrix δ_X defined by $\delta_X(x,x) = 1_k$ for $x\in X$, and $\delta_X(x',x) = 0_k$ for $x'\neq x\in X$.*

12.4.3. Proposition : *Let k be a field of characteristic p, and P be a p-group. The category $\underline{\mathsf{Perm}}_k(P)$ is equivalent to $\mathsf{Perm}_k(P)$.*

Proof: Let $e_P : \underline{\mathsf{Perm}}_k(P) \to \mathsf{Perm}_k(P)$ be the correspondence sending a finite P-set X to the permutation module kX, and $m\in \mathrm{Hom}_{\underline{\mathsf{Perm}}_k(P)}(X,Y)$, where Y is a finite P-set, to the morphism $e_P(m) : kX\to kY$ defined by the matrix m, i.e. by

$$\forall x\in X,\quad e_P(m)(x) = \sum_{y\in Y} m(y,x)y\ .$$

Then e_P is obviously a functor, and it is also clearly fully faithful. Finally e_P is essentially surjective, since any permutation module is isomorphic to kX, for some finite P-set X. □

12.4.4. Remark : Let $f : X\to Y$ be a morphism of finite P-sets. Then there is a corresponding morphism $f_* : X\to Y$ in $\underline{\mathsf{Perm}}_k(P)$ defined by $f_*(y,x) = \delta_{y,f(x)}$ (equal to 1 if $y=f(x)$ and to 0 otherwise). This yields a functor from the category of finite P-sets to $\underline{\mathsf{Perm}}_k(P)$, which is the identity on objects, and maps a morphism f to f_*.

In particular, if X and Y are isomorphic finite P-sets, then X and Y are isomorphic in $\underline{\mathsf{Perm}}_k(P)$. The converse is also true: if X and Y are isomorphic in $\underline{\mathsf{Perm}}_k(P)$, then the kP-modules kX and kY are isomorphic, and since P is a p-group, this implies that the P-sets X and Y are isomorphic.

12.4.5. Remark : The usual constructions

$$(X,Y)\mapsto X\sqcup Y \qquad\text{and}\qquad (X,Y)\mapsto X\times Y$$

on P-sets have straightforward extensions to functors $\underline{\mathsf{Perm}}_k(P)\times\underline{\mathsf{Perm}}_k(P)\to \underline{\mathsf{Perm}}_k(P)$, that will be denoted with the same symbols. If $m : X\to X'$ and $n : Y\to Y'$ are morphisms in $\underline{\mathsf{Perm}}_k(P)$, then the morphism $m\sqcup n : X\sqcup Y\to X'\sqcup Y'$ is defined by

$$\forall z, t \in (X \sqcup Y) \times (X' \sqcup Y'),\ (m \sqcup n)(t, z) = \begin{cases} m(t, z) \text{ if } t \in X' \text{ and } z \in X \\ n(t, z) \text{ if } t \in Y' \text{ and } z \in Y \\ 0 \text{ otherwise,} \end{cases}$$

and the morphism $m \times n : X \times Y \to X' \times Y'$ is defined by

$$\forall (x, y, x', y') \in X \times Y \to X' \times Y',\ \ (m \times n)\big((x', y'), (x, y)\big) = m(x', x)n(y', y)\ .$$

12.4.6. Remark : Let $\bullet$ denote a one element P-set. Then $e_P(\bullet) \cong k$. If X and Y are finite P-sets, it is straightforward to check that $e_P(X \sqcup Y) \cong e_P(X) \oplus e_P(Y)$, and that $e_P(X \times Y) \cong e_P(X) \otimes e_P(Y)$. Moreover, these isomorphisms are functorial with respect to X and Y: the following diagrams of categories and functors

$$\begin{array}{ccc} \underline{\mathsf{Perm}_k}(P) \times \underline{\mathsf{Perm}_k}(P) & \xrightarrow{\underline{*}} & \underline{\mathsf{Perm}_k}(P) \\ \Big\downarrow{\scriptstyle e_P \times e_P} & & \Big\downarrow{\scriptstyle e_P} \\ \mathsf{Perm}_k(P) \times \mathsf{Perm}_k(P) & \xrightarrow{*} & \mathsf{Perm}_k(P)\ , \end{array}$$

where the pair $(\underline{*}, *)$ is one of $(\sqcup, \oplus)$ or $(\times, \otimes)$, are "commutative": this means that the functors $e_P \circ \underline{*}$ and $* \circ (e_P \times e_P)$ are isomorphic.

12.4.7. Generalized Tensor Induction. Let P and Q be finite p-groups, and U be a finite (Q, P)-biset. If X is a finite P-set, denote by

$$T_U(X) = \mathrm{Hom}_P(U^{op}, X)$$

the set of P-equivariant maps from U^{op} to X, i.e. the set of maps $\varphi : U \to X$ such that $\varphi(ug^{-1}) = g\varphi(u)$ for any $g \in P$ and $u \in U$. Then $T_U(X)$ has a natural structure of (finite) Q-set, given by

$$\forall h \in Q,\ \forall \varphi \in T_U(X),\ \forall u \in U,\ \ (h\varphi)(u) = \varphi(h^{-1}u)\ .$$

Suppose now that Y is another finite P-set, and that $m \in \mathrm{Hom}_{\underline{\mathsf{Perm}_k}(P)}(X, Y)$. For $\varphi, \psi \in T_U(X)$, define an element $T_U(m)(\varphi, \psi)$ of k by

$$T_U(m)(\varphi, \psi) = \prod_{u \in [U/P]} m\big(\varphi(u), \psi(u)\big)\ ,$$

where $[U/P]$ is a set of representatives of U/P. This is clearly independent of the choice of such a set of representatives, since φ and ψ are P-equivariant maps, and m is a P-invariant matrix.

12.4.8. Lemma : *This definition yields a morphism $T_U(m)$ from $T_U(X)$ to $T_U(Y)$ in the category $\underline{\mathsf{Perm}}_k(Q)$.*

Proof: Let $h \in Q$. Then there exists a permutation σ_h of $[U/P]$, and, for any $u \in [U/P]$, there exists an element $g_{h,u} \in P$ such that $hu = \sigma_h(u)g_{h,u}$. It follows that

$$\begin{aligned}
T_U(m)(h^{-1}\varphi, h^{-1}\psi) &= \prod_{u\in[U/P]} m\Big((h^{-1}\varphi)(u), (h^{-1}\psi)(u)\Big)\\
&= \prod_{u\in[U/P]} m(\varphi(hu), \psi(hu))\\
&= \prod_{u\in[U/P]} m\Big(\varphi\big(\sigma_h(u)g_{h,u}\big), \psi\big(\sigma_h(u)g_{h,u}\big)\Big)\\
&= \prod_{u\in[U/P]} m\Big(g_{h,u}^{-1}\varphi\big(\sigma_h(u)\big), g_{h,u}^{-1}\psi\big(\sigma_h(u)\big)\Big)\\
&= \prod_{u\in[U/P]} m\Big(\varphi\big(\sigma_h(u)\big), \psi\big(\sigma_h(u)\big)\Big)\\
&= \prod_{u\in[U/P]} m(\varphi(u), \psi(u)) = T_U(m)(\varphi, \psi)\ .
\end{aligned}$$

So the matrix $T_U(m)$ is Q-invariant, as was to be shown. □

12.4.9. Lemma : [19, Lemma 2.2] *Let k be a field of characteristic p. Let P and Q be finite p-groups, and let U be a finite (Q,P)-biset. Then T_U is a functor from $\underline{\mathsf{Perm}}_k(P)$ to $\underline{\mathsf{Perm}}_k(Q)$.*

Proof: Let X be a finite P-set, let $\varphi, \psi \in T_U(X)$. Then

$$T_U(\delta_X)(\varphi, \psi) = \prod_{u\in[U/P]} \delta_X\big(\varphi(u), \psi(u)\big)$$

is non zero if and only if $\varphi(u) = \psi(u)$ for any $u \in [U/P]$, hence for any $u \in U$. This implies $\varphi = \psi$, and $T_U(\delta_X)(\varphi, \psi) = 1$ in this case. This shows that $T_U(\delta_X) = \delta_{T_U(X)}$, so T_U maps the identity morphism to the identity morphism.

Now let X, Y and Z be finite P-sets, let $m \in \mathrm{Hom}_{\underline{\mathsf{Perm}}_k(P)}(X, Y)$ and $n \in \mathrm{Hom}_{\underline{\mathsf{Perm}}_k(P)}(Y, Z)$. Then for $\varphi \in T_U(X)$ and $\varphi \in T_U(Z)$

$$
\begin{aligned}
T_U(n \circ m)(\varphi, \varphi) &= \prod_{u \in [U/P]} (n \circ m)\big(\varphi(u), \varphi(u)\big) \\
&= \prod_{u \in [U/P]} \Big(\sum_{y \in Y} n\big(\varphi(u), y\big) m\big(y, \varphi(u)\big)\Big) .
\end{aligned}
$$

Now for $u \in [U/P]$

$$
\begin{aligned}
\sum_{y \in Y} n\big(\varphi(u), y\big) m\big(y, \varphi(u)\big) &= \sum_{y \in [P_u \backslash Y]} \sum_{g \in P_u / P_{u,y}} n\big(\varphi(u), gy\big) m\big(gy, \varphi(u)\big) \\
&= \sum_{y \in [P_u \backslash Y]} \sum_{g \in P_u / P_{u,y}} n\big(g^{-1}\varphi(u), y\big) m\big(y, g^{-1}\varphi(u)\big) \\
&= \sum_{y \in [P_u \backslash Y]} \sum_{g \in P_u / P_{u,y}} n\big(\varphi(ug), y\big) m\big(y, \varphi(ug)\big) \\
&= \sum_{y \in [P_u \backslash Y]} |P_u / P_{u,y}| n\big(\varphi(u), y\big) m\big(y, \varphi(u)\big) ,
\end{aligned}
$$

where P_u is the (right) stabilizer of u in P, and $P_{u,y}$ is the intersection of P_u with the (left) stabilizer of y in P. Since $|P_u/P_{u,y}| = 0$ in k, unless $y \in Y^{P_u}$, this gives finally

$$
\sum_{y \in Y} n\big(\varphi(u), y\big) m\big(y, \varphi(u)\big) = \sum_{y \in Y^{P_u}} n\big(\varphi(u), y\big) m\big(y, \varphi(u)\big) .
$$

Thus

$$
T_U(n \circ m)(\varphi, \varphi) = \prod_{u \in [U/P]} \Big(\sum_{y \in Y^{P_u}} n\big(\varphi(u), y\big) m\big(y, \varphi(u)\big)\Big) .
$$

Expanding this product amounts to choosing, for each $u \in [U/P]$, an element $y_u \in Y^{P_u}$. This in turn is equivalent to choosing a P-equivariant map $\psi : U^{op} \to Y$, and setting $y_u = \psi(u)$. In other words

$$
\begin{aligned}
T_U(n \circ m)(\varphi, \varphi) &= \sum_{\psi \in T_U(Y)} \prod_{u \in [U/P]} n\big(\theta(u), \psi(u)\big) m\big(\psi(u), \varphi(u)\big) \\
&= \sum_{\psi \in T_U(Y)} \Big(\prod_{u \in [U/P]} n\big(\theta(u), \psi(u)\big)\Big)\Big(\prod_{u \in [U/P]} m\big(\psi(u), \varphi(u)\big)\Big) \\
&= \sum_{\psi \in T_U(Y)} T_U(n)(\theta, \psi) T_U(m)(\psi, \varphi) \\
&= \big(T_U(n) \circ T_U(m)\big)(\theta, \varphi) .
\end{aligned}
$$

Thus $T_U(n \circ m) = T_U(n) \circ T_U(m)$, as was to be shown. □

12.4.10. Examples : elementary bisets :

• Let Q be a subgroup of P, and let U denote the set P, viewed as a (Q,P)-biset for left and right multiplication. Then for any P-set X, the Q-set $T_U(X) = \mathrm{Hom}_P(P,X)$ is isomorphic to the restriction $\mathrm{Res}_Q^P X$. Since moreover $|U/P| = 1$ in this case, one has that $T_U(m) = m$, for any morphism $m : X \to Y$ in $\underline{\mathsf{Perm}}_k(P)$. This shows that the isomorphism $T_U(X) \cong \mathrm{Res}_Q^P X$ is functorial in X. In other words, there is a commutative diagram

$$\begin{array}{ccc} \underline{\mathsf{Perm}}_k(P) & \xrightarrow{e_P} & \mathsf{Perm}_k(P) \\ {\scriptstyle T_U}\downarrow & & \downarrow{\scriptstyle \mathrm{Res}_Q^P} \\ \underline{\mathsf{Perm}}_k(Q) & \xrightarrow{e_Q} & \mathsf{Perm}_k(Q) \end{array}$$

in this case, so T_U can be viewed as restriction from P to Q, up to the above equivalences of categories.

• In the same situation, let V denote the set P, viewed as a (P,Q)-biset for left and right multiplication. Then for any finite Q-set Y, the set $T_V(X) = \mathrm{Hom}_Q(P^{op},X)$ is in one to one correspondence with $X^{|P:Q|}$. It follows that the kP-module $e_P T_V(X)$ is isomorphic to the *tensor induced module* $\mathrm{Ten}_Q^P(kX)$. It is easy to check that there is a commutative diagram

$$\begin{array}{ccc} \underline{\mathsf{Perm}}_k(Q) & \xrightarrow{e_Q} & \mathsf{Perm}_k(Q) \\ {\scriptstyle T_V}\downarrow & & \downarrow{\scriptstyle \mathrm{Ten}_Q^P} \\ \underline{\mathsf{Perm}}_k(P) & \xrightarrow{e_P} & \mathsf{Perm}_k(P) \end{array}$$

in this case, so T_V can be viewed as tensor induction from Q to P, up to the above equivalences of categories.

• Let $N \trianglelefteq P$. Consider first the set $U = P/N$, for its natural $(P,P/N)$-biset structure. For any P/N-set X, the P-set $T_U(X) = \mathrm{Hom}_{P/N}\big((P/N)^{op},X\big)$ is isomorphic to $\mathrm{Inf}_{P/N}^P X$. Again, it is easy to check that this isomorphism is functorial in X, i.e. there is a commutative diagram

$$\begin{array}{ccc} \underline{\mathsf{Perm}}_k(P/N) & \xrightarrow{e_{P/N}} & \mathsf{Perm}_k(P/N) \\ {\scriptstyle T_U}\downarrow & & \downarrow{\scriptstyle \mathrm{Inf}_{P/N}^P} \\ \underline{\mathsf{Perm}}_k(P) & \xrightarrow{e_P} & \mathsf{Perm}_k(P) \end{array}$$

So in this case, the functor T_U can be viewed as inflation from P/N to P, up to equivalences of categories.

• In the same situation, let V denote the set P/N, for its natural $(P/N, P)$-biset structure. Then for any P-set X, the (P/N)-set $T_V(X)$=$\mathrm{Hom}_P(P/N, X)$ is isomorphic to X^N. So the module $e_{P/N}T_V(kX)$ is isomorphic to kX^N, i.e. to the Brauer quotient $(kX)[N]$. Again, one checks easily that there is a commutative diagram

$$\begin{array}{ccc}
\underline{\mathsf{Perm}}_k(P) & \xrightarrow{e_P} & \mathsf{Perm}_k(P) \\
{\scriptstyle T_V}\downarrow & & \downarrow{\scriptstyle \mathrm{Br}_N^P} \\
\underline{\mathsf{Perm}}_k(P/N) & \xrightarrow{e_{P/N}} & \mathsf{Perm}_k(P/N)
\end{array}$$

where Br_N^P is the Brauer quotient functor at N. In this case, the functor T_V can be viewed as the Brauer quotient functor from P to P/N, up to equivalences of categories.

• Finally if $f : P \to Q$ is a group isomorphism, and if X is a finite P-set, then $T_U(X)$ is the finite Q-set obtained from X by transporting the action of P via f. Similarly, the kQ-module $kT_U(X)$ is the module obtained from the kP-module kX via f.

These examples, and in particular the second one, are the motivation for the following definition:

12.4.11. Definition : *In the situation of Lemma 12.4.9, the functor T_U is called the* generalized tensor induction *associated to U.*

12.4.12. Proposition : *Let k be a field of characteristic p, let P and Q be finite p-groups, and let U be a finite (Q, P)-biset.*

1. *If X is a one element P-set, then $T_U(X)$ is a one element Q-set.*
2. *If X and Y are finite P-sets, there is an isomorphism*

$$T_U(X \times Y) \cong T_U(X) \times T_U(Y) ,$$

which is functorial in X and Y: there is a commutative diagram of categories and functors

$$\begin{array}{ccc}
\underline{\mathsf{Perm}}_k(P) \times \underline{\mathsf{Perm}}_k(P) & \xrightarrow{\times} & \underline{\mathsf{Perm}}_k(P) \\
{\scriptstyle T_u \times T_U}\downarrow & & \downarrow{\scriptstyle T_U} \\
\underline{\mathsf{Perm}}_k(Q) \times \underline{\mathsf{Perm}}_k(Q) & \xrightarrow{\times} & \underline{\mathsf{Perm}}_k(Q) .
\end{array}$$

3. *If X and Y are finite P-sets, let $\sigma : X \times Y \to Y \times X$ denote the switch morphism. Then up to the isomorphism of Assertion 2, the morphism $T_U(\sigma)$ is the switch morphism $T_U(X) \times T_U(Y) \to T_U(Y) \times T_U(X)$.*
4. *If U' is another finite (Q,P)-biset, then the functors $T_{U \sqcup U'}$ and $T_U \times T_{U'}$ are isomorphic.*

Proof: (Sketched: for details, see [19] Proposition 2.10.) Assertion 1 is obvious: if X has cardinality one, there is a unique map $U^{op} \to X$.

Assertion 2 follows from the isomorphism

$$\mathrm{Hom}_P(U^{op}, X \times Y) \cong \mathrm{Hom}_P(U^{op}, X) \times \mathrm{Hom}_P(U^{op}, Y) .$$

Assertion 3 is straightforward, and Assertion 4 follows from the isomorphism

$$\mathrm{Hom}_P(U^{op} \sqcup U'^{op}, X) \cong \mathrm{Hom}_P(U^{op}, X) \times \mathrm{Hom}_P(U'^{op}, X) .$$

□

12.4.13. Composition and Galois Twists. Let P, Q, and R be finite p-groups. Let U be a finite (Q,P)-biset, and V be a finite (R,Q)-biset. Then $T_V \circ T_U$ and $T_{V \times_Q U}$ are two functors from $\underline{\mathsf{Perm}}_k(P)$ to $\underline{\mathsf{Perm}}_k(R)$, and a natural question is to ask whether they are isomorphic or not.

It turns out that the answer is no, in general, and the difference between these two functors can be precisely expressed via *Galois twists*: to each endomorphism a of the field k is associated an endofunctor $\gamma_{a,P}$ of the category $\underline{\mathsf{Perm}}_k(P)$, called the a-Galois twist for P, defined as follows:

12.4.14. Definition : *Let a be an endomorphism of the field k. If P is a finite p-group, denote by $\gamma_{a,P}$ (or γ_a) the functor $\underline{\mathsf{Perm}}_k(P) \to \underline{\mathsf{Perm}}_k(P)$ which is the identity on objects, and sends the matrix $m \in \mathrm{Hom}_{\underline{\mathsf{Perm}}_k(P)}(X,Y)$ to the matrix $a(m) \in \mathrm{Hom}_{\underline{\mathsf{Perm}}_k(P)}(X,Y)$ defined by*

$$a(m)(y,x) = a\big(m(y,x)\big) .$$

For $n \in \mathbb{N}$, let $\gamma(p^n)$ denote the functor γ_a for the endomorphism $a : x \mapsto x^{p^n}$ of k.

It is easy to check that γ_a is indeed a functor from $\underline{\mathsf{Perm}}_k(P)$ to itself, for any endomorphism a of the field k. The functor $\gamma(p)$ is called *the Frobenius twist* functor.

12.4.15. Lemma :

1. *Let a and b be endomorphisms of k. Then for any finite p-groups P*

$$\gamma_{a,P} \circ \gamma_{b,P} \cong \gamma_{a \circ b, P} .$$

2. *Let P and Q be finite p-groups, and U be a finite (Q,P)-biset. Then*

$$T_U \circ \gamma_{a,P} = \gamma_{a,Q} \circ T_U \ .$$

Proof: Assertion 1 is obvious, from the definition. For Assertion 2, if X is a finite P-set, then

$$T_U \circ \gamma_{a,P}(X) = T_U(X) = \gamma_{a,Q} \circ T_U(X) = \mathrm{Hom}_P(U^{op}, X) \ ,$$

since both functors $\gamma_{a,P}$ and $\gamma_{a,Q}$ are the identity on objects.

And if $m \in \mathrm{Hom}_{\underline{\mathsf{Perm}}_k(P)}(X,Y)$, then for φ and ψ in $T_U(X)$

$$\begin{aligned}\big(T_U \circ \gamma_{a,P}(m)\big)(\varphi,\psi) &= \prod_{u\in[U/P]} a(m)\big(\varphi(u),\psi(u)\big)\\ &= \prod_{u\in[U/P]} a\Big(m\big(\varphi(u),\psi(u)\big)\Big)\\ &= a\Big(\prod_{u\in[U/P]} m\big(\varphi(u),\psi(u)\big)\Big)\\ &= a\Big(T_U(m)(\varphi,\psi)\Big)\\ &= \big(\gamma_{a,Q}\circ T_U(m)\big)(\varphi,\psi) \ ,\end{aligned}$$

as was to be shown. □

12.4.16. Proposition :

1. *Let P be a finite p-group. If U is the identity (P,P)-biset, then T_U is the identity functor of $\underline{\mathsf{Perm}}_k(P)$.*
2. *Let P, Q, and R be finite p-groups, let U be a finite (Q,P)-biset, and let V be a finite (R,Q)-biset. Then*

$$\begin{aligned}T_{V\times_Q U} &\cong \prod_{(v,_Q u)\in[R\backslash(V\times_Q U)/P]} T_{R(v,_Q u)P}\\ T_V\circ T_U &\cong \prod_{(v,_Q u)\in[R\backslash(V\times_Q U)/P]} \gamma\big(|\mathbf{1}^v : \mathbf{1}^v\cap {}^uP|\big)\circ T_{R(v,_Q u)P}\end{aligned}$$

where $(v,_Q u)$ is the image of $(v,u)\in V\times U$ in $V\times_Q U$.

Proof: Assertion 1 follows from the fact that if $U = \mathrm{Id}_P$ is the set P, for its natural structure of (P,P)-biset, then for any P-set X, there is an isomorphism of P-sets

$$\mathrm{Hom}_P(U^{op}, X) = \mathrm{Hom}_P(P, X) \cong X \ .$$

Moreover U is a transitive P-set, and it follows easily that T_U is also the identity on morphisms in $\underline{\mathsf{Perm}}_k(P)$.

The first isomorphism of functors in Assertion 2 is a consequence of Assertion 4 of Proposition 12.4.12, after decomposition of $V \times_Q U$ as a disjoint union of transitive (R, P)-bisets. For the second one, let X be a finite P-set. Then

$$\begin{aligned} T_V \circ T_U(X) &= \mathrm{Hom}_Q\big(V^{op}, \mathrm{Hom}_P(U^{op}, X)\big) \\ &\cong \mathrm{Hom}_P(U^{op} \times_Q V^{op}, X) \\ &\cong \mathrm{Hom}_P\big((V \times_Q U)^{op}, X\big) \\ &= T_{V \times_Q U}(X) \ . \end{aligned}$$

This is also isomorphic to

$$\prod_{(v,_Q u) \in [R\backslash (V \times_Q U)/P]} \gamma\big(|\mathbf{1}^v : \mathbf{1}^v \cap {}^uP|\big) \circ T_{R(v,_Q u)P}(X) \ ,$$

since every functor $\gamma\big(|\mathbf{1}^v : \mathbf{1}^v \cap {}^uP|\big)$ is the identity on objects.

Now let X and Y be finite P-sets, and $m \in \mathrm{Hom}_{\underline{\mathsf{Perm}}_k(P)}(X, Y)$. Then, for $\varphi, \psi \in T_{V \times_Q U}(X)$

$$T_{V \times_Q U}(m)(\varphi, \psi) = \prod_{(v,_Q u) \in [(V \times_Q U)/P]} m\big(\varphi(v,_Q u), \psi(v,_Q u)\big) \ .$$

On the other hand, after identification of $T_{V \times_Q U}(X)$ with $T_V \circ T_U(X)$

$$\begin{aligned} (T_V \circ T_U)(m)(\varphi, \psi) &= \prod_{v \in [V/Q]} T_U(m)\big(\varphi(v), \psi(u)\big) \\ &= \prod_{v \in [V/Q]} \prod_{u \in [U/P]} m\big(\varphi(v)(u), \psi(v)(u)\big) \\ &= \prod_{v \in [V/Q]} \prod_{u \in [U/P]} m\big(\varphi(v,_Q u), \psi(v,_Q u)\big) \ . \end{aligned}$$

Let $\theta : [V/Q] \times [U/P] \to (V \times_Q U)/P$ be the map defined by $\theta(v, u) = (v,_Q u)P$. This map is surjective, since for $(v,_Q u)P \in (V \times_Q U)/P$, there exists a unique $v_0 \in [V/Q]$ and $h \in Q$ such that $v = v_0 h$. Now there exists $u_0 \in [U/P]$ and $g \in P$ such that $hu = u_0 g$. Thus

$$(v,_Q u)P = (v_0 h,_Q u)P = (v_0,_Q hu)P = (v_0,_Q u_0 g)P = (v_0,_Q u_0)P = \theta(v_0, u_0) \ .$$

Now two pairs (v_0, u_0) and (v_1, u_1) have the same image under θ if and only if there exists $h \in H$ and $g \in G$ such that $v_1 = v_0 h$ and $hu_1 = u_0 g$. The first condition implies $v_1 = v_0$, since v_1 and v_0 both belong to $[V/Q]$. In this case $v_0 = v_0 h$, thus $h \in \mathbf{1}^{v_0}$.

The second condition is now equivalent to saying that $u_1 P$ belongs to the orbit of $u_0 P$ under the group $\mathbf{1}^{v_0}$. The left stabilizer of $u_0 P$ in $\mathbf{1}^{v_0}$ is equal to $\mathbf{1}^{v_0} \cap {}^{u_0}P$, so this orbit has length $|\mathbf{1}^{v_0} : \mathbf{1}^{v_0} \cap {}^{u_0}P|$. So the inverse image $\theta^{-1}\big((v,_{Q} u)\big)$ has cardinality $|\mathbf{1}^v : \mathbf{1}^v \cap {}^u P|$. Moreover this index depends only on the orbit $R(v,_{Q} u)P$ of $(v,_{Q} u)$. It follows that

$$(T_V \circ T_U)(m)(\varphi, \psi) = \prod_{(v,_{Q} u) \in [(V \times_Q U)/P]} m\big(\varphi(v,_{Q} u), \psi(v,_{Q} u)\big)^{|\mathbf{1}^v : \mathbf{1}^v \cap {}^u P|} .$$

This is also equal to

$$\prod_{(v',_{Q} u') \in [R \backslash (V \times_Q U)/P]} \Big(\prod_{(v,_{Q} u) \in [R(v',_{Q} u')P/P]} m\big(\varphi(v,_{Q} u), \psi(v,_{Q} u)\big)\Big)^{|\mathbf{1}^{v'} : \mathbf{1}^{v'} \cap {}^{u'} P|},$$

i.e. to

$$\prod_{(v',_{Q} u') \in [R \backslash (V \times_Q U)/P]} \Big(\big(\gamma(|\mathbf{1}^{v'} : \mathbf{1}^{v'} \cap {}^{u'} P|) \circ T_{R(v',_{Q} u')P}\big)(m)\Big)(\varphi, \psi)$$

as was to be shown. □

12.4.17. Corollary : *If V is a free right Q-set, or if U is a transitive right P-set, then $T_V \circ T_U \cong T_{V \times_Q U}$.*

Proof: Indeed in either case, all the indices $|\mathbf{1}^v : \mathbf{1}^v \cap {}^u P|$ are equal to 1: if V is free, then $\mathbf{1}^v = \mathbf{1}$, for any $v \in V$. And if U is transitive, then ${}^u P = P$, for any $u \in U$. □

12.5. Bisets and Permutation Algebras

12.5.1. Permutation Algebras Revisited. There is an interpretation of the notion of algebra purely in terms of categories of modules: a P-algebra over k is *a monoid* in the monoidal category kP-Mod of kP-modules (see [39] VII.3): a P-algebra over k is a triple (A, μ, η), where A is a kP-module, where $\mu : A \otimes A \to A$ is a multiplication morphism, and $\eta : k \to A$ is a unit morphism, such that the diagrams

$$\begin{array}{ccccc} A \otimes (A \otimes A) & \xrightarrow{\alpha} & & & (A \otimes A) \otimes A \\ \Big\downarrow{\scriptstyle \mathrm{Id} \otimes \mu} & & & & \Big\downarrow{\scriptstyle \mu \otimes \mathrm{Id}} \\ A \otimes A & \xrightarrow{\mu} & A & \xleftarrow{\mu} & A \otimes A \end{array}$$

and

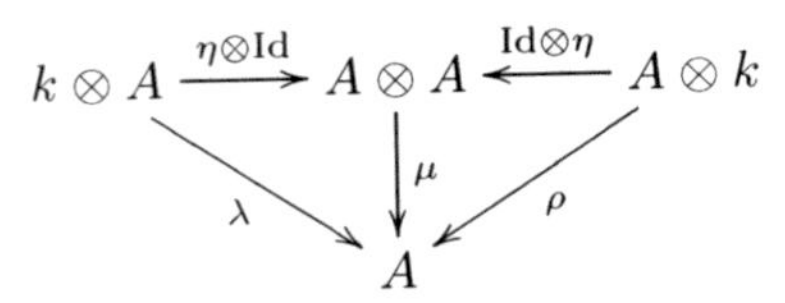

are commutative. Here α is the associativity isomorphism, and λ and ρ are the kP-module isomorphisms $k \otimes A \cong A \cong A \otimes k$.

A morphism of P-algebras over k from (A, μ, η) to (A', μ', η') is a morphism of kP-modules $f : A \to A'$ such that the diagrams

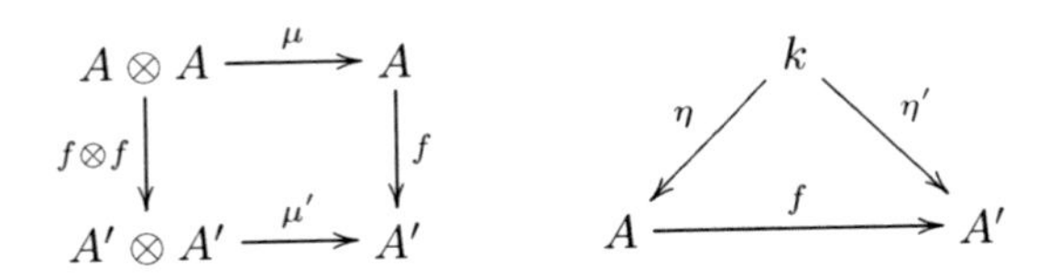

are commutative.

Similarly, a finite dimensional permutation P-algebra over k is a monoid in the monoidal subcategory $\mathsf{Perm}_k(P)$ of kP-Mod. Permutation P-algebras over k, and morphism of P-algebras over k, form a category, denoted by $\mathsf{PermAlg}_k(P)$.

By Remark 12.4.6, the equivalence of categories e_P of Proposition 12.4.3 is an equivalence of monoidal categories. This yields the following equivalent category $\underline{\mathsf{PermAlg}}_k(P)$ of monoids in $\underline{\mathsf{Perm}}_k(P)$: its objects are triples $(X, \underline{\mu}, \underline{\eta})$, where X is a finite P-set, and where $\underline{\mu} : X \times X \to X$ and $\underline{\eta} : \bullet \to X$ are morphisms in the category $\underline{\mathsf{Perm}}_k(P)$, such that the diagrams

(12.5.2)
$$\begin{array}{ccccc}
X \times (X \times X) & & \xrightarrow{\underline{\alpha}} & & (X \times X) \times X \\
\downarrow{\scriptstyle \mathrm{Id} \times \underline{\mu}} & & & & \downarrow{\scriptstyle \underline{\mu} \times \mathrm{Id}} \\
X \times X & \xrightarrow{\underline{\mu}} & X & \xleftarrow{\underline{\mu}} & X \times X
\end{array}$$

and

(12.5.3)
$$\begin{array}{ccccc}
\bullet \times X & \xrightarrow{\underline{\eta} \times \mathrm{Id}} & X \times X & \xleftarrow{\mathrm{Id} \times \underline{\eta}} & X \times \bullet \\
& {\scriptstyle \underline{\lambda}} \searrow & \downarrow{\scriptstyle \underline{\mu}} & \swarrow {\scriptstyle \underline{\rho}} & \\
& & X & &
\end{array}$$

are commutative. Here $\underline{\alpha}$ is the associativity isomorphism in $\underline{\mathsf{Perm}}_k(P)$, and $\underline{\lambda}$ and $\underline{\mu}$ are the isomorphisms $\bullet \times X \cong X \cong X \times \bullet$ in $\underline{\mathsf{Perm}}_k(P)$.

A morphism from $(X, \underline{\mu}, \underline{\eta})$ to $(X', \underline{\mu}', \underline{\eta}')$ in the category $\underline{\mathsf{PermAlg}}_k(P)$ is a morphism $f : X \to X'$ in $\underline{\mathsf{Perm}}_k(P)$ such that the diagrams

(12.5.4)

$$\begin{array}{ccc} X \times X & \xrightarrow{\underline{\mu}} & X \\ {\scriptstyle f\times f}\downarrow & & \downarrow{\scriptstyle f} \\ X' \times X' & \xrightarrow{\underline{\mu}'} & X' \end{array} \qquad \begin{array}{c} \bullet \\ {\scriptstyle \underline{\eta}}\swarrow \quad \searrow{\scriptstyle \underline{\eta}'} \\ X \xrightarrow{\quad f \quad} X' \end{array}$$

are commutative.

12.5.5. Generalized Tensor Induction of Algebras. Let P and Q be finite p-groups, and U be a finite (Q,P)-biset. If $(X,\underline{\mu},\underline{\eta})$ is an algebra in $\underline{\mathsf{Perm}}_k(P)$, then in particular $\underline{\mu}$ is a morphism from $X\times X$ to X. Applying the functor T_U, and using the isomorphism $T_U(X\times X)\cong T_U(X)\times T_U(X)$ of Proposition 12.4.12 yields a morphism $T_U(\underline{\mu}) : T_U(X)\times T_U(X)\to T_U(X)$ in $\underline{\mathsf{Perm}}_k(Q)$. Similarly $T_U(\underline{\eta})$ is a morphism from $T_U(\bullet) = \bullet$ to $T_U(X)$ in $\underline{\mathsf{Perm}}_k(Q)$. These two morphisms give an algebra structure on $T_U(X)$:

12.5.6. Proposition : *Let P and Q be finite p-groups, and U be a finite (Q,P)-biset.*

1. *If $(X,\underline{\mu},\underline{\eta})$ is an object in $\underline{\mathsf{PermAlg}}_k(P)$, then $\big(T_U(X),T_U(\underline{\mu}),T_U(\underline{\eta})\big)$ is an object $\underline{\mathsf{PermAlg}}_k(Q)$.*
2. *If $f:(X,\underline{\mu},\underline{\eta})\to(X',\underline{\mu}',\underline{\eta}')$ is a morphism in $\underline{\mathsf{PermAlg}}_k(P)$, then $T_U(f)$ is a morphism from $\big(T_U(X),T_U(\underline{\mu}),T_U(\underline{\eta})\big)$ to $\big(T_U(X'),T_U(\underline{\mu}'),T_U(\underline{\eta}')\big)$ in $\underline{\mathsf{PermAlg}}_k(Q)$.*
3. *These correspondences yield a functor, still denoted by T_U, from $\underline{\mathsf{PermAlg}}_k(P)$ to $\underline{\mathsf{PermAlg}}_k(Q)$.*

Proof: Applying the functor T_U to the diagrams 12.5.2 and 12.5.3, and using the isomorphisms of Proposition 12.4.12 gives the following diagrams

$$\begin{array}{ccccc} T_U(X)\times\big(T_U(X)\times T_U(X)\big) & & \xrightarrow{\quad T_U(\underline{\alpha})\quad} & & \big(T_U(X)\times T_U(X)\big)\times T_U(X) \\ {\scriptstyle \mathrm{Id}\times T_U(\underline{\mu})}\downarrow & & & & \downarrow{\scriptstyle T_U(\underline{\mu})\times\mathrm{Id}} \\ T_U(X)\times T_U(X) & \xrightarrow{T_U(\underline{\mu})} & T_U(X) & \xleftarrow{T_U(\underline{\mu})} & T_U(X)\times T_U(X) \end{array}$$

and

$$\begin{array}{ccccc} \bullet\times T_U(X) & \xrightarrow{T_U(\underline{\eta})\times\mathrm{Id}} & T_U(X)\times T_U(X) & \xleftarrow{\mathrm{Id}\times T_U(\underline{\eta})} & T_U(X)\times\bullet \\ & {\scriptstyle T_U(\underline{\lambda})}\searrow & \downarrow{\scriptstyle T_U(\underline{\mu})} & \swarrow{\scriptstyle T_U(\underline{\rho})} & \\ & & T_U(X) & & \end{array}$$

One checks easily that $T_U(\underline{\alpha})$, $T_U(\underline{\lambda})$ and $T_U(\underline{\rho})$ are respectively equal to the corresponding morphisms $\underline{\alpha}$, $\underline{\lambda}$ and $\underline{\rho}$ for the Q-set $T_U(X)$. This shows Assertion 1.

The proof of Assertion 2 is similar, applying T_U to the diagrams 12.5.4. Assertion 3 is a straightforward consequence of Assertion 1 and 2, and Lemma 12.4.9. □

12.5.7. Corollary : *If A is a permutation P-algebra, and if X and X' are P-invariant k-bases of A, then there is an isomorphism of P-algebras over k from $kT_U(X) = e_Q\big(T_U(X)\big)$ to $kT_U(X') = e_Q\big(T_U(X)\big)$.*

Proof: Indeed if $\underline{\mu}$ and $\underline{\eta}$ (resp. $\underline{\mu}'$ and $\underline{\eta}'$) are the multiplication morphism and unit morphism of the algebra A corresponding to X (resp. X'), there is an isomorphism $\varphi : (X, \underline{\mu}, \underline{\eta}) \to (X', \underline{\mu}', \underline{\eta}')$ in $\underline{\mathsf{Perm}}_k(P)$. Hence $T_U(\varphi)$ is an isomorphism from $\big(T_U(X), T_U(\underline{\mu}), T_U(\underline{\eta})\big)$ to $\big(T_U(X'), T_U(\underline{\mu}'), T_U(\underline{\eta}')\big)$ in $\underline{\mathsf{Perm}}_k(Q)$, and $e_Q(\varphi) : kT_U(X) \to kT_U(X')$ is an isomorphism of P-algebras over k. □

12.5.8. Notation : *If A is a permutation P-algebra over k, let $T_U(A)$ denote the permutation Q-algebra $kT_U(X)$, where X is a P-invariant k-basis of A. The algebra $T_U(A)$ is well defined up to isomorphism of Q-algebras over k.*

12.5.9. Remark : The algebra $T_U(A)$ can be described in terms of *structure constants*, as follows: in the basis X, the multiplication in A is given by

$$\forall x, y \in X, \quad x \cdot y = \sum_{z \in X} c_{x,y}^{z}\, z \;,$$

where the scalars $c_{x,y}^{z} \in k$ are the structure constants of A. Then the multiplication in $T_U(A)$ is given by

$$\forall \varphi, \psi \in T_U(X), \quad \varphi \cdot \psi = \sum_{\theta \in T_U(X)} C_{\varphi,\psi}^{\theta}\, \theta \;,$$

where

$$C_{\varphi,\psi}^{\theta} = \prod_{u \in [U/P]} c_{\varphi(u),\psi(u)}^{\theta(u)} \;.$$

12.5.10. Examples: elementary bisets : It is easy to check that if U is an elementary biset (see 12.4.10), corresponding to one of the operations of restriction, induction, inflation, deflation, or transport by isomorphism, then $T_U(A)$ is obtained from A by the corresponding natural construction. More precisely, suppose that P is a finite p-group and A is a permutation P-algebra over k. Then:

- Let Q be a subgroup of P, and $U = \mathrm{Res}_Q^P$. Then the algebra $T_U(A)$ is the algebra A, with the Q-action obtained by restriction from P.

- Let P be a subgroup of Q, and $U = \mathrm{Ind}_P^Q$. Then the algebra $T_U(A)$ is isomorphic to the tensor induced algebra $\mathrm{Ten}_P^Q(A)$.
- Suppose that $P = Q/M$, for some finite p-group Q and some normal subgroup M of Q. Then $T_U(A)$ is isomorphic to A as an algebra, and the action of Q is obtained by inflation from P.
- Let N be a normal subgroup of P. Let $Q = P/N$, and let $U = \mathrm{Def}_Q^P$. Then by the fourth example in 12.4.10, the functor T_U corresponds to the Brauer quotient by N, and $T_U(A)$ is isomorphic to the Brauer quotient algebra $A[N]$.
- Finally, if $f : P \to Q$ is a group isomorphism, and if $U = \mathrm{Iso}(f)$, then the algebra $T_U(A)$ is isomorphic to A as an algebra, and the action of Q is obtained by transport via f.

12.5.11. Lemma : *With the same notation:*

1. *If A is a permutation P-algebra over k, then there is an isomorphism $T_U(A^{op}) \cong T_U(A)^{op}$ of permutation Q-algebras over k.*
2. *If A and A' are permutation P-algebras over k, then there is an isomorphism $T_U(A \otimes A') \cong T_U(A) \otimes T_U(A')$ of permutation Q-algebras over k.*
3. *If A is a permutation P-algebra over k, and if U and U' are finite (Q, P)-bisets, then there is an isomorphism $T_{U \sqcup U'}(A) \cong T_U(A) \otimes T_{U'}(A)$ of permutation Q-algebras over k.*
4. *If X is a finite P-set, then there is an isomorphism $T_U\big(\mathrm{End}_k kX)\big) \cong \mathrm{End}_k\big(kT_U(X)\big)$ of permutation Q-algebras over k.*

Proof: Assertion 1, Assertion 2, and Assertion 3 follow respectively from Assertion 3, Assertion 2 and Assertion 4 of Proposition 12.4.12. The proof of Assertion 3 is straightforward (see Lemma 2.11 of [19] for details). □

12.5.12. Theorem : [Bouc–Thévenaz [19]] *Let k be a field of characteristic p, let P and Q be finite p-groups, and let U be a finite (Q, P)-biset.*

1. *If A is a Dade P-algebra over k, then $T_U(A)$ is a Dade Q-algebra over k.*
2. *If A and A' are equivalent Dade P-algebras over k, then $T_U(A)$ and $T_U(A')$ are equivalent Dade Q-algebras over k.*
3. *The correspondence $A \mapsto T_U(A)$ induces a group homomorphism $D_k(P) \to D_k(Q)$, denoted by $D_k(U)$.*
4. *If U' is another finite (Q, P)-biset, then $D_k(U \sqcup U') = D_k(U) + D_k(U')$. Hence the correspondence $U \mapsto D_k(U)$ extends to a group homomorphism $D_k : B(Q, P) \to \mathrm{Hom}\big(D_k(P), D_k(Q)\big)$.*

Proof: For Assertion 1, let X be a P-invariant k-basis of A. Then the image of the set X^P in the Brauer quotient $A[P]$ is a k-basis of $A[P]$. Thus, saying that $A[P]$ is non zero is equivalent to saying that X^P is non empty. If $x \in X^P$, then the constant map $c_x : U^{op} \to X$, defined by $c_x(u) = x$ for all $u \in U$, is an element of $T_U(X) = \mathrm{Hom}_P(U^{op}, X)$, and it is clearly fixed by Q. So $T_U(X)^Q \neq \emptyset$, i.e. $T_U(A)[Q] \neq \{0\}$.

It remains to show that $T_U(A)$ is isomorphic to a full matrix algebra over k. This is independent of the action of Q: in other words, it is enough to show that $\mathrm{Res}_{\mathbf{1}}^Q T_U(A)$ is isomorphic to a full matrix algebra over k. Denote by V the $(\mathbf{1}, Q)$-biset $\mathrm{Res}_{\mathbf{1}}^Q$, i.e. the set Q, with right Q-action by multiplication. Then $\mathrm{Res}_{\mathbf{1}}^Q T_U(A) = T_V \circ T_U(A) = T_{V \times_Q U}(A)$, by Corollary 12.4.17, since V is a free right Q-set. This shows that one can suppose $Q = \mathbf{1}$ here.

Now U is a $(\mathbf{1}, P)$-biset, i.e. a right P-set. It decomposes as a disjoint union of transitive right P-sets U_i, for $i \in \{1, \ldots, n\}$, and the algebra $T_U(A)$ is isomorphic to the tensor product of the algebras $T_{U_i}(A)$. Since a tensor product of matrix algebras is a matrix algebra, it suffices to consider the case where U is transitive, i.e. of the form $U = S\backslash P$, for some subgroup S of P. In this case $T_U(A) = \mathrm{Defres}_{S/S}^P A$ is isomorphic to the Brauer quotient $A[S]$, which is isomorphic to a full matrix algebra over k by Theorem 12.3.3.

For Assertion 2, if A and A' are equivalent, then there is a finite P-set X and an isomorphism

$$A' \otimes A^{op} \cong \mathrm{End}_k(kX)$$

of permutation P-algebras over k. Then by Lemma 12.5.11, there are isomorphisms

$$\begin{aligned} T_U(A') \otimes T_U(A)^{op} &\cong T_U(A') \otimes T_U(A^{op}) \cong T_U(A' \otimes A^{op}) \\ &\cong T_U\big(\mathrm{End}_k(kX)\big) \cong \mathrm{End}_k\big(kT_U(X)\big) \end{aligned}$$

of permutation Q-algebras over k, so $T_U(A)$ and $T_U(A')$ are equivalent.

It follows that the correspondence $A \mapsto T_U(A)$ induces a well defined map $D_k(U) : D_k(P) \to D_k(Q)$, which is a group homomorphism by Assertions 1 and 2 of Lemma 12.5.11. This proves Assertion 3.

Finally, Assertion 4 follows from Assertion 3 of Lemma 12.5.11. □

12.5.13. Galois Twists and Permutation Algebras. The functors T_U yield a correspondence $A \mapsto T_U(A)$ of Dade algebras. The Galois twist functors γ_a defined in 12.4.14 yield similar constructions:

12.5.14. Proposition : *Let a be an endomorphism of k, and P be a finite p-group.*

1. *The functor $\gamma_a : \underline{\mathsf{Perm}}_k(P) \to \underline{\mathsf{Perm}}_k(P)$ extends to a functor $\underline{\mathsf{PermAlg}}_k(P) \to \underline{\mathsf{PermAlg}}_k(P)$, still denoted by γ_a, defined by $\gamma_a\big((X, \underline{\mu}, \underline{\eta})\big) = \big(X, \gamma_a(\underline{\mu}), \gamma_a(\underline{\eta})\big)$.*
2. *Let A be a permutation P-algebra over k. Let X be a P-invariant k-basis of A, with corresponding multiplication morphism $\underline{\mu} : X \times X \to X$ and unit morphism $\eta : \bullet \to X$ in $\underline{\mathsf{Perm}}_k(P)$. Then the algebra $\gamma_a(A) = e_P \circ$*

$\gamma_a\big((X,\underline{\mu},\underline{\eta})\big)$ does not depend on the choice of the P-invariant basis X, up to isomorphism of permutation P-algebras over k.

3. *If A is a permutation P-algebra over k, then there is an isomorphism $\gamma_a(A^{op}) \cong \gamma_a(A)^{op}$ of permutation P-algebras over k.*
4. *If A and A' are permutation P-algebras over k, then there is an isomorphism $\gamma_a(A \otimes A') \cong \gamma_a(A) \otimes \gamma_a(A')$ of permutation P-algebras over k.*
5. *If A is a Dade P-algebra over k, then $\gamma_a(A)$ is a Dade P-algebra over k.*
6. *If A and A' are equivalent Dade P-algebras over k, then $\gamma_a(A)$ and $\gamma_a(A')$ are equivalent Dade P-algebras over k.*
7. *The correspondence $A \mapsto \gamma_a(A)$ on the class of Dade P-algebras over k induces a group homomorphism $D_k(P) \to D_k(P)$, still denoted by γ_a, or $\gamma(p^n)$ when a is the endomorphism $x \mapsto x^{p^n}$, for $n \in \mathbb{N}$.*

Proof: For Assertion 1, let $(X,\underline{\mu},\underline{\eta})$ be an object in $\underline{\mathsf{PermAlg}}_k(P)$. Applying the functor γ_a to the diagrams 12.5.2 and 12.5.3 gives the following diagrams

$$\begin{array}{ccccc}
X\times(X\times X) & & \xrightarrow{\gamma_a(\underline{\alpha})} & & (X\times X)\times X \\
\downarrow{\scriptstyle \gamma_a(\mathrm{Id}\times\underline{\mu})} & & & & \downarrow{\scriptstyle \gamma_a(\underline{\mu}\times\mathrm{Id})} \\
X\times X & \xrightarrow{\gamma_a(\underline{\mu})} & X & \xleftarrow{\gamma_a(\underline{\mu})} & X\times X
\end{array}$$

and

$$\begin{array}{ccccc}
\bullet\times X & \xrightarrow{\gamma_a(\underline{\eta}\times\mathrm{Id})} & X\times X & \xleftarrow{\gamma_a(\mathrm{Id}\times\underline{\eta})} & X\times\bullet \\
& {\scriptstyle \gamma_a(\underline{\lambda})}\searrow & \downarrow{\scriptstyle \gamma_a(\underline{\mu})} & \swarrow{\scriptstyle \gamma_a(\underline{\rho})} & \\
& & X & &
\end{array}$$

since the functor γ_a is the identity on objects. Now $\gamma_a(\underline{\alpha}) = \underline{\alpha}$, since the coefficients of the associativity isomorphism are equal to 0 or 1. Similarly $\gamma_a(\underline{\lambda}) = \underline{\lambda}$ and $\gamma_a(\underline{\rho}) = \underline{\rho}$. Now obviously $\gamma_a(f \times g) = \gamma_a(f) \times \gamma_a(g)$ for any two morphisms f and g in the category $\underline{\mathsf{Perm}}_k(P)$. The above diagrams become

$$\begin{array}{ccccc}
X\times(X\times X) & & \xrightarrow{\underline{\alpha}} & & (X\times X)\times X \\
\downarrow{\scriptstyle \mathrm{Id}\times\gamma_a(\underline{\mu})} & & & & \downarrow{\scriptstyle \gamma_a(\underline{\mu})\times\mathrm{Id}} \\
X\times X & \xrightarrow{\gamma_a(\underline{\mu})} & X & \xleftarrow{\gamma_a(\underline{\mu})} & X\times X
\end{array}$$

and

$$\begin{array}{ccccc}
\bullet\times X & \xrightarrow{\gamma_a(\underline{\eta})\times\mathrm{Id}} & X\times X & \xleftarrow{\mathrm{Id}\times\gamma_a(\underline{\eta})} & X\times\bullet\ , \\
& {\scriptstyle \gamma_a(\underline{\lambda})}\searrow & \downarrow{\scriptstyle \gamma_a(\underline{\mu})} & \swarrow{\scriptstyle \gamma_a(\underline{\rho})} & \\
& & X & &
\end{array}$$

showing that $\big(X,\gamma_a(\underline{\mu}),\gamma_a(\underline{\eta})\big)$ is an object of $\underline{\mathsf{PermAlg}}_k(P)$.

Similarly, if $f : (X, \underline{\mu}, \underline{\eta}) \to (X', \underline{\mu}', \underline{\eta}')$ is a morphism in $\underline{\mathsf{PermAlg}_k}(P)$, then applying γ_a to the diagrams 12.5.4 gives the diagrams

$$\begin{array}{ccc} X \times X & \xrightarrow{\gamma_a(\underline{\mu})} & X \\ {\scriptstyle \gamma_a(f)\times\gamma_a(f)}\downarrow & & \downarrow{\scriptstyle \gamma_a(f)} \\ X' \times X' & \xrightarrow{\gamma_a(\underline{\mu}')} & X' \end{array} \qquad \begin{array}{ccc} & \bullet & \\ {\scriptstyle \gamma_a(\underline{\eta})}\swarrow & & \searrow{\scriptstyle \gamma_a(\underline{\eta}')} \\ X & \xrightarrow{\gamma_a(f)} & X' \end{array} ,$$

showing that $\gamma_a(f)$ is a morphism in $\underline{\mathsf{PermAlg}_k}(P)$. This completes the proof of Assertion 1.

Assertion 2 follows: if A is a permutation P-algebra over k, and if X and X' are two P-invariant k-bases of A, then there is an isomorphism $f : (X, \underline{\mu}, \underline{\eta}) \to (X', \underline{\mu}', \underline{\eta}')$ between the corresponding objects of $\underline{\mathsf{PermAlg}_k}(P)$. Then $e_P\big(\gamma_a(f)\big)$ is an isomorphism of P-algebras over k from $e_P \circ \gamma_a\big((X, \underline{\mu}, \underline{\eta})\big)$ to $e_P \circ \gamma_a\big((X', \underline{\mu}', \underline{\eta}')\big)$.

Assertion 3 and 4 are straightforward.

For Assertion 5, since the algebras A and $\gamma_a(A)$ have the same P-invariant k-bases, saying that $A[P] \neq \{0\}$ is equivalent to saying that $\gamma_a(A)[P] \neq \{0\}$, i.e. that P has a fixed point on some invariant basis of A. To check that $\gamma_a(A)$ is isomorphic to a full matrix algebra over k, it suffices to check that $\mathrm{Res}^P_{\mathbf{1}} \gamma_a(A)$ is. Now $\mathrm{Res}^P_{\mathbf{1}} = T_U$, where U is the set P, viewed as an $(\mathbf{1}, P)$-biset in the obvious way. Since $T_U \circ \gamma_a = \gamma_a \circ T_U$ by Lemma 12.4.15, and since $T_U(A)$ is a Dade $\mathbf{1}$-algebra over k (i.e. a full matrix algebra over k), it is enough to suppose $P = \mathbf{1}$, and that A is a matrix algebra $M_n(k)$. In the canonical basis of A, the structure constants are equal to 0 or 1, hence they are invariant by a. So the multiplication morphism $\underline{\mu}$ is invariant by a, and $\gamma_a(A) \cong A$ in this case. This completes the proof of Assertion 5.

For Assertion 6, if A and A' are equivalent Dade P-algebras over k, there exists a finite P-set X and an isomorphism $A' \otimes A^{op} \cong \mathrm{End}_k(kX)$ of P-algebras over k. It follows from Assertions 3 and 4 that

$$\gamma_a(A') \otimes \gamma_a(A)^{op} \cong \gamma_a(A' \otimes A^{op}) \cong \gamma_a\big(\mathrm{End}_k(kX)\big) .$$

Moreover the canonical k-basis of the matrix algebra $\mathrm{End}_k(kX)$ is invariant by P, and the structure constants in this basis are equal to 0 or 1. It follows that $\gamma_a\big(\mathrm{End}_k(kX)\big) \cong \mathrm{End}_k(kX)$, so there is an isomorphism

$$\gamma_a(A') \otimes \gamma_a(A) \cong \mathrm{End}_k(kX)$$

of P-algebras over k, and the Dade P-algebras $\gamma_a(A)$ and $\gamma_a(A')$ are equivalent.

It follows that the correspondence $A \mapsto \gamma_a(A)$ induces a well defined map from $D_k(P)$ to itself, and this map is a group homomorphism by Assertions 3 and 4. This proves Assertion 7, and completes the proof of Proposition 12.5.14. □

12.5.15. Remark : If A is a permutation P-algebra over k, with structure constants $c_{x,y}^z$ in the P-invariant k-basis X, then the algebra $\gamma_a(A)$ is the k-vector space with the same basis X, and multiplication obtained by twisting by a the structure constants of A, i.e. by setting

$$\forall x, y \in X, \quad x \cdot y = \sum_{z \in X} a(c_{x,y}^z)\, z \ .$$

12.5.16. Composition of Biset Operations on Dade Groups. Recall from Theorem 12.5.12 that when P and Q are finite p-groups, to each finite (Q,P)-biset U is associated a group homomorphism

$$D_k(U) : D_k(P) \to D_k(Q) \ .$$

This correspondence $U \mapsto D_k(U)$ is additive with respect to U. However, in general, this does not give a biset functor structure on the correspondence $P \mapsto D_k(P)$, since the maps $D_k(U)$ are not well behaved for composition:

12.5.17. Proposition : [Bouc–Thévenaz [19]]

1. *For every finite p-group P, the morphism $D_k(\mathrm{Id}_P)$ associated to the identity (P,P)-biset is equal to the identity map of $D_k(P)$.*
2. *If P, Q and R are finite p-groups, if U is a finite (Q,P)-biset and V is a finite (R,Q)-biset, then*

$$D_k(V \times_Q U) = \sum_{(v,_Q u) \in [R \backslash (V \times_Q U)/P]} D_k\big(R(v,_Q u)P\big) \ ,$$

$$D_k(V) \circ D_k(U) = \sum_{(v,_Q u) \in [R \backslash (V \times_Q U)/P]} \gamma\big(|\mathbf{1}^v : \mathbf{1}^v \cap {}^uP|\big) \circ D_k\big(R(v,_Q u)P\big) \ .$$

Proof: Each assertion is a straightforward consequence of the corresponding assertion of Proposition 12.4.16, switching from functors to the group homomorphisms induced on Dade groups, and to the corresponding additive notation. □

12.5.18. Corollary : *If V is a free right Q-set, or if U is a transitive right P-set, then $D_k(V) \circ D_k(U) = D_k(V \times_Q U)$.*

Proof: This follows from Corollary 12.4.17. □

12.6. Relative Syzygies

Though the correspondence sending a finite p-group P to its Dade group $D_k(P)$ is generally not a biset functor, because of Frobenius twists appearing in Proposition 12.5.17, there is an important biset functor closely related to the Dade group, namely *the functor of relative syzygies*: this functor sends a finite p-group P to the group $D^\Omega(P)$ to be defined now:

12.6.1. Definition : *Let P be a finite p-group. If X is a non-empty finite P-set, let $\epsilon_X : kX \to k$ denote the augmentation map, defined by $\epsilon_X(x) = 1$, for $x \in X$. The kernel $\Omega_X(k)$ of ϵ_X is called the X-relative syzygy of the trivial module.*

12.6.2. Theorem : [Alperin [2]] *Let P be a finite p-group, and X be a non-empty finite P-set. Then $\Omega_X(k)$ is an endo-permutation kP-module. Moreover:*

1. *If $X^P \neq \emptyset$, then $\Omega_X(k)$ is a permutation kP-module.*
2. *If $|X^P| \neq 1$, then $\Omega_X(k)$ is a capped endo-permutation kP-module.*

Proof: (Sketched: for details see [2], or Lemma 2.3.3 of [10]) If $X^P \neq \emptyset$, then the short exact sequence

$$0 \longrightarrow \Omega_X(k) \longrightarrow kX \xrightarrow{\epsilon_X} k \longrightarrow 0$$

is split, so $\Omega_X(k)$ is a permutation module. And if $|X^P| > 1$, then $\Omega_X(k)$ admits a direct summand isomorphic to the trivial module, with multiplicity $|X^P| - 1$, so $\Omega_X(k)$ is a capped (endo-)permutation kP-module.

The proof that $\Omega_X(k)$ is an endo-permutation kP-module uses a relative version of Shanuel's Lemma, to show that there is an isomorphism of kP-modules

$$\mathrm{End}_k\big(\Omega_X(k)\big) \oplus kX \oplus kX \cong k \oplus (kX \otimes kX) ,$$

which also shows that $\mathrm{End}_k\big(\Omega_X(k)\big)[P] \cong k$ if $X^P = \emptyset$. □

12.6.3. Definition : *Let P be a finite p-group. If X is a finite P-set, let Ω_X denote the element of $D_k(P)$ defined by*

$$\Omega_X = \begin{cases} [\Omega_X(k)] & \text{if } X \neq \emptyset \text{ and } X^P = \emptyset \\ 0 & \text{otherwise.} \end{cases}$$

Let $D_k^\Omega(P)$ denote the subgroup of $D_k(P)$ generated by the elements Ω_X, where X is a finite P-set.

12.6.4. Example: Dade's theorem : The Dade group of abelian p-groups was determined by Dade in 1978 [31, 32]. Using the above notation, Dade's theorem can be stated as follows:

Theorem : *Let k be a field of characteristic p, and P be a finite abelian p-group. Then $D_k(P) = D_k^{\Omega}(P)$. More precisely: let $NC(P)$ denote the set of subgroups Q of P such that P/Q is non cyclic, and $C'(P)$ denote the set of subgroups Q of P such that P/Q is cyclic of order at least equal to 3. Then*

1. *If $Q \leq P$ and $|P:Q| \leq 2$, then $\Omega_{P/Q} = 0$.*
2. *The elements $\Omega_{P/Q}$, for $Q \in C'(P)$, have order 2, and they are linearly independent over $\mathbb{F}_2$. They generate the torsion subgroup $D_k(P)_{tors}$ of $D_k(P)$.*
3. *The elements $\Omega_{P/Q}$, for $Q \in NC(P)$, have infinite order, and they are linearly independent over $\mathbb{Z}$. They generate a (free) complement to $D_k(P)_{tors}$ in $D_k(P)$.*

12.6.5. Proposition : *Let k be a field of characteristic p.*

1. *If P is a finite p-group, if Q is a subgroup of P, and if X is a finite P-set, then $\mathrm{Res}_Q^P \Omega_X = \Omega_{\mathrm{Res}_Q^P X}$ in $D_k(Q)$.*
2. *If P is a finite p-group, if $N \trianglelefteq P$, and if Y is a finite (P/N)-set, then $\mathrm{Inf}_{P/N}^P \Omega_Y = \Omega_{\mathrm{Inf}_{P/N}^P Y}$ in $D_k(P)$.*
3. *If $\varphi : P \to Q$ is a group isomorphism between finite p-groups P and Q, and if X is a finite P-set, then $\mathrm{Iso}(\varphi)(\Omega_X) = \Omega_{^{\varphi}X}$, where $^{\varphi}X$ is the Q-set obtained from X by transport via φ.*
4. *If P is a finite p-group, if $N \trianglelefteq P$, and if X is a finite P-set, then $\mathrm{Def}_{P/N}^P \Omega_X = \Omega_{X^N}$ in $D_k(P/N)$.*

Proof: All the assertions are straightforward. For details, see Corollary 4.1.2 and Lemma 4.2.1 of [10]. □

The previous proposition shows that all elementary bisets except tensor induction preserve the group of relative syzygies. The following theorem handles the remaining case:

12.6.6. Theorem : [10, Theorem 5.1.2] *Let k be a field of characteristic p, let P be a finite p-group, let Q be a subgroup of P, and let X be a finite Q-set. Then in $D_k(P)$*

$$\mathrm{Ten}_Q^P \Omega_X = \sum_{\substack{S,T \in [s_P] \\ S \leq_P T}} \mu_P(S,T) \, |\{a \in T\backslash P/Q \mid X^{T^a \cap Q} \neq \emptyset\}| \, \Omega_{P/S} ,$$

where $[s_P]$ is a set of representatives of conjugacy classes of subgroups of P, ordered by the relation $S \leq_P T$ if and only if some P-conjugate of S is a subgroup of T, and μ_P is the Möbius function of this poset.

Proof: At the time, no easy proof of this formula is known. The original proof is *highly* technical: it uses induction on the index $|P:Q|$, the crucial (and really hard) case being the case $|P:Q| = p$. For details, see Sect. 5 of [10]. □

12.6.7. Corollary : *Let X be a finite P-set. Then in $D_k^\Omega(P)$*

$$\Omega_X = \sum_{\substack{S,T \in [s_P] \\ S \leq_P T,\ X^T \neq \emptyset}} \mu_P(S,T)\Omega_{P/S} \ ,$$

and in particular the elements $\Omega_{P/S}$, for $S \in [s_P]$, generate $D_k(\Omega)$.

Proof: This is not really a corollary of Theorem 12.6.6, but rather the initial step $Q = P$ in the inductive proof of this theorem. See Lemma 5.2.3 of [10] for details. □

12.6.8. Theorem : *Let k be a field of characteristic p.*

1. *If P and Q are finite p-groups, and if U is a finite (Q,P)-biset, then*

$$D_k(U)\big(D_k^\Omega(P)\big) \subseteq D_k^\Omega(Q) \ .$$

2. *The correspondence $P \mapsto D_k^\Omega(P)$ is a p-biset functor.*

Proof: For Assertion 1, since the correspondence $U \mapsto D_k(U)$ is additive in U, it suffices to consider the case where U is a transitive (Q,P)-biset. In this case, by Lemma 2.3.26, the biset U can be factored as

$$U = \mathrm{Ind}_T^Q \circ \mathrm{Inf}_{T/S}^T \circ \mathrm{Iso}(\varphi) \circ \mathrm{Def}_{B/A}^B \circ \mathrm{Res}_B^P \ ,$$

where (B,A) is a section of P, where (T,S) is a section of Q, and $\varphi : B/A \to T/S$ is a group isomorphism. Repeated application of both cases of Corollary 12.5.18 show that

$$D_k(U) = \mathrm{Ten}_T^Q \circ \mathrm{Inf}_{T/S}^T \circ \mathrm{Iso}(\varphi) \circ \mathrm{Def}_{B/A}^B \circ \mathrm{Res}_B^P \ ,$$

so $D_k(U)\big(D_k^\Omega(P)\big) \subseteq D^\Omega(Q)$, by Proposition 12.6.5 and Theorem 12.6.6.

Now for any finite P-set X, the module $\Omega_X(k)$ is clearly defined over the prime field $\mathbb{F}_p$, i.e. $\Omega_X(k) \cong k \otimes_{\mathbb{F}_p} \Omega_X(\mathbb{F}_p)$. It follows that the structure constants of the algebra $A = \mathrm{End}_k\big(\Omega_X(k)\big)$ are all in $\mathbb{F}_p$, so they are fixed by any field endomorphism of k. This means in particular that $\gamma(p^n)([A]) = [A]$ in $D_k(P)$, for any $n \in \mathbb{N}$.

In other words, all the Frobenius twists $\gamma(p^n)$ act as the identity on $D_k^\Omega(P)$, for any finite p-group P. So if P, Q, and R are finite p-groups, if U is a finite (Q,P)-biset and V is a finite (R,Q)-biset, and if $D_k^\Omega(U)$ denotes the restriction of the map $D_k(U)$ to the group $D_k^\Omega(P)$ (so $D_k^\Omega(U)$ maps $D_k^\Omega(P)$ inside $D_k^\Omega(Q)$, by Assertion 1), it follows from Proposition 12.5.17 that

$$D_k^\Omega(V) \circ D_k^\Omega(U) = D_k^\Omega(V \times_Q U) ,$$

which proves Assertion 2. □

12.7. A Short Exact Sequence of Biset Functors

The functor of relative syzygies D_k^Ω is closely related to the dual functor B^* of the Burnside functor, and also to the functor $R_\mathbb{Q}$ of rational representations. More precisely there exist a surjective morphism of p-biset functors from B^* to D_k^Ω, defined hereafter:

12.7.1. Notation : *Let P be a finite p-group. If X is a finite P-set, denote by ω_X the element of $B^*(P) = \mathrm{Hom}_\mathbb{Z}\big(B(P),\mathbb{Z}\big)$ defined by*

$$\forall Q \leq P,\ \ \omega_X(P/Q) = \begin{cases} 1 \text{ if } X^Q \neq \emptyset \\ 0 \text{ if } X^Q = \emptyset \,. \end{cases}$$

If $S \leq P$, denote by $\delta_{P/S}$ the element of $B^*(P)$ defined by

$$\forall Q \leq P,\ \ \delta_{P/S}(P/Q) = \begin{cases} 1 \text{ if } Q =_P S \\ 0 \text{ otherwise.} \end{cases}$$

The set of elements $\delta_{P/S}$, for $S \in [s_P]$, is the dual basis of the canonical $\mathbb{Z}$-basis of $B(P)$.

12.7.2. Lemma : *Let P be a finite p-group. Then the elements $\omega_{P/S}$, for $S \in [s_P]$, form a $\mathbb{Z}$-basis of $B^*(P)$. More precisely, for $S \leq P$*

$$\omega_{P/S} = \sum_{\substack{T \in [s_P] \\ T \leq_P S}} \delta_{P/T} \,,$$

$$\delta_{P/S} = \sum_{\substack{T \in [s_P] \\ T \leq_P S}} \mu_P(T,S)\omega_{P/T} \,.$$

Proof: The first equality follows from the fact that for $T \leq P$, the set $(P/S)^T$ is non empty if and only if $T \leq_P S$. The second formula follows by Möbius inversion. So the set $\{\omega_{P/S} \mid S \in [s_P]\}$ is a basis of $B^*(P)$, since it is obtained from the basis $\{\delta_{P/S} \mid S \in [s_P]\}$ by a linear transformation whose matrix is triangular with 1's on the diagonal. □

12.7.3. Theorem : [13, Theorem 1.7] *Let k be a field of characteristic p. There is a unique morphism of biset functors $\Theta : B^* \to D_k^\Omega$ such that*

$$\Theta_P(\omega_X) = \Omega_X$$

for any finite p-group P and any finite P-set X. Moreover Θ is surjective.

Proof: (Sketched: see [13] for details) The uniqueness part of Assertion 1 follows from Lemma 12.7.2: for any finite p-group P, the map Θ_P is uniquely defined by $\Theta_P(\omega_{P/Q}) = \Omega_{P/Q}$, for $Q \in [s_P]$.

This implies that $\Theta_P(\omega_X) = \Omega_X$ for any finite P-set X, because the expression of ω_X in the basis $\{\omega_{P/Q} \mid Q \in [s_P]\}$ is the same as the linear combination of Corollary 12.6.7.

Now the fact that Θ is a morphism of biset functors is essentially equivalent to Theorem 12.6.6. Moreover Θ is clearly surjective, by definition of D_k^Ω. □

12.7.4. Example : Let $E \cong (\mathbb{Z}/p\mathbb{Z})^2$. Then by Dade's theorem 12.6.4, if $p = 2$, the group $D_k(E)$ is free of rank one, generated by $\Omega_{E/\mathbf{1}}$. And if $p > 2$, then $D_k(E) \cong \mathbb{Z} \oplus (\mathbb{Z}/2\mathbb{Z})^{p+1}$, generated by $\Omega_{E/\mathbf{1}}$, which has infinite order, and the $p+1$ elements $\Omega_{E/F}$, for $|E : F| = p$, which have order 2, and are linearly independent over $\mathbb{F}_2$.

This example will be an important special case in the proof of the next theorem, so it is worth considering it in full details: let $\chi_E^* : R_\mathbb{Q}^*(E) \to B^*(E)$ denote the transposed map of $\chi_E : B(E) \to R_\mathbb{Q}(E)$. Then χ_E^* is injective, since χ_E is surjective by the Ritter–Segal theorem. The rational irreducible representations of E are the trivial representation $\mathbb{Q}$, and the representations $J_{E/F} = \mathrm{Ker}(\mathbb{Q}E/F \xrightarrow{\epsilon} \mathbb{Q})$, for $|E : F| = p$, where ϵ is the augmentation map. The group $R_\mathbb{Q}^*(E)$ has a $\mathbb{Z}$-basis consisting of the elements V^*, for $V \in \mathrm{Irr}_\mathbb{Q}(E)$. The element $\chi_E^*(V^*)$ of $B^*(E)$ is the linear form on $B(E)$ sending E/X, for $X \leq E$, to the multiplicity of V in $\mathbb{Q}(E/X)$.

It follows easily that $\chi_E^*(\mathbb{Q}^*)(E/X) = 1$ for any $X \leq E$. In other words $\chi_E^*(\mathbb{Q}^*) = \omega_{E/E}$. Similarly, if $|E : F| = p$, then $\chi_E^*(J_{E/F}^*)(E/X)$ is equal to 1 if $X \leq F$, and to zero otherwise. Equivalently $\chi_E^*(J_{E/F}^*) = \omega_{E/F}$.

In particular, the image of the map $\Theta_E \circ \chi_E^*$ is the subgroup of $D_k(E)$ generated by $\Omega_{E/E} = 0$, and the elements $\Omega_{E/F}$, for $|E : F| = p$. If $p > 2$, these elements generate the torsion subgroup of $D_k(E)$. And if $p = 2$, these elements are all equal to 0 in $D_k(E)$. Thus in any case, the image of $\Theta_E \circ \chi_E^*$ is equal to the torsion subgroup of $D_k(E) = D_k^\Omega(E)$.

12.7.5. Theorem : [13, Theorem 1.8] *Let $\chi^* : R_{\mathbb{Q}}^* \to B^*$ denote the transpose of $\chi : B \to R_{\mathbb{Q}}$. Then, with the notation of Theorem 12.7.3, the image of $\Theta \circ \chi^*$ is equal to the torsion part $(D_k^{\Omega})_{tors}$ of D_k^{Ω}. In other words, if $\overline{\Theta}$ denotes the composition $\pi \circ \Theta$, where $\pi : D_k^{\Omega} \to D_k^{\Omega}/(D_k^{\Omega})_{tors}$ is the projection map, there is a short exact sequence*

$$0 \longrightarrow R_{\mathbb{Q}}^* \xrightarrow{\chi^*} B^* \xrightarrow{\overline{\Theta}} D_k^{\Omega}/(D_k^{\Omega})_{tors} \longrightarrow 0$$

of p-biset functors.

Proof: Let P be a finite p-group, and let $u \in D_k^{\Omega}(P)$. Choose $f \in B^*(P)$ such that $\Theta_P(f) = u$. Then u is a torsion element of $D_k^{\Omega}(P)$ if and only if u is a torsion element of $D_k(P)$, and by Theorem 1.6 of [19], this is equivalent to saying that $\mathrm{Defres}_{T/S}^P u$ is a torsion element in $D_k(T/S)$, for any section (T,S) of P such that T/S is elementary abelian of rank 2.

Now $\mathrm{Im}(\Theta_E \circ \chi_E^*) = (D_k^{\Omega})_{tors}(E)$ when $E \cong (\mathbb{Z}/p\mathbb{Z})^2$, by Example 12.7.4. Then, since $\mathrm{Defres}_{T/S}^P \circ \Theta_P = \Theta_{T/S} \circ \mathrm{Defres}_{T/S}^P$ by Theorem 12.7.3, saying that $u \in \mathrm{Im}(\Theta_P \circ \chi_P^*)$ is equivalent to saying that $\mathrm{Defres}_{T/S}^P f \in \mathrm{Im}\chi_{T/S}^*$, for any section (T,S) of P such that $T/S \cong (\mathbb{Z}/p\mathbb{Z})^2$. This in turn is equivalent to saying that $f \in \mathrm{Im}\chi_P^*$, as was to be shown, by the following lemma. □

12.7.6. Lemma : [Bouc–Yalçın [22] Lemma 4.2] *Let P be a finite p-group, and $f \in B^*(P)$. Then $f \in \mathrm{Im}\chi_P^*$ if and only if $\mathrm{Defres}_{T/S}^P f \in \mathrm{Im}\chi_{T/S}^*$, for any section (T,S) of P such that $T/S \cong (\mathbb{Z}/p\mathbb{Z})^2$. Equivalently, for any such section*

$$f(P/S) - \sum_{S<Y<T} f(P/Y) + pf(P/T) = 0 \; . \tag{12.7.7}$$

Proof: (Sketched: see [22] for details) One direction of the implication is obvious. For the converse, it is enough to show that $f \in \mathbb{Q}\mathrm{Im}(\chi_P^*)$, assuming that Condition 12.7.7 holds for any section (T,S) of P such that $T/S \cong (\mathbb{Z}/p\mathbb{Z})^2$. This is equivalent to show that $f(e_Q^P) = 0$ for any non cyclic subgroup Q of P, and it can be proved by induction on the order of Q, Condition 12.7.7 allowing to start induction in the case $Q = P \cong (\mathbb{Z}/p\mathbb{Z})^2$. □

12.8. Borel–Smith Functions

It follows from Theorem 12.7.5 that there is an exact sequence of p-biset functors

$$0 \longrightarrow \mathrm{Ker}\,\Theta \longrightarrow \chi^*(R_{\mathbb{Q}}^*) \xrightarrow{\Theta_|} (D_k^{\Omega})_{tors} \longrightarrow 0 \; ,$$

where $\Theta_|$ is the restriction of Θ to $\chi^*(R_\mathbb{Q}^*)$. But $\chi^*(R_\mathbb{Q}^*) \cong R_\mathbb{Q}^*$, since χ^* is injective, and this yields the following exact sequence of p-biset functors

$$0 \longrightarrow H \longrightarrow R_\mathbb{Q}^* \xrightarrow{\theta} (D_k^\Omega)_{tors} \longrightarrow 0 \; ,$$

where $H = (\chi^*)^{-1}(\mathrm{Ker}\;\Theta)$, and $\theta = \Theta \circ \chi^*$.

12.8.1. Lemma : *With the notation of Theorem 12.7.3, the functors* Ker Θ *and* $(D_k^\Omega)_{tors}$ *are rational p-biset functors.*

Proof: Indeed Ker $\Theta \cong H$, and H is a subfunctor of $R_\mathbb{Q}^*$, so H is rational. Similarly $(D_k^\Omega)_{tors}$ is isomorphic to a quotient of $R_\mathbb{Q}^*$, so it is rational. □

It follows from Proposition 10.2.1 that the functor H is entirely determined by its values at p-groups of normal p-rank 1: more precisely, if N is such a p-group, there exists an integer a_N such that $\partial H(N) = \langle a_N \Phi_N^* \rangle$, and for any finite p-group P, the subgroup $H(P)$ of $R_\mathbb{Q}^*(P)$ is equal to the set of elements $h \in R_\mathbb{Q}^*(P)$ such that

$$f_{\mathbf{1}}^{T/S} \mathrm{Defres}_{T/S}^P h \in \langle a_{T/S} \Phi_{T/S}^* \rangle \; , \qquad \textbf{(12.8.2)}$$

for every section (T, S) of P such that $T/S \in \mathcal{N}$.

12.8.3. The first thing to do is to find the values of the integers a_N. So let $N \in \mathcal{N}$. The following sequence

$$0 \longrightarrow \partial H(N) \longrightarrow \partial R_\mathbb{Q}^*(N) \longrightarrow \partial (D_k^\Omega)_{tors}(N) \longrightarrow 0$$

is exact, since it is the image of a short exact sequence by an idempotent endomorphism.

- If $|N| \leq 2$, then $D_k(N) = \{0\}$, so $H(N) = R_\mathbb{Q}^*(N)$, and $a_N = 1$.
- If N is cyclic and $|N| \geq 3$, then by Dade's theorem 12.6.4, the group $D_k(P) = D_k^\Omega(P)$ is an elementary abelian 2-group, with an $\mathbb{F}_2$-basis consisting of the elements $\Omega_{N/M}$, where $M < N$. If $M \neq \mathbf{1}$, then $\Omega_{N/M}$ is inflated from the proper quotient N/M, so $f_\mathbf{1}^N \Omega_{N/M} = 0$. And if $M = \mathbf{1}$, then any proper deflation of $\Omega_{N/M}$ is equal to 0, by Proposition 12.6.5, so $\Omega_{N/M} = f_\mathbf{1}^N \Omega_{N/M}$. It follows that $\partial D_k^\Omega(N)$ has order 2, generated by $\Omega_{N/\mathbf{1}}$. This shows that $a_N = 2$.
- The remaining cases occur for $p = 2$, when N is generalized quaternion, dihedral, or semi-dihedral. In these cases, the Dade group $D_k(N)$ has been determined by Dade [30] and Carlson-Thévenaz [25]. It follows that:
  - ⋆ If $N \cong Q_{2^n}$, for $n \geq 3$, then $D_k^\Omega(N)_{tors} \cong \mathbb{Z}/4\mathbb{Z}$, generated by $\Omega_{N/\mathbf{1}}$. Thus $a_N = 4$ in this case.
  - ⋆ If $N \cong D_{2^n}$, for $n \geq 4$, then $D_k(N)$ is torsion free, and $a_N = 1$ in this case.

- ⋆ If $N \cong SD_{2^n}$, for $n \geq 4$, then $D_k^\Omega(N)_{tors} \cong \mathbb{Z}/2\mathbb{Z}$, generated by the element $\Omega_{N/1} + \Omega_{N/M}$, where M is a non central subgroup of order 2 of N. So $a_N = 2$ in this case.

12.8.4. The next step it to interpret Condition 12.8.2, knowing the integers a_N. First, the condition is void if $a_{T/S} = 1$, so in particular if $T = S$. And if T/S is a non-trivial group in $\mathcal{N}$, then there exists a unique subgroup $\hat{S}$ such that $\hat{S}/S$ is a central subgroup of order p in T/S, and $f_1^{T/S} = \mathrm{Id}_{T/S} - \mathrm{Inf}_{T/\hat{S}}^{T/S}\mathrm{Def}_{T/\hat{S}}^{T/S}$. For any $h \in R_\mathbb{Q}^*(P)$, there is an integer n such that

$$f_1^{T/S}\mathrm{Defres}_{T/S}^P h = n\,\Phi_{T/S}^* \; .$$

The value of n can be found by computing the value of both sides at the free module $\mathbb{Q}T/S$. This gives

$$h(\mathrm{Indinf}_{T/S}^P f_1^{T/S}\mathbb{Q}T/S) = n\,m(\Phi_{T/S}, \mathbb{Q}T/S) = n\,\mathsf{d}_{T/S} \; ,$$

by Lemma 9.5.4, where the integer $\mathsf{d}_{T/S}$ was defined in 9.4.4. Recall that $\mathsf{d}_{T/S} = 2$ if T/S is dihedral or semi-dihedral, and $\mathsf{d}_{T/S} = 1$ otherwise. Equivalently

$$h(\mathbb{Q}P/S - \mathbb{Q}P/\hat{S}) = n\,\mathsf{d}_{\mathsf{T/S}} \; ,$$

thus $n = \dfrac{h(\mathbb{Q}P/S) - h(\mathbb{Q}P/\hat{S})}{\mathsf{d}_{T/S}}$.

It follows that Condition 12.8.2 for the section (T,S) of P is equivalent to

$$h(\mathbb{Q}P/S) \equiv h(\mathbb{Q}P/\hat{S}) \;(\mathrm{mod.}\; a_{T/S}\mathsf{d}_{T/S}) \; .$$

This gives the following conditions:

- If T/S is cyclic with $|T/S| \geq 3$, then $h(\mathbb{Q}P/S) \equiv h(\mathbb{Q}P/\hat{S})$ (mod. 2). Since T/S contains a unique subgroup (T'/S) of order p if $p > 2$, or of order 4 if $p = 2$, this condition follows from the condition obtained for the section (T', S) of P. In other words, it is enough to consider sections (T,S) for which T/S is cyclic of order p if $p > 2$, or of order 4 if $p = 2$.
- If T/S is generalized quaternion, then $h(\mathbb{Q}P/S) \equiv h(\mathbb{Q}P/\hat{S})$ (mod. 4). Since T/S contains a quaternion subgroup (T', S) of order 8, which contains $\hat{S}/S$, this condition follows from the condition obtained for the section (T', S) of P. In other words, it is enough to consider sections (T,S) for which T/S is a quaternion group of order 8.
- If T/S is semi-dihedral, then $h(\mathbb{Q}P/S) \equiv h(\mathbb{Q}P/\hat{S})$ (mod. 4), since $a_{T/S} = 2$ and $\mathsf{d}_{T/S} = 2$ in this case. As T/S contains a quaternion group T'/S of order 8, which contains $\hat{S}/S$, this condition follows from condition obtained for the section (T', S). So the semi-dihedral subquotient do not yield new conditions.

Finally, this shows the following:

12.8.5. Proposition : *With the above notation, if P is a finite p-group, then the subgroup $H(P)$ of $R_{\mathbb{Q}}^*(P)$ consists of the elements h fulfilling the following conditions:*

- $h(\mathbb{Q}P/S) \equiv h(\mathbb{Q}P/\hat{S})$ (mod. 2) *for any section (T,S) of P such that T/S is cyclic of order p if $p > 2$, or of order 4 if $p = 2$.*
- $h(\mathbb{Q}P/S) \equiv h(\mathbb{Q}P/\hat{S})$ (mod. 4) *for any section (T,S) of P for which T/S is a quaternion group of order 8.*

It turns out that this result has a natural interpretation in terms of *Borel-Smith functions*, which have been introduced by tom Dieck ([53] page 210) for studying homotopy representations of finite groups:

12.8.6. Definition : *Let G be a finite group. A* superclass function *is a function from the set of subgroups of G to $\mathbb{Z}$, which is constant on each conjugacy class of subgroups of G.*

Such a function h is called a Borel-Smith function *if it satisfies moreover the following conditions, called* the Borel-Smith conditions*:*

- *If (T,S) is a section of G such that $T/S \cong (\mathbb{Z}/p\mathbb{Z})^2$, for some prime number p, then*

$$h(S) - \sum_{S<Y<T} h(Y) + ph(T) = 0 \ .$$

- *If (T,S) is a section of G such that T/S is cyclic of order p, for an odd prime number p, or cyclic of order 4, then $h(S) \equiv h(\hat{S})$ (mod. 2), where $\hat{S}/S$ is the unique subgroup of prime order of T/S.*
- *If (T,S) is a section of G such that T/S is a quaternion group of order 8, then $h(S) \equiv h(\hat{S})$ (mod. 4), where $\hat{S}/S$ is the unique subgroup of order 2 of T/S.*

The Borel-Smith functions for the group G form a additive group, denoted by $C_b(G)$.

The superclass functions for the group G also form a group, that can be (and *will be*, from now on) identified with the group $B^*(G)$: any superclass function h defines a linear form h' on $B(G)$, by $h'(G/H) = h(H)$, for $H \leq G$, and this correspondence is clearly bijective.

12.8.7. Theorem : [Bouc–Yalçın [22] Theorem 1.2] *The correspondence sending a finite p-group P to the group of Borel–Smith functions $C_b(P)$ is a rational p-biset functor, which identifies with the kernel of $\Theta : B^* \to D_k^\Omega$. In other words, there is a short exact sequence of p-biset functors*

$$0 \longrightarrow C_b \longrightarrow B^* \xrightarrow{\Theta} D_k^\Omega \longrightarrow 0 \ .$$

Proof: Let P be a finite p-group, and $l \in B^*(P)$. First $\mathrm{Ker}\,\Theta_P$ is contained in $\chi_P^*\big(R_{\mathbb{Q}}^*(P)\big)$, by Theorem 12.7.5. Now Lemma 12.7.6 shows that there exists $h \in R_{\mathbb{Q}}^*(P)$ such that $l = \chi_P^*(h)$ if and only if l fulfills the first Borel–Smith condition. By Proposition 12.8.5, the two other Borel–Smith conditions are then equivalent to $h \in H(P)$, i.e. to $l \in \mathrm{Ker}\,\Theta_P$, by definition of H. Thus $\mathrm{Ker}\,\Theta_P = C_b(P)$, and this shows that the correspondence $P \mapsto C_b(P)$ is a subfunctor of B^*, and this subfunctor is rational by Lemma 12.8.1. □

12.8.8. Generators and Relations. Theorem 12.8.7 shows in particular that the functor D_k^Ω does not depend on k. For this reason, it will simply be denoted by D^Ω, from now on.

Another consequence of Theorem 12.8.7 is a presentation of the group $D^\Omega(P)$, for a finite p-group P, by generators and relations. First some notation:

12.8.9. Notation : *Let P be a finite p-group.*

- *If $S \leq P$, set $\Delta_{P/S} = \Theta_P(\delta_{P/S})$, i.e.*

$$\Delta_{P/S} = \sum_{\substack{T \in [s_P] \\ T \leq_P S}} \mu_P(T,S)\,\Omega_{P/T}\ .$$

- *If S and Q are subgroups of P, set*

$$\begin{aligned}
i_P(Q,S) &= |\{x \in Q\backslash P/N_P(S) \mid Q^x \cap N_P(S) \leq S\}| \\
j_P(Q,S) &= |\{x \in Q\backslash P/N_P(S) \mid |J_x(Q,S)| = p,\ J_x(Q,S) \not\leq Z\big(N_P(S)/S\big)\}|
\end{aligned}$$

where $J_x(Q,S) = \big(Q^x \cap N_P(S)\big)S/S$.

12.8.10. Theorem : [15, Theorem 9.5] *Let P be a finite p-group, let $[s_P]$ be a set of representatives of conjugacy classes of subgroups of P, and $\mathcal{G}$ be a genetic basis of P.*

Then $D^\Omega(P)$ is generated by the elements $\Delta_{P/Q}$, for $Q \in [s_P]$. These generators are subject to the following relations:

$$\forall S \in \mathcal{G},\ \ \tau_S \sum_{Q \in [s_P]} \big(\mathsf{d}_S i_P(Q,S) + j_P(Q,S)\big)\Delta_{P/Q} = 0\ ,$$

where $\mathsf{d}_S = 1$ if $N_P(S)/S$ is cyclic or generalized quaternion, and $\mathsf{d}_S = 2$ otherwise, and

$$\tau_S = \begin{cases} 1 \text{ if } |N_P(S)/S| \leq 2 \text{ or if } N_P(S)/S \text{ is dihedral} \\ 2 \text{ if } N_P(S)/S \text{ is cyclic of order at least 3 or semi-dihedral} \\ 4 \text{ if } N_P(S)/S \text{ is generalized quaternion.} \end{cases}$$

Proof: The morphism $\Theta : B^* \to D^\Omega$ is surjective, and its kernel C_b is a rational p-biset functor, by Lemma 12.8.1. Since the elements $\delta_{P/Q}$, for $Q \in [s_P]$, form a basis of $B^*(P)$, the elements $\Delta_{P/Q}$, for $Q \in [s_P]$, generate $D^\Omega(P)$. The relations between these elements are given by the group $C_b(P)$, and the map

$$\mathcal{I}_\mathcal{G} : \bigoplus_{S\in\mathcal{G}} \partial C_b\big(N_P(S)/S\big) \to C_b(P)$$

given by $\mathcal{I}_\mathcal{G} = \bigoplus_{S\in\mathcal{G}} \mathrm{Indinf}_{N_P(S)/S}^P$, is an isomorphism. Moreover, if $S \in \mathcal{G}$, the group $N_S = N_P(S)/S$ is in $\mathcal{N}$, and $\partial C_b(N_S)$ is free of rank 1, generated by $a_{N_S}\chi^*_{N_S}(\Phi^*_{N_S})$, where

$$a_{N_S} = \begin{cases} 1 \text{ if } |N_S| \leq 2 \text{ or if } N_S \text{ is dihedral} \\ 2 \text{ if } N_S \text{ is cyclic of order at least 3 or semi-dihedral} \\ 4 \text{ if } N_S \text{ is generalized quaternion.} \end{cases}$$

Thus with the present notation $a_{N_S} = \tau_S$, and the group $C_b(P)$ has a $\mathbb{Z}$-basis consisting of the elements $\tau_S l_S$, for $S \in \mathcal{G}$, where $l_S = \mathrm{Indinf}_{N_S}^P \chi^*_{N_S}(\Phi^*_{N_S})$.

Now the linear form l_S is such that

$$\begin{aligned} \forall Q \leq P,\ \ l_S(P/Q) &= \Phi^*_{N_S}(\mathrm{Defres}^P_{N_S}\mathbb{Q}P/Q) \\ &= m(\Phi_{N_S}, \mathrm{Defres}^P_{N_S}\mathbb{Q}P/Q) \\ &= m\big(V(S), \mathbb{Q}P/Q\big)\ , \end{aligned}$$

by Frobenius reciprocity, since $\langle \Phi_{N_S}, \Phi_{N_S}\rangle_{N_S} = \langle V(S), V(S)\rangle_P$, where $V(S) = \mathrm{Indinf}_{N_S}^P \Phi_{N_S}$ is the irreducible $\mathbb{Q}P$-module associated to the genetic subgroup S of P.

By Lemma 9.5.3, with Notation 12.8.9,

$$m(\Phi_{N_S}, \mathrm{Defres}^P_{N_S}\mathbb{Q}P/Q) = \mathsf{d}_S i_P(Q,S) + j_P(Q,S)\ .$$

It follows that

$$l_S = \sum_{Q\in[s_P]} \big(\mathsf{d}_S i_P(Q,S) + j_P(Q,S)\big)\delta_{P/Q}\ ,$$

and Theorem 12.8.10 follows. □

12.8.11. Corollary : *Let P be a finite p-group, and $\mathcal{G}$ be a genetic basis of P. Then*

$$D^\Omega(P) \cong \mathbb{Z}^{nc(P)} \oplus (\mathbb{Z}/4\mathbb{Z})^{q(P)} \oplus (\mathbb{Z}/2\mathbb{Z})^{c'(P)+sd(P)}\ ,$$

where

$$\begin{aligned} nc(P) &= |\{S \in \mathcal{G} \mid N_P(S)/S \textit{ is non cyclic}\}| \\ q(P) &= |\{S \in \mathcal{G} \mid N_P(S)/S \textit{ is generalized quaternion}\}| \\ c'(P) &= |\{S \in \mathcal{G} \mid N_P(S)/S \textit{ is cyclic of order at least } 3\}| \\ sd(P) &= |\{S \in \mathcal{G} \mid N_P(S)/S \textit{ is semi-dihedral}\}| \ . \end{aligned}$$

12.9. The Dade Group up to Relative Syzygies

This section is devoted to showing that if k is a field of characteristic p, and P is a finite p-group, the group $D_k(P)/D^{\Omega}(P)$ is finite. This relies on a detection result proved by Puig, and on the following lemma, inspired by Sect. 7.4 of [10], and proved in [20]. First some notation:

12.9.1. Notation : *If P is a finite p-group, denote by $\mathcal{E}(P)$ the set of sections (T,S) of P such that T/S is elementary abelian. If F is a p-biset functor, let $\varprojlim_{(T,S)\in\mathcal{E}(P)} F(T/S)$ be the set of sequences $u_{T,S} \in F(T/S)$, indexed by $\mathcal{E}(P)$, fulfilling the following two conditions:*

1. *If $x \in P$, and $(T,S) \in \mathcal{E}(P)$, then ${}^x u_{T,S} = u_{{}^xT,{}^xS}$.*
2. *If $(T,S), (T',S') \in \mathcal{E}(P)$ and $S \le S' \le T' \le T$, then*

$$\mathrm{Defres}^{T/S}_{T'/S'} u_{T,S} = u_{T',S'} \ .$$

12.9.2. Lemma : *Let F be a p-biset functor, and P be a finite p-group. Let $d_P : F(P) \to \varprojlim_{(T,S)\in\mathcal{E}(P)} F(T/S)$ be the map defined by*

$$d_P(v)_{T,S} = \mathrm{Defres}^P_{T/S} v \ ,$$

and $s_P : \varprojlim_{(T,S)\in\mathcal{E}(P)} F(T/S) \to F(P)$ be the map defined by

$$s_P\big((u_{T,S})_{(T,S)\in\mathcal{E}(P)}\big) = \sum_{(T,S)\in\mathcal{E}(P)} |S|\mu(S,T)\mathrm{Indinf}^P_{S/\Phi(T)} u_{S,\Phi(T)} \ .$$

Then $s_P \circ d_P = |P|\,\mathrm{Id}$, and in particular $|P|\,\mathrm{Coker}\, d_P = \{0\}$.

Proof: First note that if (T,S) is a section of P, then $(T,S)\in\mathcal{E}(P)$ if and only if $S\geq\Phi(T)$, so the map s_P is well defined.

Let $u=(u_{T,S})_{(T,S)\in\mathcal{E}(P)}\in\varprojlim_{(T,S)\in\mathcal{E}(P)}F(T/S)$, and $(B,A)\in\mathcal{E}(P)$. Then

$$\mathrm{Defres}^P_{B/A}s_P(u)=\sum_{(T,S)\in\mathcal{E}(P)}|S|\mu(S,T)\big(A\backslash P/\Phi(T)\big)(u_{S,\Phi(T)})\ ,$$

where $A\backslash P/\Phi(T)$ is a $\big(B/A,S/\Phi(T)\big)$-biset. This biset is a disjoint union of orbits

$$A\backslash P/\Phi(T)=\bigsqcup_{x\in[B\backslash P/S]}A\backslash BxS/\Phi(T)\ .$$

Setting $\Phi=\Phi(T)$, by Remark 4.3.14, this is isomorphic to

$$A\backslash P/\Phi(T)\cong\bigsqcup_{x\in[B\backslash P/S]}\mathrm{Indinf}^{B/A}_{(B\cap{}^xS)A/(B\cap{}^x\Phi)A}\mathrm{Iso}(x)c_x\mathrm{Defres}^{S/\Phi}_{(S\cap B^x)\Phi/(S\cap A^x)\Phi},$$

where $\mathrm{Iso}(x):({}^xS\cap B)^x\Phi/({}^xS\cap A)^x\Phi\to(B\cap{}^xS)A/(B\cap{}^x\Phi)A$ is the group isomorphism sending $z({}^xS\cap A)^x\Phi$ to $z(B\cap{}^x\Phi)A$, for $z\in{}^xS\cap B$, and c_x is the conjugation $y\mapsto{}^xy$. Since $u\in\varprojlim_{(T,S)\in\mathcal{E}(P)}F(T/S)$, it follows that

$$\begin{aligned}\mathrm{Iso}(x)c_x\mathrm{Defres}^{S/\Phi}_{(S\cap B^x)\Phi/(S\cap A^x)\Phi}(u_{S,\Phi})&=\mathrm{Iso}(x)c_x(u_{(S\cap B^x)\Phi,(S\cap A^x)\Phi})\\&=\mathrm{Iso}(x)(u_{({}^xS\cap B)^x\Phi,({}^xS\cap A)^x\Phi})\ ,\end{aligned}$$

Denoting by u_x this element, for short, this gives

$$\begin{aligned}\mathrm{Defres}^P_{B/A}(u)&=\sum_{(T,S)\in\mathcal{E}(P)}\sum_{x\in[B\backslash P/S]}|S|\mu(S,T)\mathrm{Indinf}^{B/A}_{(B\cap{}^xS)A/(B\cap{}^x\Phi)A}(u_x)\\&=\sum_{(T,S)\in\mathcal{E}(P)}\sum_{x\in[B\backslash P]}|B\cap{}^xS|\mu(S,T)\mathrm{Indinf}^{B/A}_{(B\cap{}^xS)A/(B\cap{}^x\Phi)A}(u_x)\ ,\end{aligned}$$

using the fact that $u_{xs}=u_x$, for any $x\in P$ and $s\in S$. Exchanging the order of summation gives

$$\begin{aligned}\mathrm{Defres}^P_{B/A}(u)&=\sum_{x\in[B\backslash P]}\sum_{(T,S)\in\mathcal{E}(P)}|B\cap{}^xS|\mu(S,T)\mathrm{Indinf}^{B/A}_{(B\cap{}^xS)A/(B\cap{}^x\Phi)A}(u_x)\\&=\sum_{x\in[B\backslash P]}\sum_{({}^xT,{}^xS)\in\mathcal{E}(P)}|B\cap{}^xS|\mu({}^xS,{}^xT)\mathrm{Indinf}^{B/A}_{(B\cap{}^xS)A/(B\cap{}^x\Phi)A}(u_x)\\&=\sum_{x\in[B\backslash P]}\sum_{(T,S)\in\mathcal{E}(P)}|B\cap S|\mu(S,T)\mathrm{Indinf}^{B/A}_{(B\cap S)A/(B\cap\Phi)A}(u_1)\end{aligned}$$

since $\mu({}^xS, {}^xT) = \mu(S,T)$ and ${}^x\Phi(T) = \Phi({}^xT)$, and since u_x only depends on the pair $({}^xT, {}^xS)$. This gives finally

$$\mathrm{Defres}^P_{B/A}(u) = |P:B| \sum_{(T,S)\in\mathcal{E}(P)} |B\cap S|\mu(S,T)\mathrm{Indinf}^{B/A}_{(B\cap S)A/(B\cap\Phi)A}(u_1)\ .$$

This summation can be made by summing first on T, and then on any subgroup S of T: indeed, the coefficient $\mu(S,T)$ is non zero if and only if $S \geq \Phi(T)$, i.e. if $(T,S) \in \mathcal{E}(P)$. Grouping together the terms for which $B\cap S$ is a given subgroup R of T, this becomes

$$\mathrm{Defres}^P_{B/A}(u) = |P:B| \sum_{R\leq T\leq P} |R|\Big(\sum_{\substack{S\leq T\\ B\cap S=R}} \mu(S,T)\Big)\mathrm{Indinf}^{B/A}_{RA/(R\cap\Phi)A}(u_1)\ ,$$

where $u_1 = \mathrm{Iso}(1)(u_{R\Phi,(R\cap A)\Phi})$ only depends on R, since $\mathrm{Iso}(1)$ is the isomorphism $(R\cap B)\Phi/(R\cap A)\Phi \to RA/(R\cap\Phi)A$ sending $z(R\cap A)\Phi$ to $z(R\cap\Phi)A$, for $z\in R$.

Now by a classical combinatorial result ([46] Corollary 3.9.3), the sum

$$\sum_{\substack{S\leq T\\ B\cap S=R}} \mu(S,T)$$

is equal to 0, unless $T\leq B$. As B is a p-group, this implies $\Phi = \Phi(T) \leq \Phi(B)$, and moreover $\Phi(B)\leq A$ since $(B,A)\in\mathcal{E}(P)$. This gives

$$\mathrm{Defres}^P_{B/A}(u) = |P:B| \sum_{R\leq T\leq B} |R|\mu(R,T)\mathrm{Indinf}^{B/A}_{RA/A}(u_1)\ ,$$

where $u_1 = \mathrm{Iso}^{RA/A}_{R/(R\cap A)}(u_{RA/A})$. Now for a fixed R, the sum $\sum\limits_{R\leq T\leq B} \mu(R,T)$ is equal to 0 if $R\neq B$,and to 1 if $R=B$. In this case $u_1 = u_{B,A}$ and finally

$$\mathrm{Defres}^P_{B/A}(u) = |P:B||B|\mathrm{Indinf}^{B/A}_{B/A}u_{B/A} = |P|u_{B,A}\ ,$$

as was to be shown. □

12.9.3. Remark : It is shown in [21] that the correspondence

$$P \mapsto \varprojlim_{(T,S)\in\mathcal{E}(P)} F(T/S)$$

has a natural structure of p-biset functor: let $\mathcal{E}_p$ denote the full subcategory of the biset category $\mathcal{C}_p$ whose objects are elementary abelian p-groups. If F is a p-biset functor and P is a finite p-group, then using Notation 3.3.2

$$\lim_{\longleftarrow \atop (T,S)\in\mathcal{E}(P)} F(T/S) \cong ({}^r\mathcal{I}nd_{\mathcal{E}_p}^{\mathcal{C}_p}\mathcal{R}es_{\mathcal{E}_p}^{\mathcal{C}_p}F)(P) \ .$$

Moreover, the map d_P defined in Lemma 12.9.2 is the evaluation at P of the unit morphism $F \to {}^r\mathcal{I}nd_{\mathcal{E}_p}^{\mathcal{C}_p}\mathcal{R}es_{\mathcal{E}_p}^{\mathcal{C}_p}F$ for the adjoint pair of functors $(\mathcal{R}es_{\mathcal{E}_p}^{\mathcal{C}_p}, {}^r\mathcal{I}nd_{\mathcal{E}_p}^{\mathcal{C}_p})$.

12.9.4. Theorem : *Let k be a field of characteristic p, and P be a finite p-group. Set $\mathbb{Q}D_k(P) = \mathbb{Q} \otimes_{\mathbb{Z}} D_k(P)$.*

1. *The map $\mathbb{Q}d_{k,P} : \mathbb{Q}D_k(P) \to \prod_{(T,S)\in\mathcal{E}(P)} \mathbb{Q}D_k(T/S)$ defined by $(\mathbb{Q}d_{k,P})(u)_{T,S} = \mathrm{Defres}_{T/S}^P(u)$, is injective.*
2. *The correspondence $P \mapsto \mathbb{Q}D_k(P)$ is a p-biset functor.*
3. *The group $D_k(P)/D^\Omega(P)$ is a torsion group. In other words*

$$\mathbb{Q}D_k = \mathbb{Q}D^\Omega \ .$$

Proof: Assertion 1 is an easy consequence (see [19] Theorem 1.6 for details) of a result of Puig [40] on the detection of endo-trivial modules by restriction to elementary abelian subgroups.

Assertion 2 is essentially equivalent to showing that the Galois twists act trivially on $\mathbb{Q}D_k(P)$, and this follows from Assertion 1: if a is an endomorphism of k, and if $u \in \mathbb{Q}D_k(P)$, then for any $(T,S) \in \mathcal{E}(P)$

$$\mathrm{Defres}_{T/S}^P\gamma_a(u) = \gamma_a\mathrm{Defres}_{T/S}^P(u) = \mathrm{Defres}_{T/S}^P(u) \ ,$$

since γ_a commutes with $\mathrm{Defres}_{T/S}^P$ by Lemma 12.4.15, and since it acts trivially on $D_k(T/S) = D^\Omega(T/S)$, by Dade's Theorem 12.6.4.

Thus $(\mathbb{Q}d_{k,P})\big(\gamma_a(u) - u\big) = 0$, and $\gamma_a(u) = u$, by Assertion 1.

For Assertion 3, there is a commutative diagram

$$\begin{array}{ccccc}
0 \longrightarrow & \mathbb{Q}D_k(P) & \xrightarrow{\mathbb{Q}d_{k,P}} & \lim\limits_{\longleftarrow \atop (T,S)\in\mathcal{E}(P)} & \mathbb{Q}D_k(T/S) \\
& \uparrow & & & \| \\
0 \longrightarrow & \mathbb{Q}D^\Omega(P) & \xrightarrow{\mathbb{Q}d_P} & \lim\limits_{\longleftarrow \atop (T,S)\in\mathcal{E}(P)} & \mathbb{Q}D^\Omega(T/S)
\end{array}$$

where the map d_P is the map defined in Lemma 12.9.2. The inclusion on the right is an equality, because $D_k(T/S) = D^\Omega(T/S)$ by Dade's Theorem, since T/S is abelian. Now the bottom horizontal map $\mathbb{Q}d_P$ is an isomorphism, since it is injective, and since its cokernel is zero, for it is a both a $\mathbb{Q}$-vector space and a torsion group by Lemma 12.9.2. Hence the top horizontal map is surjective, so it is an isomorphism, too. It follows that the inclusion on the left is an isomorphism, hence an equality. This shows that $\mathbb{Q}D_k = \mathbb{Q}D^\Omega$. Equivalently, for any P, the group $D_k(P)/D^\Omega(P)$ is a torsion group. □

12.9.5. Theorem : [15, Theorem 7.1] *Let k be a field of characteristic p, and n be a positive integer.*

1. *The correspondence $nD \cap D^\Omega$ sending a p-group P to $nD_k(P) \cap D^\Omega(P)$ is a p-biset subfunctor of D^Ω.*
2. *Moreover*
$$(nD_k \cap D^\Omega) + D^\Omega_{tors} = nD^\Omega + D^\Omega_{tors} .$$

Proof: Let P and Q be finite p-groups, and U be a finite (Q,P)-biset. Then the map $D_k(U) : D_k(P) \to D_k(Q)$ restricts to $D^\Omega(U) : D^\Omega(P) \to D^\Omega(Q)$, hence to a map $D_n(U) : nD_k(P) \cap D^\Omega(P) \to nD_k(Q) \cap D^\Omega(Q)$. Moreover, since for any finite p-group P, the Galois twists act trivially on $D^\Omega(P)$, it follows that if R is another finite p-group and V is a finite (R,Q)-biset, then $D_n(V) \circ D_n(U) = D_n(V \times_Q U)$, by Proposition 12.5.17. This proves Assertion 1.

For Assertion 2, if $n = 1$, there is nothing to prove, so assume $n \geq 2$. By Theorem 12.7.5, there is an exact sequence of p-biset functors

$$0 \to R^*_{\mathbb{Q}} \to B^* \to D^\Omega/D^\Omega_{tors} \to 0 .$$

If P is a finite p-group, evaluation at P gives the exact sequence of abelian groups

$$0 \to R^*_{\mathbb{Q}}(P) \to B^*(P) \to (D^\Omega/D^\Omega_{tors})(P) \to 0 ,$$

and this sequence is actually split exact, since $(D^\Omega/D^\Omega_{tors})(P)$ is a free abelian group. In particular, this sequence remains exact after tensoring (over $\mathbb{Z}$) with $\Gamma_n = \mathbb{Z}/n\mathbb{Z}$, and this shows that the sequence

$$0 \to \Gamma_n R^*_{\mathbb{Q}} \to \Gamma_n B^* \to \Gamma_n(D^\Omega/D^\Omega_{tors}) \to 0$$

is an exact sequence of p-biset functors, where the $\otimes_{\mathbb{Z}}$ symbols have been dropped (so, e.g. $\Gamma_n R^*_{\mathbb{Q}}$ denotes $\Gamma_n \otimes_{\mathbb{Z}} R^*_{\mathbb{Q}}$). Now there are canonical isomorphisms

$$\Gamma_n R^*_{\mathbb{Q}} \cong \mathrm{Hom}_{\Gamma_n}(\Gamma_n R_{\mathbb{Q}}, \Gamma_n)$$
$$\Gamma_n B^* \cong \mathrm{Hom}_{\Gamma_n}(\Gamma_n B, \Gamma_n) .$$

Moreover

$$\begin{aligned}\Gamma_n(D^\Omega/D^\Omega_{tors}) &\cong (D^\Omega/D^\Omega_{tors})\big/n(D^\Omega/D^\Omega_{tors})\\ &\cong (D^\Omega/D^\Omega_{tors})\Big/\Big((nD^\Omega+D^\Omega_{tors})/D^\Omega_{tors}\Big)\\ &\cong D^\Omega/(nD^\Omega+D^\Omega_{tors})\ .\end{aligned}$$

Setting $T_n = D^\Omega/(nD^\Omega + D^\Omega_{tors})$, this gives the exact sequence

(12.9.6) $$0 \to \mathrm{Hom}_{\Gamma_n}(\Gamma_n R_{\mathbb{Q}}, \Gamma_n) \to \mathrm{Hom}_{\Gamma_n}(\Gamma_n B, \Gamma_n) \to T_n \to 0\ .$$

Note that $T_n(P)$ is a free Γ_n-module, for any p-group P.

Now on the other hand the exact sequence of p-biset functors

$$0 \to K \to B \to R_{\mathbb{Q}} \to 0$$

remains exact after tensoring with Γ_n, since every evaluation of this sequence is a split exact sequence of abelian groups. Hence there is an exact sequence

$$0 \to \Gamma_n K \to \Gamma_n B \to \Gamma_n R_{\mathbb{Q}} \to 0\ ,$$

and every evaluation of this sequence is a split exact sequence of Γ_n-modules. Now taking Γ_n-duals yields the exact sequence

(12.9.7)
$$0 \to \mathrm{Hom}_{\Gamma_n}(\Gamma_n R_{\mathbb{Q}}, \Gamma_n) \to \mathrm{Hom}_{\Gamma_n}(\Gamma_n B, \Gamma_n) \to \mathrm{Hom}_{\Gamma_n}(\Gamma_n K, \Gamma_n) \to 0\ .$$

Comparing the sequences 12.9.6 and 12.9.7 gives the following natural isomorphism of biset functors

(12.9.8) $$T_n \cong \mathrm{Hom}_{\Gamma_n}(\Gamma_n K, \Gamma_n)\ .$$

This means that for any p-group P, there is a non-degenerate bilinear form

$$(\ ,\ )_P : T_n(P) \times \Gamma_n K(P) \to \Gamma_n$$

with the property that for any p-group Q and any $\varphi \in \mathrm{Hom}_{\mathcal{C}_p}(Q,P)$, one has that

(12.9.9) $$\forall u \in T_n(P),\ \forall v \in \Gamma_n K(Q),\ \big(u, \varphi(v)\big)_P = \big(\varphi^{op}(u), v\big)_Q\ .$$

Now set

$$F = \big((nD \cap D^\Omega) + D^\Omega_{tors}\big)\big/(nD^\Omega + D^\Omega_{tors})\ .$$

Then F is a biset subfunctor of T_n. Suppose that $F \neq \{0\}$, and let P be a p-group of minimal order such that $F(P) \neq \{0\}$. Then $F(P)$ is a subset of

$$\underline{T}_n(P) \subseteq \{u \in T_n(P) \mid \mathrm{Defres}^P_{T/S}(u) = 0,\ \forall S \trianglelefteq T \leq P,\ |T/S| < |P|\} .$$

By the above duality 12.9.9, it follows that if $u \in \underline{T_n}(P)$, then for any proper section T/S of P, and any $v \in \Gamma_n K(P)$

$$\big(u, \mathrm{Indinf}^P_{T/S}(v)\big)_P = 0 .$$

Hence $(u, w)_P = 0$ for any $w \in \sum_{\substack{S \trianglelefteq T \subseteq P \\ |T/S| < |P|}} \mathrm{Indinf}^P_{T/S} \Gamma_n K(T/S)$. By Corollary 11.1.16, this is the whole of $\Gamma_n K(P)$, unless P is isomorphic to X_{p^3} or E_{p^2} if $p \neq 2$, or if P is dihedral of order at least 4 if $p = 2$. Thus if P is not isomorphic to one of the groups in this list, then $(u, v)_P = 0$ for any $v \in \Gamma_n K(P)$, hence $u = 0$ since the bilinear form $(\ ,\ )_P$ is non degenerate. Thus $\underline{T}_n(P) = \{0\}$.

Hence the minimal group P is one of the groups X_{p^3}, E_{p^2} or D_{2^n} (note that $E_{2^2} \cong D_{2^2}$). If $F(P) = \{0\}$ in each of these cases, this gives a contradiction, proving that $F = \{0\}$. But if P is one of the groups X_{p^3}, E_{p^2} or D_{2^n}, then $D(P) = D^{\Omega}(P)$: for $P = X_{p^3}$, this follows from Theorem 10.2 of [18], for $P = E_{p^2}$, this follows from Theorem 12.6.4, and for $P = D_{2^n}$, this follows from Theorem 10.3 of [25]. Hence $nD(P) \cap D^{\Omega}(P) = nD^{\Omega}(P)$, and $F(P) = \{0\}$. This completes the proof of the theorem. □

12.9.10. Theorem : *Let k be a field of characteristic p, and P be a p-group.*

1. $D_k(P) = D^{\Omega}(P) + \big(D_k(P)\big)_{tors}$.
2. *If $p > 2$, then* $D_k(P) = D^{\Omega}(P)$.

Proof: Let $u \in D_k(P)$. By Theorem 12.9.4, there exists a positive integer n such that $nu \in D^{\Omega}(P)$. Thus $nu \in nD_k(P) \cap D^{\Omega}(P)$. By Theorem 12.9.5, there exists $v \in D^{\Omega}(P)$ and $t \in D^{\Omega}_{tors}(P)$ such that $nu = nv + t$. Thus $n(u-v)$ is a torsion element of $D_k(P)$, and so is $u-v$. Hence $u \in D^{\Omega}(P) + \big(D_k(P)\big)_{tors}$, proving Assertion 1.

For Assertion 2, consider the commutative diagram

$$\begin{array}{ccc} D_k(P) & \xrightarrow{d_{k,P}} & \varprojlim_{(T,S)\in\mathcal{E}(P)} D_k(T/S) \\ \uparrow & & \| \\ D^{\Omega}(P) & \xrightarrow{d_P} & \varprojlim_{(T,S)\in\mathcal{E}(P)} D^{\Omega}(T/S) . \end{array}$$

It has been shown by Carlson and Thévenaz (see [26] Theorem 13.1) that the map $d_{k,P}$ is injective when p is odd. On the other hand, the cokernel of d_P is a finite p-group, by Lemma 12.9.2. This shows that in the previous argument, the integer n is a power of p, say $n = p^m$, for some $m \in \mathbb{N}$. Thus $w = p^m(u - v) \in D^{\Omega}_{tors}(P)$. But by Corollary 12.8.11, the group $D^{\Omega}_{tors}(P)$ is a 2-group of exponent 2. Thus $p^m w = w$, hence $p^m(u - v - w) = 0$. Now by Corollary 13.2 of [26], there is no p-torsion in $D_k(P)$, when p is odd. Thus $u = v + w$, so $u \in D^{\Omega}(P)$. Hence $D_k(P) = D^{\Omega}(P)$ in this case, as was to be shown. □

12.9.11. Corollary :

1. *If P is a finite p-group, then the inclusion $D^{\Omega}(P) \subseteq D_k(P)$ induces an isomorphism $D^{\Omega}(P)/D^{\Omega}_{tors}(P) \cong D_k(P)/D_k(P)_{tors}$.*
2. *The correspondence $P \mapsto D_k(P)/D_k(P)_{tors}$ is a p-biset functor $D_k/(D_k)_{tors}$, isomorphic to $D^{\Omega}/D^{\Omega}_{tors}$.*
3. *There are exact sequences of p-bisets functors*

$$0 \longrightarrow R^*_{\mathbb{Q}} \to B^* \longrightarrow D_k/(D_k)_{tors} \longrightarrow 0\ .$$

$$0 \longrightarrow \mathbb{Q}D_k \to \mathbb{Q}B \xrightarrow{\mathbb{Q}\chi} \mathbb{Q}R_{\mathbb{Q}} \longrightarrow 0\ .$$

4. *The functor $\mathbb{Q}D_k$ is simple, isomorphic to the functor $S_{E,\mathbb{Q}}$, where $E \cong (\mathbb{Z}/p\mathbb{Z})^2$.*
5. *If P is a finite p-group, then $D_k(P)/\big(D_k(P)\big)_{tors} \cong \mathbb{Z}^{nc(P)}$, where $nc(P)$ is the number of conjugacy classes of non cyclic subgroups of P.*

Proof: Assertions 1 and 2 follow from the equality $D_k = D^{\Omega} + (D_k)_{tors}$. The first exact sequence in Assertion 3 follows from Theorem 12.7.5. The second exact sequence follows by applying the functor $\mathrm{Hom}_{\mathbb{Z}}(-, \mathbb{Q})$. Assertion 4 follows from Sect. 5.6.9: the Burnside functor has a unique proper non zero subfunctor, isomorphic to $S_{E,\mathbb{Q}}$. Assertion 5 follows from Assertion 3, since the rank of $B^*(P)$ is the number of conjugacy classes of subgroups of P, and the rank of $R^*_{\mathbb{Q}}(P)$ is the number of conjugacy classes of cyclic subgroups of P. □

12.9.12. Remark : The exposition given here is far from chronological: Assertion 4, Assertion 5 and the second exact sequence in Assertion 3 were first proved in [19] (Theorems 10.1, 4.1, and 10.4, respectively). It was the first important application of the formalism of biset functors, and a major motivation for developing it.

12.10. The Torsion Subgroup of the Dade Group

In order to complete the description of the Dade group $D_k(P)$ of a finite p-group P, it remains to find the structure of its torsion part $\big(D_k(P)\big)_{tors}$. Though the correspondence $P \mapsto \big(D_k(P)\big)_{tors}$ is not a biset functor in general, it is close enough to being a rational p-biset functor. In particular, if P is a finite p-group, the map

$$f_{\mathbf{1}}^P = \sum_{Z \leq \Omega_1 Z(P)} \mu(\mathbf{1}, Z)\mathrm{Inf}_{P/Z}^P \mathrm{Def}_{P/Z}^P : D_k(P) \to D_k(P)$$

is well-defined, and it is an idempotent endomorphism. Thus one can set

$$\partial D_k(P) = f_{\mathbf{1}}^P D_k(P) \ ,$$

that may be called the subgroup of *faithful elements* of $D_k(P)$.

With this notation, the correspondence $P \mapsto D_k(P)_{tors}$ fulfills the conditions of Theorem 10.6.1:

12.10.1. Theorem : [15, Proposition 8.1] *Let k be a field of characteristic p, and P be a finite p-group. Then:*

1. *If $Z(P)$ is non cyclic, then $\partial D_k(P)_{tors} = \{0\}$.*
2. *If $E \trianglelefteq P$ is a normal elementary abelian subgroup of rank 2, if Z is a central subgroup of order p of P contained in E, then the map*

$$\mathrm{Res}_{C_P(E)}^P \oplus \mathrm{Def}_{P/Z}^P : D_k(P)_{tors} \longrightarrow D_k\big(C_P(E)\big)_{tors} \oplus D_k(P/Z)_{tors}$$

is injective.

Proof: Both assertions are proved in [15] by induction on $|P|$. The key ingredients are, on the one hand, the detection theorems of Carlson and Thévenaz ([26] Theorems 13.1 and 13.4), and, on the other hand, the structure of the Dade group of some (almost) extra-special p-groups, stated in a joint paper with N. Mazza ([18] Theorems 9.1 and 10.1). □

12.10.2. Theorem : [15, Theorem 8.2] *Let k be a field of characteristic p, let P be a finite p-group, and $\mathcal{G}$ be a genetic basis of P. Then the map*

$$\mathcal{I}_{\mathcal{G}} = \bigoplus_{S \in \mathcal{G}} \mathrm{Teninf}_{N_P(S)/S}^P : \bigoplus_{S \in \mathcal{G}} \partial D_k\big(N_P(S)/S\big)_{tors} \to D_k(P)_{tors}$$

is an isomorphism.

Proof: Though the correspondence $P \mapsto D_k(P)_{tors}$ is not a biset functor, because of the Galois twists, one can show that the map $\mathcal{I}_{\mathcal{G}}$ is always split injective (see [13] Theorem 6.1), with explicit section. Now the proof of Theorem 10.6.1 goes through without change. □

12.10.3. Corollary : *Let P be a finite p-group, and $\mathcal{G}$ be a genetic basis of P. Then*

$$D_k(P)_{tors} \cong (\mathbb{Z}/4\mathbb{Z})^{q(P)} \oplus (\mathbb{Z}/2\mathbb{Z})^{c'(P)+sd(P)+q'(P)} ,$$

where

$$\begin{aligned} c'(P) &= |\{S \in \mathcal{G} \mid N_P(S)/S \cong C_{p^n},\ p^n \geq 3\}|, \\ q(P) &= |\{S \in \mathcal{G} \mid N_P(S)/S \cong Q_{2^n},\ n \geq 3\}|, \\ sd(P) &= |\{S \in \mathcal{G} \mid N_P(S)/S \cong SD_{2^n},\ n \geq 4\}|, \\ q'(P) &= |\{S \in \mathcal{G} \mid N_P(S)/S \cong Q_{p^n},\ n \geq 4\}|,\ \textit{if}\ k \not\supseteq \mathbb{F}_4 \\ &= q(P),\ \textit{if}\ k \supseteq \mathbb{F}_4 . \end{aligned}$$

Proof: This follows from the structure of $D_k(Q)_{tors}$, when Q is a group of normal p-rank 1: the case of cyclic groups follows from Dade's Theorem, the case of generalized quaternion groups follows from an earlier result of Dade [30], and the cases of dihedral and semi-dihedral 2-groups have been settled by Carlson and Thévenaz [25]. More precisely:

- If $Q \cong C_{p^n}$, then $\partial D_k(Q)_{tors} \cong \mathbb{Z}/2\mathbb{Z}$ if $p^n \geq 3$, and $\partial D_k(Q)_{tors} = \{0\}$ otherwise.
- If $Q \cong Q_{2^n}$, then $D_k(Q)_{tors} = \partial D_k(Q)_{tors}$. Moreover $D^\Omega(Q)_{tors} \cong \mathbb{Z}/4\mathbb{Z}$, and $D_k(Q)_{tors} = D^\Omega(Q)_{tors} \oplus \mathbb{Z}/2\mathbb{Z}$, except if $n = 3$ and $k \not\supseteq \mathbb{F}_4$, where $D_k(Q)_{tors} = D^\Omega(Q)_{tors}$. So apart from this case, there are exactly two elements η_Q and η'_Q of order 2 in $D_k(Q) - D^\Omega(Q)$.
- If $Q \cong D_{2^n}$, then $D_k(Q)_{tors} = \{0\}$.
- If $Q \cong SD_{2^n}$, then $D_k(Q)_{tors} = \partial D_k(Q)_{tors} = D^\Omega(Q)_{tors} \cong \mathbb{Z}/2\mathbb{Z}$. □

12.10.4. Theorem : *Let P be a finite p-group, and $\mathcal{G}$ be a genetic basis of P. Denote by $\mathcal{Q}$ the set of $S \in \mathcal{G}$ for which $N_P(S)/S$ is generalized quaternion, and, in the case $k \not\supseteq \mathbb{F}_4$, of order at least 16. For each $S \in \mathcal{Q}$, choose an element $\eta_{N_P(S)/S}$ of order 2 in $D_k\big(N_P(S)/S\big) - D^\Omega\big(N_P(S)/S\big)$, and set*

$$\Lambda_S = \mathrm{Teninf}_{N_P(S)/S}^P \eta_{N_P(S)/S} .$$

Then $2\Lambda_S = 0$, and the set $\mathcal{E}_{\mathcal{Q}} = \{\Lambda_S \mid S \in \mathcal{Q}\}$ is linearly independent over $\mathbb{F}_2$. The subgroup $D_k^{\mathcal{Q}}(P)$ of $D_k(P)$ generated by $\mathcal{E}_{\mathcal{Q}}$ is an elementary abelian 2-group of rank $|\mathcal{E}_{\mathcal{Q}}| = q'(P)$, and

$$D_k(P) = D^\Omega(P) \oplus D_k^{\mathcal{Q}}(P) .$$

References

1. J. Adams, J. H. Gunawardena, and H. Miller. The Segal conjecture for elementary abelian p-groups. *Topology*, 24(4):435–460, 1885.
2. J. L. Alperin. A construction of endo-permutation modules. *J. Group Theory*, 4:3–10, 2001.
3. F. W. Andersen and K. R. Fuller. *Rings and Categories of Modules*, volume 13 of *Graduate texts in Mathematics*. Springer, 1992. Second edition.
4. L. Barker. Rhetorical biset functors, rational p-biset functors, and their semisimplicity in characteristic zero. *J. Algebra*, 319(9):3810–3853, 2008.
5. S. Bouc. Homologie de certains ensembles ordonnés. *C. R. Acad. Sc. Paris*, t.299(2): 49–52, 1984. Série I.
6. S. Bouc. Foncteurs d'ensembles munis d'une double action. *J. of Algebra*, 183(0238): 664–736, 1996.
7. S. Bouc. *Green-functors and G-sets*, volume 1671 of *Lecture Notes in Mathematics*. Springer, October 1997.
8. S. Bouc. Burnside rings. In *Handbook of Algebra*, volume 2, chapter 6E, pages 739–804. Elsevier, 2000.
9. S. Bouc. *Non-additive exact functors and tensor induction for Mackey functors*, volume 144 of *Memoirs*. A.M.S., 2000. no. 683.
10. S. Bouc. Tensor induction of relative syzygies. *J. reine angew. Math.*, 523:113–171, 2000.
11. S. Bouc. A remark on a theorem of Ritter and Segal. *J. of Group Theory*, 4:11–18, 2001.
12. S. Bouc. The functor of rational representations for p-groups. *Advances in Mathematics*, 186:267–306, 2004.
13. S. Bouc. A remark on the Dade group and the Burnside group. *J. of Algebra*, 279(1):180–190, 2004.
14. S. Bouc. Biset functors and genetic sections for p-groups. *Journal of Algebra*, 284(1):179–202, 2005.
15. S. Bouc. The Dade group of a p-group. *Inv. Math.*, 164:189–231, 2006.
16. S. Bouc. The functor of units of Burnside rings for p-groups. *Comm. Math. Helv.*, 82:583–615, 2007.
17. S. Bouc. Rational p-biset functors. *Journal of Algebra*, 319:1776–1800, 2008.
18. S. Bouc and N. Mazza. The Dade group of (almost) extraspecial p-groups. *J. Pure and Applied Algebra*, 192:21–51, 2004.
19. S. Bouc and J. Thévenaz. The group of endo-permutation modules. *Invent. Math.*, 139:275–349, 2000.
20. S. Bouc and J. Thévenaz. Gluing torsion endo-permutation modules. *J. London Math. Soc.*, 78(2):477–501, 2008.
21. S. Bouc and J. Thévenaz. A sectional characterization of the Dade group. *Journal of Group Theory*, 11(2):155–298, 2008.

22. S. Bouc and E. Yalçın. Borel–Smith functions and the Dade group. *Journal of Algebra*, 311:821–839, 2007.
23. M. Broué and L. Puig. Characters and local structures in G-algebras. *J. of Algebra*, 63:306–317, 1980.
24. W. Burnside. *Theory of groups of finite order*. Cambridge University Press, second edition, 1911.
25. J. Carlson and J. Thévenaz. Torsion endo-trivial modules. *Algebras and Representation Theory*, 3:303–335, 2000.
26. J. Carlson and J. Thévenaz. The classification of torsion endo-trivial modules. *Ann. of Math.*, 162(2):823–883, September 2005.
27. J. F. Carlson and J. Thévenaz. The classification of endo-trivial modules. *Invent. Math.*, 158(2):389–411, 2004.
28. O. Coşkun. Alcahestic subalgebras of the alchemic algebra and a correspondence of simple modules. *J. Algebra*, 320:2422–2450, 2008.
29. C. Curtis and I. Reiner. *Methods of representation theory with applications to finite groups and orders*, volume 1 of *Wiley classics library*. Wiley, 1990.
30. E. Dade. Une extension de la théorie de Hall et Higman,. *J. Algebra*, 20:570–609, 1972.
31. E. Dade. Endo-permutation modules over p-groups I. *Ann. of Math.*, 107:459–494, 1978.
32. E. Dade. Endo-permutation modules over p-groups II. *Ann. of Math.*, 108:317–346, 1978.
33. A. Dress. A characterization of solvable groups. *Math. Zeit.*, 110:213–217, 1969.
34. A. Dress. *Contributions to the theory of induced representations*, volume 342 of *Lecture Notes in Mathematics*, pages 183–240. Springer-Verlag, 1973.
35. D. Gluck. Idempotent formula for the Burnside ring with applications to the p-subgroup simplicial complex. *Illinois J. Math.*, 25:63–67, 1981.
36. W. Gorenstein. *Finite groups*. Chelsea, 1968.
37. I. Hambleton, L. Taylor, and B. Williams. Detection theorems for K-theory and L-theory. *J. Pure Appl. Algebra*, 63:247–299, 1990.
38. C. Kratzer and J. Thévenaz. Fonction de Möbius et anneau de Burnside. *Comment. Math. Helv.*, 59:425–438, 1984.
39. S. Mac Lane. *Categories for the working mathematician*, volume 5 of *Graduate texts in Mathematics*. Springer, 1971.
40. L. Puig. Affirmative answer to a question of Feit. *J. of Algebra*, 131:513–526, 1990.
41. J. Ritter. Ein Induktionssatz für rational Charaktere von nilpotenten Gruppen. *J. f. reine u. angew. Math.*, 254:133–151, 1972.
42. P. Roquette. Realisierung von Darstellungen endlicher nilpotente Gruppen. *Arch. Math.*, 9:224–250, 1958.
43. G. Segal. Permutation representations of finite p-groups. *Quart. J. Math. Oxford*, 23:375–381, 1972.
44. J.-P. Serre. *Corps locaux*, volume VIII/1296 of *Publications de l'Institut de Mathématique de l'Université de Nancago*. Hermann, 1962.
45. J.-P. Serre. *Représentations linéaires des groupes finis*. Collection Méthodes. Hermann, 1971.
46. R. P. Stanley. *Enumerative combinatorics*, volume 1. Wadsworth & Brooks/Cole, Monterey, 1986.
47. P. Symonds. A splitting principle for group representations. *Comment. Math. Helv.*, 66(2):169–184, 1991.
48. J. Thévenaz. Permutation representation arising from simplicial complexes. *J. Combin. Theory*, 46:122–155, 1987. Ser. A.
49. J. Thévenaz. *G-algebras and modular representation theory*. Clarendon Press, Oxford, 1995.
50. J. Thévenaz. Endo-permutation modules, a guided tour. In *Group representation theory*. EPFL Press Lausanne, 2007. Edited by M. Geck, D. Testerman, J. Thévenaz.

51. J. Thévenaz and P. Webb. The structure of Mackey functors. *Trans. Amer. Math. Soc.*, 347(6):1865–1961, June 1995.
52. T. tom Dieck. *Transformation groups and representation theory*, volume 766 of *Lecture Notes in Mathematics*. Springer-Verlag, 1979.
53. T. tom Dieck. *Transformation groups*, volume 8 of *De Gruyter series in mathematics*. De Gruyter, 1987.
54. P. Webb. Two classification of simple Mackey functors with application to group cohomology and the decomposition of classifying spaces. *J. of Pure and App. Alg.*, 88:265–304, 1993.
55. P. Webb. A guide to Mackey functors. In *Handbook of Algebra*, volume 2, chapter 6E, pages 805–836. Elsevier, 2000.
56. P. Webb. Stratification of globally defined Mackey functors. http://www.math.umn.edu/~webb/Talks/GDMFatMSRI.pdf, 2008.
57. P. Webb. Stratifications and Mackey functors II: globally defined Mackey functors. Preprint, 2008.
58. E. Yalçın. An induction theorem for the unit groups of Burnside rings of 2-groups. *J. of Algebra*, 289:105–127, 2005.
59. E. Yaraneri. A filtration of the modular representation functor. *J. Algebra*, 318(1): 140–179, 2007.
60. T. Yoshida. Idempotents in Burnside rings and Dress induction theorem. *J. Algebra*, 80:90–105, 1983.

Index

N

O

P

R

S

T

W

Y

Lecture Notes in Mathematics

For information about earlier volumes
please contact your bookseller or Springer
LNM Online archive: springerlink.com

Vol. 1807: V. D. Milman, G. Schechtman (Eds.), Geometric Aspects of Functional Analysis. Israel Seminar 2000-2002 (2003)

Vol. 1808: W. Schindler, Measures with Symmetry Properties (2003)

Vol. 1809: O. Steinbach, Stability Estimates for Hybrid Coupled Domain Decomposition Methods (2003)

Vol. 1810: J. Wengenroth, Derived Functors in Functional Analysis (2003)

Vol. 1811: J. Stevens, Deformations of Singularities (2003)

Vol. 1812: L. Ambrosio, K. Deckelnick, G. Dziuk, M. Mimura, V. A. Solonnikov, H. M. Soner, Mathematical Aspects of Evolving Interfaces. Madeira, Funchal, Portugal 2000. Editors: P. Colli, J. F. Rodrigues (2003)

Vol. 1813: L. Ambrosio, L. A. Caffarelli, Y. Brenier, G. Buttazzo, C. Villani, Optimal Transportation and its Applications. Martina Franca, Italy 2001. Editors: L. A. Caffarelli, S. Salsa (2003)

Vol. 1814: P. Bank, F. Baudoin, H. Föllmer, L.C.G. Rogers, M. Soner, N. Touzi, Paris-Princeton Lectures on Mathematical Finance 2002 (2003)

Vol. 1815: A. M. Vershik (Ed.), Asymptotic Combinatorics with Applications to Mathematical Physics. St. Petersburg, Russia 2001 (2003)

Vol. 1816: S. Albeverio, W. Schachermayer, M. Talagrand, Lectures on Probability Theory and Statistics. Ecole d'Eté de Probabilités de Saint-Flour XXX-2000. Editor: P. Bernard (2003)

Vol. 1817: E. Koelink, W. Van Assche (Eds.), Orthogonal Polynomials and Special Functions. Leuven 2002 (2003)

Vol. 1818: M. Bildhauer, Convex Variational Problems with Linear, nearly Linear and/or Anisotropic Growth Conditions (2003)

Vol. 1819: D. Masser, Yu. V. Nesterenko, H. P. Schlickewei, W. M. Schmidt, M. Waldschmidt, Diophantine Approximation. Cetraro, Italy 2000. Editors: F. Amoroso, U. Zannier (2003)

Vol. 1820: F. Hiai, H. Kosaki, Means of Hilbert Space Operators (2003)

Vol. 1821: S. Teufel, Adiabatic Perturbation Theory in Quantum Dynamics (2003)

Vol. 1822: S.-N. Chow, R. Conti, R. Johnson, J. Mallet-Paret, R. Nussbaum, Dynamical Systems. Cetraro, Italy 2000. Editors: J. W. Macki, P. Zecca (2003)

Vol. 1823: A. M. Anile, W. Allegretto, C. Ringhofer, Mathematical Problems in Semiconductor Physics. Cetraro, Italy 1998. Editor: A. M. Anile (2003)

Vol. 1824: J. A. Navarro González, J. B. Sancho de Salas, $\mathcal{C}^\infty$ – Differentiable Spaces (2003)

Vol. 1825: J. H. Bramble, A. Cohen, W. Dahmen, Multiscale Problems and Methods in Numerical Simulations, Martina Franca, Italy 2001. Editor: C. Canuto (2003)

Vol. 1826: K. Dohmen, Improved Bonferroni Inequalities via Abstract Tubes. Inequalities and Identities of Inclusion-Exclusion Type. VIII, 113 p, 2003.

Vol. 1827: K. M. Pilgrim, Combinations of Complex Dynamical Systems. IX, 118 p, 2003.

Vol. 1828: D. J. Green, Gröbner Bases and the Computation of Group Cohomology. XII, 138 p, 2003.

Vol. 1829: E. Altman, B. Gaujal, A. Hordijk, Discrete-Event Control of Stochastic Networks: Multimodularity and Regularity. XIV, 313 p, 2003.

Vol. 1830: M. I. Gil', Operator Functions and Localization of Spectra. XIV, 256 p, 2003.

Vol. 1831: A. Connes, J. Cuntz, E. Guentner, N. Higson, J. E. Kaminker, Noncommutative Geometry, Martina Franca, Italy 2002. Editors: S. Doplicher, L. Longo (2004)

Vol. 1832: J. Azéma, M. Émery, M. Ledoux, M. Yor (Eds.), Séminaire de Probabilités XXXVII (2003)

Vol. 1833: D.-Q. Jiang, M. Qian, M.-P. Qian, Mathematical Theory of Nonequilibrium Steady States. On the Frontier of Probability and Dynamical Systems. IX, 280 p, 2004.

Vol. 1834: Yo. Yomdin, G. Comte, Tame Geometry with Application in Smooth Analysis. VIII, 186 p, 2004.

Vol. 1835: O.T. Izhboldin, B. Kahn, N.A. Karpenko, A. Vishik, Geometric Methods in the Algebraic Theory of Quadratic Forms. Summer School, Lens, 2000. Editor: J.-P. Tignol (2004)

Vol. 1836: C. Năstăsescu, F. Van Oystaeyen, Methods of Graded Rings. XIII, 304 p, 2004.

Vol. 1837: S. Tavaré, O. Zeitouni, Lectures on Probability Theory and Statistics. Ecole d'Eté de Probabilités de Saint-Flour XXXI-2001. Editor: J. Picard (2004)

Vol. 1838: A.J. Ganesh, N.W. O'Connell, D.J. Wischik, Big Queues. XII, 254 p, 2004.

Vol. 1839: R. Gohm, Noncommutative Stationary Processes. VIII, 170 p, 2004.

Vol. 1840: B. Tsirelson, W. Werner, Lectures on Probability Theory and Statistics. Ecole d'Eté de Probabilités de Saint-Flour XXXII-2002. Editor: J. Picard (2004)

Vol. 1841: W. Reichel, Uniqueness Theorems for Variational Problems by the Method of Transformation Groups (2004)

Vol. 1842: T. Johnsen, A. L. Knutsen, K_3 Projective Models in Scrolls (2004)

Vol. 1843: B. Jefferies, Spectral Properties of Noncommuting Operators (2004)

Vol. 1844: K.F. Siburg, The Principle of Least Action in Geometry and Dynamics (2004)

Vol. 1845: Min Ho Lee, Mixed Automorphic Forms, Torus Bundles, and Jacobi Forms (2004)

Vol. 1846: H. Ammari, H. Kang, Reconstruction of Small Inhomogeneities from Boundary Measurements (2004)

Vol. 1847: T.R. Bielecki, T. Björk, M. Jeanblanc, M. Rutkowski, J.A. Scheinkman, W. Xiong, Paris-Princeton Lectures on Mathematical Finance 2003 (2004)
Vol. 1848: M. Abate, J. E. Fornaess, X. Huang, J. P. Rosay, A. Tumanov, Real Methods in Complex and CR Geometry, Martina Franca, Italy 2002. Editors: D. Zaitsev, G. Zampieri (2004)
Vol. 1849: Martin L. Brown, Heegner Modules and Elliptic Curves (2004)
Vol. 1850: V. D. Milman, G. Schechtman (Eds.), Geometric Aspects of Functional Analysis. Israel Seminar 2002-2003 (2004)
Vol. 1851: O. Catoni, Statistical Learning Theory and Stochastic Optimization (2004)
Vol. 1852: A.S. Kechris, B.D. Miller, Topics in Orbit Equivalence (2004)
Vol. 1853: Ch. Favre, M. Jonsson, The Valuative Tree (2004)
Vol. 1854: O. Saeki, Topology of Singular Fibers of Differential Maps (2004)
Vol. 1855: G. Da Prato, P.C. Kunstmann, I. Lasiecka, A. Lunardi, R. Schnaubelt, L. Weis, Functional Analytic Methods for Evolution Equations. Editors: M. Iannelli, R. Nagel, S. Piazzera (2004)
Vol. 1856: K. Back, T.R. Bielecki, C. Hipp, S. Peng, W. Schachermayer, Stochastic Methods in Finance, Bressanone/Brixen, Italy, 2003. Editors: M. Fritelli, W. Runggaldier (2004)
Vol. 1857: M. Émery, M. Ledoux, M. Yor (Eds.), Séminaire de Probabilités XXXVIII (2005)
Vol. 1858: A.S. Cherny, H.-J. Engelbert, Singular Stochastic Differential Equations (2005)
Vol. 1859: E. Letellier, Fourier Transforms of Invariant Functions on Finite Reductive Lie Algebras (2005)
Vol. 1860: A. Borisyuk, G.B. Ermentrout, A. Friedman, D. Terman, Tutorials in Mathematical Biosciences I. Mathematical Neurosciences (2005)
Vol. 1861: G. Benettin, J. Henrard, S. Kuksin, Hamiltonian Dynamics – Theory and Applications, Cetraro, Italy, 1999. Editor: A. Giorgilli (2005)
Vol. 1862: B. Helffer, F. Nier, Hypoelliptic Estimates and Spectral Theory for Fokker-Planck Operators and Witten Laplacians (2005)
Vol. 1863: H. Führ, Abstract Harmonic Analysis of Continuous Wavelet Transforms (2005)
Vol. 1864: K. Efstathiou, Metamorphoses of Hamiltonian Systems with Symmetries (2005)
Vol. 1865: D. Applebaum, B.V. R. Bhat, J. Kustermans, J. M. Lindsay, Quantum Independent Increment Processes I. From Classical Probability to Quantum Stochastic Calculus. Editors: M. Schürmann, U. Franz (2005)
Vol. 1866: O.E. Barndorff-Nielsen, U. Franz, R. Gohm, B. Kümmerer, S. Thorbjønsen, Quantum Independent Increment Processes II. Structure of Quantum Lévy Processes, Classical Probability, and Physics. Editors: M. Schürmann, U. Franz, (2005)
Vol. 1867: J. Sneyd (Ed.), Tutorials in Mathematical Biosciences II. Mathematical Modeling of Calcium Dynamics and Signal Transduction. (2005)
Vol. 1868: J. Jorgenson, S. Lang, $\mathrm{Pos}_n(\mathrm{R})$ and Eisenstein Series. (2005)
Vol. 1869: A. Dembo, T. Funaki, Lectures on Probability Theory and Statistics. Ecole d'Eté de Probabilités de Saint-Flour XXXIII-2003. Editor: J. Picard (2005)
Vol. 1870: V.I. Gurariy, W. Lusky, Geometry of Müntz Spaces and Related Questions. (2005)
Vol. 1871: P. Constantin, G. Gallavotti, A.V. Kazhikhov, Y. Meyer, S. Ukai, Mathematical Foundation of Turbulent Viscous Flows, Martina Franca, Italy, 2003. Editors: M. Cannone, T. Miyakawa (2006)
Vol. 1872: A. Friedman (Ed.), Tutorials in Mathematical Biosciences III. Cell Cycle, Proliferation, and Cancer (2006)
Vol. 1873: R. Mansuy, M. Yor, Random Times and Enlargements of Filtrations in a Brownian Setting (2006)
Vol. 1874: M. Yor, M. Émery (Eds.), In Memoriam Paul-André Meyer - Séminaire de Probabilités XXXIX (2006)
Vol. 1875: J. Pitman, Combinatorial Stochastic Processes. Ecole d'Eté de Probabilités de Saint-Flour XXXII-2002. Editor: J. Picard (2006)
Vol. 1876: H. Herrlich, Axiom of Choice (2006)
Vol. 1877: J. Steuding, Value Distributions of L-Functions (2007)
Vol. 1878: R. Cerf, The Wulff Crystal in Ising and Percolation Models, Ecole d'Eté de Probabilités de Saint-Flour XXXIV-2004. Editor: Jean Picard (2006)
Vol. 1879: G. Slade, The Lace Expansion and its Applications, Ecole d'Eté de Probabilités de Saint-Flour XXXIV-2004. Editor: Jean Picard (2006)
Vol. 1880: S. Attal, A. Joye, C.-A. Pillet, Open Quantum Systems I, The Hamiltonian Approach (2006)
Vol. 1881: S. Attal, A. Joye, C.-A. Pillet, Open Quantum Systems II, The Markovian Approach (2006)
Vol. 1882: S. Attal, A. Joye, C.-A. Pillet, Open Quantum Systems III, Recent Developments (2006)
Vol. 1883: W. Van Assche, F. Marcellàn (Eds.), Orthogonal Polynomials and Special Functions, Computation and Application (2006)
Vol. 1884: N. Hayashi, E.I. Kaikina, P.I. Naumkin, I.A. Shishmarev, Asymptotics for Dissipative Nonlinear Equations (2006)
Vol. 1885: A. Telcs, The Art of Random Walks (2006)
Vol. 1886: S. Takamura, Splitting Deformations of Degenerations of Complex Curves (2006)
Vol. 1887: K. Habermann, L. Habermann, Introduction to Symplectic Dirac Operators (2006)
Vol. 1888: J. van der Hoeven, Transseries and Real Differential Algebra (2006)
Vol. 1889: G. Osipenko, Dynamical Systems, Graphs, and Algorithms (2006)
Vol. 1890: M. Bunge, J. Funk, Singular Coverings of Toposes (2006)
Vol. 1891: J.B. Friedlander, D.R. Heath-Brown, H. Iwaniec, J. Kaczorowski, Analytic Number Theory, Cetraro, Italy, 2002. Editors: A. Perelli, C. Viola (2006)
Vol. 1892: A. Baddeley, I. Bárány, R. Schneider, W. Weil, Stochastic Geometry, Martina Franca, Italy, 2004. Editor: W. Weil (2007)
Vol. 1893: H. Hanßmann, Local and Semi-Local Bifurcations in Hamiltonian Dynamical Systems, Results and Examples (2007)
Vol. 1894: C.W. Groetsch, Stable Approximate Evaluation of Unbounded Operators (2007)
Vol. 1895: L. Molnár, Selected Preserver Problems on Algebraic Structures of Linear Operators and on Function Spaces (2007)
Vol. 1896: P. Massart, Concentration Inequalities and Model Selection, Ecole d'Été de Probabilités de Saint-Flour XXXIII-2003. Editor: J. Picard (2007)
Vol. 1897: R. Doney, Fluctuation Theory for Lévy Processes, Ecole d'Été de Probabilités de Saint-Flour XXXV-2005. Editor: J. Picard (2007)

Vol. 1898: H.R. Beyer, Beyond Partial Differential Equations, On linear and Quasi-Linear Abstract Hyperbolic Evolution Equations (2007)

Vol. 1899: Séminaire de Probabilités XL. Editors: C. Donati-Martin, M. Émery, A. Rouault, C. Stricker (2007)

Vol. 1900: E. Bolthausen, A. Bovier (Eds.), Spin Glasses (2007)

Vol. 1901: O. Wittenberg, Intersections de deux quadriques et pinceaux de courbes de genre 1, Intersections of Two Quadrics and Pencils of Curves of Genus 1 (2007)

Vol. 1902: A. Isaev, Lectures on the Automorphism Groups of Kobayashi-Hyperbolic Manifolds (2007)

Vol. 1903: G. Kresin, V. Maz'ya, Sharp Real-Part Theorems (2007)

Vol. 1904: P. Giesl, Construction of Global Lyapunov Functions Using Radial Basis Functions (2007)

Vol. 1905: C. Prévôt, M. Röckner, A Concise Course on Stochastic Partial Differential Equations (2007)

Vol. 1906: T. Schuster, The Method of Approximate Inverse: Theory and Applications (2007)

Vol. 1907: M. Rasmussen, Attractivity and Bifurcation for Nonautonomous Dynamical Systems (2007)

Vol. 1908: T.J. Lyons, M. Caruana, T. Lévy, Differential Equations Driven by Rough Paths, Ecole d'Été de Probabilités de Saint-Flour XXXIV-2004 (2007)

Vol. 1909: H. Akiyoshi, M. Sakuma, M. Wada, Y. Yamashita, Punctured Torus Groups and 2-Bridge Knot Groups (I) (2007)

Vol. 1910: V.D. Milman, G. Schechtman (Eds.), Geometric Aspects of Functional Analysis. Israel Seminar 2004-2005 (2007)

Vol. 1911: A. Bressan, D. Serre, M. Williams, K. Zumbrun, Hyperbolic Systems of Balance Laws. Cetraro, Italy 2003. Editor: P. Marcati (2007)

Vol. 1912: V. Berinde, Iterative Approximation of Fixed Points (2007)

Vol. 1913: J.E. Marsden, G. Misiołek, J.-P. Ortega, M. Perlmutter, T.S. Ratiu, Hamiltonian Reduction by Stages (2007)

Vol. 1914: G. Kutyniok, Affine Density in Wavelet Analysis (2007)

Vol. 1915: T. Bıyıkoğlu, J. Leydold, P.F. Stadler, Laplacian Eigenvectors of Graphs. Perron-Frobenius and Faber-Krahn Type Theorems (2007)

Vol. 1916: C. Villani, F. Rezakhanlou, Entropy Methods for the Boltzmann Equation. Editors: F. Golse, S. Olla (2008)

Vol. 1917: I. Veselić, Existence and Regularity Properties of the Integrated Density of States of Random Schrödinger (2008)

Vol. 1918: B. Roberts, R. Schmidt, Local Newforms for GSp(4) (2007)

Vol. 1919: R.A. Carmona, I. Ekeland, A. Kohatsu-Higa, J.-M. Lasry, P.-L. Lions, H. Pham, E. Taflin, Paris-Princeton Lectures on Mathematical Finance 2004. Editors: R.A. Carmona, E. Çinlar, I. Ekeland, E. Jouini, J.A. Scheinkman, N. Touzi (2007)

Vol. 1920: S.N. Evans, Probability and Real Trees. Ecole d'Été de Probabilités de Saint-Flour XXXV-2005 (2008)

Vol. 1921: J.P. Tian, Evolution Algebras and their Applications (2008)

Vol. 1922: A. Friedman (Ed.), Tutorials in Mathematical BioSciences IV. Evolution and Ecology (2008)

Vol. 1923: J.P.N. Bishwal, Parameter Estimation in Stochastic Differential Equations (2008)

Vol. 1924: M. Wilson, Littlewood-Paley Theory and Exponential-Square Integrability (2008)

Vol. 1925: M. du Sautoy, L. Woodward, Zeta Functions of Groups and Rings (2008)

Vol. 1926: L. Barreira, V. Claudia, Stability of Nonautonomous Differential Equations (2008)

Vol. 1927: L. Ambrosio, L. Caffarelli, M.G. Crandall, L.C. Evans, N. Fusco, Calculus of Variations and Non-Linear Partial Differential Equations. Cetraro, Italy 2005. Editors: B. Dacorogna, P. Marcellini (2008)

Vol. 1928: J. Jonsson, Simplicial Complexes of Graphs (2008)

Vol. 1929: Y. Mishura, Stochastic Calculus for Fractional Brownian Motion and Related Processes (2008)

Vol. 1930: J.M. Urbano, The Method of Intrinsic Scaling. A Systematic Approach to Regularity for Degenerate and Singular PDEs (2008)

Vol. 1931: M. Cowling, E. Frenkel, M. Kashiwara, A. Valette, D.A. Vogan, Jr., N.R. Wallach, Representation Theory and Complex Analysis. Venice, Italy 2004. Editors: E.C. Tarabusi, A. D'Agnolo, M. Picardello (2008)

Vol. 1932: A.A. Agrachev, A.S. Morse, E.D. Sontag, H.J. Sussmann, V.I. Utkin, Nonlinear and Optimal Control Theory. Cetraro, Italy 2004. Editors: P. Nistri, G. Stefani (2008)

Vol. 1933: M. Petkovic, Point Estimation of Root Finding Methods (2008)

Vol. 1934: C. Donati-Martin, M. Émery, A. Rouault, C. Stricker (Eds.), Séminaire de Probabilités XLI (2008)

Vol. 1935: A. Unterberger, Alternative Pseudodifferential Analysis (2008)

Vol. 1936: P. Magal, S. Ruan (Eds.), Structured Population Models in Biology and Epidemiology (2008)

Vol. 1937: G. Capriz, P. Giovine, P.M. Mariano (Eds.), Mathematical Models of Granular Matter (2008)

Vol. 1938: D. Auroux, F. Catanese, M. Manetti, P. Seidel, B. Siebert, I. Smith, G. Tian, Symplectic 4-Manifolds and Algebraic Surfaces. Cetraro, Italy 2003. Editors: F. Catanese, G. Tian (2008)

Vol. 1939: D. Boffi, F. Brezzi, L. Demkowicz, R.G. Durán, R.S. Falk, M. Fortin, Mixed Finite Elements, Compatibility Conditions, and Applications. Cetraro, Italy 2006. Editors: D. Boffi, L. Gastaldi (2008)

Vol. 1940: J. Banasiak, V. Capasso, M.A.J. Chaplain, M. Lachowicz, J. Miękisz, Multiscale Problems in the Life Sciences. From Microscopic to Macroscopic. Będlewo, Poland 2006. Editors: V. Capasso, M. Lachowicz (2008)

Vol. 1941: S.M.J. Haran, Arithmetical Investigations. Representation Theory, Orthogonal Polynomials, and Quantum Interpolations (2008)

Vol. 1942: S. Albeverio, F. Flandoli, Y.G. Sinai, SPDE in Hydrodynamic. Recent Progress and Prospects. Cetraro, Italy 2005. Editors: G. Da Prato, M. Röckner (2008)

Vol. 1943: L.L. Bonilla (Ed.), Inverse Problems and Imaging. Martina Franca, Italy 2002 (2008)

Vol. 1944: A. Di Bartolo, G. Falcone, P. Plaumann, K. Strambach, Algebraic Groups and Lie Groups with Few Factors (2008)

Vol. 1945: F. Brauer, P. van den Driessche, J. Wu (Eds.), Mathematical Epidemiology (2008)

Vol. 1946: G. Allaire, A. Arnold, P. Degond, T.Y. Hou, Quantum Transport. Modelling, Analysis and Asymptotics. Cetraro, Italy 2006. Editors: N.B. Abdallah, G. Frosali (2008)

Vol. 1947: D. Abramovich, M. Mariño, M. Thaddeus, R. Vakil, Enumerative Invariants in Algebraic Geometry and String Theory. Cetraro, Italy 2005. Editors: K. Behrend, M. Manetti (2008)
Vol. 1948: F. Cao, J-L. Lisani, J-M. Morel, P. Musé, F. Sur, A Theory of Shape Identification (2008)
Vol. 1949: H.G. Feichtinger, B. Helffer, M.P. Lamoureux, N. Lerner, J. Toft, Pseudo-Differential Operators. Quantization and Signals. Cetraro, Italy 2006. Editors: L. Rodino, M.W. Wong (2008)
Vol. 1950: M. Bramson, Stability of Queueing Networks, Ecole d'Eté de Probabilités de Saint-Flour XXXVI-2006 (2008)
Vol. 1951: A. Moltó, J. Orihuela, S. Troyanski, M. Valdivia, A Non Linear Transfer Technique for Renorming (2009)
Vol. 1952: R. Mikhailov, I.B.S. Passi, Lower Central and Dimension Series of Groups (2009)
Vol. 1953: K. Arwini, C.T.J. Dodson, Information Geometry (2008)
Vol. 1954: P. Biane, L. Bouten, F. Cipriani, N. Konno, N. Privault, Q. Xu, Quantum Potential Theory. Editors: U. Franz, M. Schuermann (2008)
Vol. 1955: M. Bernot, V. Caselles, J.-M. Morel, Optimal Transportation Networks (2008)
Vol. 1956: C.H. Chu, Matrix Convolution Operators on Groups (2008)
Vol. 1957: A. Guionnet, On Random Matrices: Macroscopic Asymptotics, Ecole d'Eté de Probabilités de Saint-Flour XXXVI-2006 (2009)
Vol. 1958: M.C. Olsson, Compactifying Moduli Spaces for Abelian Varieties (2008)
Vol. 1959: Y. Nakkajima, A. Shiho, Weight Filtrations on Log Crystalline Cohomologies of Families of Open Smooth Varieties (2008)
Vol. 1960: J. Lipman, M. Hashimoto, Foundations of Grothendieck Duality for Diagrams of Schemes (2009)
Vol. 1961: G. Buttazzo, A. Pratelli, S. Solimini, E. Stepanov, Optimal Urban Networks via Mass Transportation (2009)
Vol. 1962: R. Dalang, D. Khoshnevisan, C. Mueller, D. Nualart, Y. Xiao, A Minicourse on Stochastic Partial Differential Equations (2009)
Vol. 1963: W. Siegert, Local Lyapunov Exponents (2009)
Vol. 1964: W. Roth, Operator-valued Measures and Integrals for Cone-valued Functions and Integrals for Cone-valued Functions (2009)
Vol. 1965: C. Chidume, Geometric Properties of Banach Spaces and Nonlinear Iterations (2009)
Vol. 1966: D. Deng, Y. Han, Harmonic Analysis on Spaces of Homogeneous Type (2009)
Vol. 1967: B. Fresse, Modules over Operads and Functors (2009)
Vol. 1968: R. Weissauer, Endoscopy for GSP(4) and the Cohomology of Siegel Modular Threefolds (2009)
Vol. 1969: B. Roynette, M. Yor, Penalising Brownian Paths (2009)
Vol. 1970: M. Biskup, A. Bovier, F. den Hollander, D. Ioffe, F. Martinelli, K. Netočný, F. Toninelli, Methods of Contemporary Mathematical Statistical Physics. Editor: R. Kotecký (2009)
Vol. 1971: L. Saint-Raymond, Hydrodynamic Limits of the Boltzmann Equation (2009)
Vol. 1972: T. Mochizuki, Donaldson Type Invariants for Algebraic Surfaces (2009)
Vol. 1973: M.A. Berger, L.H. Kauffmann, B. Khesin, H.K. Moffatt, R.L. Ricca, De W. Sumners, Lectures on Topological Fluid Mechanics. Cetraro, Italy 2001. Editor: R.L. Ricca (2009)
Vol. 1974: F. den Hollander, Random Polymers: École d'Été de Probabilités de Saint-Flour XXXVII – 2007 (2009)
Vol. 1975: J.C. Rohde, Cyclic Coverings, Calabi-Yau Manifolds and Complex Multiplication (2009)
Vol. 1976: N. Ginoux, The Dirac Spectrum (2009)
Vol. 1977: M.J. Gursky, E. Lanconelli, A. Malchiodi, G. Tarantello, X.-J. Wang, P.C. Yang, Geometric Analysis and PDEs. Cetraro, Italy 2001. Editors: A. Ambrosetti, S.-Y.A. Chang, A. Malchiodi (2009)
Vol. 1978: M. Qian, J.-S. Xie, S. Zhu, Smooth Ergodic Theory for Endomorphisms (2009)
Vol. 1979: C. Donati-Martin, M. Émery, A. Rouault, C. Stricker (Eds.), Śeminaire de Probablitiés XLII (2009)
Vol. 1980: P. Graczyk, A. Stos (Eds.), Potential Analysis of Stable Processes and its Extensions (2009)
Vol. 1981: M. Chlouveraki, Blocks and Families for Cyclotomic Hecke Algebras (2009)
Vol. 1982: N. Privault, Stochastic Analysis in Discrete and Continuous Settings. With Normal Martingales (2009)
Vol. 1983: H. Ammari (Ed.), Mathematical Modeling in Biomedical Imaging I. Electrical and Ultrasound Tomographies, Anomaly Detection, and Brain Imaging (2009)
Vol. 1984: V. Caselles, P. Monasse, Geometric Description of Images as Topographic Maps (2010)
Vol. 1985: T. Linß, Layer-Adapted Meshes for Reaction-Convection-Diffusion Problems (2010)
Vol. 1986: J.-P. Antoine, C. Trapani, Partial Inner Product Spaces. Theory and Applications (2009)
Vol. 1987: J.-P. Brasselet, J. Seade, T. Suwa, Vector Fields on Singular Varieties (2010)
Vol. 1988: M. Broué, Introduction to Complex Reflection Groups and Their Braid Groups. Cetraro, Italy 2007. Editors: G. di Pillo, F. Schoen (2010)
Vol. 1989: I.M. Bomze, V. Demyanov, Nonlinear Optimization (2010)
Vol. 1990: S. Bouc, Biset Functors for Finite Groups (2010)

Recent Reprints and New Editions

Vol. 1702: J. Ma, J. Yong, Forward-Backward Stochastic Differential Equations and their Applications. 1999 – Corr. 3rd printing (2007)
Vol. 830: J.A. Green, Polynomial Representations of GL_n, with an Appendix on Schensted Correspondence and Littelmann Paths by K. Erdmann, J.A. Green and M. Schoker 1980 – 2nd corr. and augmented edition (2007)
Vol. 1693: S. Simons, From Hahn-Banach to Monotonicity (Minimax and Monotonicity 1998) – 2nd exp. edition (2008)
Vol. 470: R.E. Bowen, Equilibrium States and the Ergodic Theory of Anosov Diffeomorphisms. With a preface by D. Ruelle. Edited by J.-R. Chazottes. 1975 – 2nd rev. edition (2008)
Vol. 523: S.A. Albeverio, R.J. Høegh-Krohn, S. Mazzucchi, Mathematical Theory of Feynman Path Integral. 1976 – 2nd corr. and enlarged edition (2008)
Vol. 1764: A. Cannas da Silva, Lectures on Symplectic Geometry 2001 – Corr. 2nd printing (2008)

LECTURE NOTES IN MATHEMATICS

Editorial Policy (for the publication of monographs)

1. Lecture Notes aim to report new developments in all areas of mathematics and their applications - quickly, informally and at a high level. Mathematical texts analysing new developments in modelling and numerical simulation are welcome.

   Monograph manuscripts should be reasonably self-contained and rounded off. Thus they may, and often will, present not only results of the author but also related work by other people. They may be based on specialised lecture courses. Furthermore, the manuscripts should provide sufficient motivation, examples and applications. This clearly distinguishes Lecture Notes from journal articles or technical reports which normally are very concise. Articles intended for a journal but too long to be accepted by most journals, usually do not have this "lecture notes" character. For similar reasons it is unusual for doctoral theses to be accepted for the Lecture Notes series, though habilitation theses may be appropriate.

2. Manuscripts should be submitted either online at www.editorialmanager.com/lnm to Springer's mathematics editorial in Heidelberg, or to one of the series editors. In general, manuscripts will be sent out to 2 external referees for evaluation. If a decision cannot yet be reached on the basis of the first 2 reports, further referees may be contacted: The author will be informed of this. A final decision to publish can be made only on the basis of the complete manuscript, however a refereeing process leading to a preliminary decision can be based on a pre-final or incomplete manuscript. The strict minimum amount of material that will be considered should include a detailed outline describing the planned contents of each chapter, a bibliography and several sample chapters.

   Authors should be aware that incomplete or insufficiently close to final manuscripts almost always result in longer refereeing times and nevertheless unclear referees' recommendations, making further refereeing of a final draft necessary.

   Authors should also be aware that parallel submission of their manuscript to another publisher while under consideration for LNM will in general lead to immediate rejection.

3. Manuscripts should in general be submitted in English. Final manuscripts should contain at least 100 pages of mathematical text and should always include

   - a table of contents;
   - an informative introduction, with adequate motivation and perhaps some historical remarks: it should be accessible to a reader not intimately familiar with the topic treated;
   - a subject index: as a rule this is genuinely helpful for the reader.

   For evaluation purposes, manuscripts may be submitted in print or electronic form (print form is still preferred by most referees), in the latter case preferably as pdf- or zipped ps-files. Lecture Notes volumes are, as a rule, printed digitally from the authors' files. To ensure best results, authors are asked to use the LaTeX2e style files available from Springer's web-server at:

   ftp://ftp.springer.de/pub/tex/latex/svmonot1/ (for monographs) and
   ftp://ftp.springer.de/pub/tex/latex/svmultt1/ (for summer schools/tutorials).

Additional technical instructions, if necessary, are available on request from: lnm@springer.com.

4. Careful preparation of the manuscripts will help keep production time short besides ensuring satisfactory appearance of the finished book in print and online. After acceptance of the manuscript authors will be asked to prepare the final LaTeX source files and also the corresponding dvi-, pdf- or zipped ps-file. The LaTeX source files are essential for producing the full-text online version of the book (see http://www.springerlink.com/openurl.asp?genre=journal&issn=0075-8434 for the existing online volumes of LNM).

   The actual production of a Lecture Notes volume takes approximately 12 weeks.

5. Authors receive a total of 50 free copies of their volume, but no royalties. They are entitled to a discount of 33.3% on the price of Springer books purchased for their personal use, if ordering directly from Springer.

6. Commitment to publish is made by letter of intent rather than by signing a formal contract. Springer-Verlag secures the copyright for each volume. Authors are free to reuse material contained in their LNM volumes in later publications: a brief written (or e-mail) request for formal permission is sufficient.

Addresses:

Professor J.-M. Morel, CMLA,
École Normale Supérieure de Cachan,
61 Avenue du Président Wilson, 94235 Cachan Cedex, France
E-mail: Jean-Michel.Morel@cmla.ens-cachan.fr

Professor F. Takens, Mathematisch Instituut,
Rijksuniversiteit Groningen, Postbus 800,
9700 AV Groningen, The Netherlands
E-mail: F.Takens@rug.nl

Professor B. Teissier, Institut Mathématique de Jussieu,
UMR 7586 du CNRS, Équipe "Géométrie et Dynamique",
175 rue du Chevaleret,
75013 Paris, France
E-mail: teissier@math.jussieu.fr

For the "Mathematical Biosciences Subseries" of LNM:

Professor P.K. Maini, Center for Mathematical Biology,
Mathematical Institute, 24-29 St Giles,
Oxford OX1 3LP, UK
E-mail: maini@maths.ox.ac.uk

Springer, Mathematics Editorial, Tiergartenstr. 17,
69121 Heidelberg, Germany,
Tel.: +49 (6221) 487-259
Fax: +49 (6221) 4876-8259
E-mail: lnm@springer.com

Lightning Source UK Ltd.
Milton Keynes UK
02 June 2010

154976UK00003B/34/P